A Dictionary of the Space Age

New Series in NASA History
Steven J. Dick, Series Editor

A Dictionary of the Space Age

Paul Dickson

The Johns Hopkins University Press   Baltimore

This book has been brought to publication with the generous assistance of the J. G. Goellner Endowment of the Johns Hopkins University Press.

9 8 7 6 5 4 3 2 1

This volume was produced under contract with the National Aeronautics and Space Administration.

The Johns Hopkins University Press
2715 North Charles Street
Baltimore, Maryland 21218-4363
www.press.jhu.edu

Library of Congress Cataloging-in-Publication Data

Dickson, Paul.
A dictionary of the space age / Paul Dickson.
    p. cm. — (New series in NASA history)
Includes bibliographical references.
ISBN-13: 978-0-8018-9115-1 (hardcover : alk. paper)
ISBN-10: 0-8018-9115-9 (hardcover : alk. paper)
1. Aeronautics—Dictionaries. I. Title.
TL509.D475 2009
629.403—dc22
                                2008022679

A catalog record for this book is available from the British Library.

Special discounts are available for bulk purchases of this book. For more information, please contact Special Sales at 410-516-6936 or specialsales@press.jhu.edu.

In time of
rapid change, few things
change more rapidly than language
itself. Newly discovered phenomena, technologies that until lately have not even existed—each
of these demands marked growth in the little conceptual
handles that we call words, and that we use as tools to communicate and to record. • Sometimes this growth takes the form
of new meanings added precariously to the top of loads carried by
existing words, thus setting traps of incomprehension for the unwary.
The growth of language is a process that races past generally unperceived
by most of us. Yet today surely millions of people must have at least a
nodding acquaintance with words and terms that, a few years ago, would
have quite baffled them in their current contexts, e.g. ablation, staging,
mid-course correction, and hold, to cite only a few. At times we
are afforded a momentary glimpse of the phenomenon of
language change: today we used the word astronaut
with casual ease, but only a few years ago as it first
crept into the language it sounded bizarre
and even pretentious to
many ears.

—Melvin S. Day, Director,

NASA Scientific and Technical

Information Division, foreword to

Dictionary of Technical Terms for

Aerospace Use (NASA SP-7, 1965)

Contents

Foreword

More than three decades have passed since *Origins of NASA Names* was published in the NASA History Series in 1976. As that volume rolled off the press during the nation's bicentennial year, the final remnants of the Apollo program had been played out, with three crews having visited Skylab in 1973–74 and the Apollo-Soyuz Test Project having come to a successful completion in July 1975. The Space Shuttle was still five years from its first launch in 1981, and the initial assembly of the International Space Station was more than two decades in the future. Two Viking spacecraft would make first landfall on Mars in 1976, but the Voyagers had not yet made their tours of the outer planets, and Pluto was still a planet (rather than a dwarf planet, as redefined by the International Astronomical Union in 2006—an example of the importance of definition). The inspiring images of the Hubble Space Telescope were still 15 years away, and the other Great Observatories were little more than a gleam in astronomers' imaginations.

In the intervening decades since *Origins* was first published, the lexicon of aerospace (see its entry!) has increased substantially. Although this compilation is by no means comprehensive, it is a fascinating commentary on how one technological discipline has affected the meaning and use of words and language, both within the aerospace community and in the broader world.

As a longtime space reporter, author, lexicographer, and space enthusiast, Paul Dickson is well qualified to undertake this update of *Origins*. His knowledge and love of the subject shine through in each entry. It is appropriate, too, that this volume should appear on the occasion of the 50th anniversaries of NASA and of the Space Age. Space exploration has arguably added much to history and culture, and as language evolves over the coming decades, one anticipates that a next edition of this volume will reflect the evolution of both the Space Age and the culture in which it is embedded.

Steven J. Dick
NASA Chief Historian
Washington, D.C.

Introduction and Notes on Method

It was silver in color, about the size of a beach ball, and weighed a mere 184 pounds (83 kg). Yet for all its simplicity, small size, and inability to do more than orbit the Earth and transmit seemingly meaningless radio blips, the influence of Sputnik on America and the world was enormous and totally unpredicted.

The reaction to Sputnik, including early attempts by America to get into orbit, gave rise to a popular new vocabulary within months of its launch on October 4, 1957. The process began the next day when newspapers all over the world proclaimed the dawning of the Space Age, an abstract term rendered real in a matter of 24 hours.

Sputnik became a name, a word, and a metaphor overnight. The Russian name for the satellite was Sputnik Zemlyi, a term that means "traveling companion of the world," or Earth satellite. Almost at once, Sputnik Zemlyi was shortened to Sputnik, and in that form it entered the languages of the world.[1] The first syllable of the word was hardly ever pronounced as it would have been in Russian. "Spootnik! Spootnik!" said a Russian woman I interviewed when working on a book on the subject, "Sputt-nick is an American word."

Indeed, it was an American word. "Some new words take years to get into the language," said lexicographer Clarence L. Barnhart at the time. "She's a record breaking word." So sure was he that the day after the launch of Sputnik 1, Barnhart called his printer to dictate a definition, and it was included in the next *Thorndike-Barnhart* (1958).[2]

A new subset of English language—called variously space-Speak, Nasan, NASA-speak, or NASAese—was created. It was, to one writer, composed of "a vocabulary as new as fresh paint" that generated more than a score of official and unofficial dictionaries and glossaries during the first decade of the Space Age.[3]

1. "The News of the Week in Review," *New York Times,* November 10, 1957, p. E1.
2. John G. Rogers, "Sputnik Soars into Type in a U.S. Dictionary," *New York Herald Tribune,* December 18, 1957, p. 17.
3. William B. Stapleton, "Space Age Sign Language," *Houston Post,* July 12, 1964, p. 6.

Suddenly, words that had belonged to the worlds of science, aviation, and science fiction were part of the larger language appearing in newspapers and television. The world was learning a host of new proper nouns and names—some of which, like Gemini and Mercury, were ancient allusions while others, like Vanguard and Explorer, evoked visions of adventure and the future. The nature of this high-speed endeavor was such that it created its own newspeak of verbal shortcuts, abbreviations, acronyms, and initialisms.

In the rush to create a vocabulary to fit the new age of satellites and rocketships, there were lexical glitches—mostly minor and obscured by time. For instance, the hurry to describe the behavior of a satellite led to the use of two *t*'s in the word orbiting in both official documents and the press.[4]

With the beginnings of NASA and the Mercury, Gemini, and Apollo programs, the process was accelerated. In *Moonport: A History of Apollo Launch Facilities and Operations,* Charles D. Benson and William Barnaby Faherty wrote, "Apollo scientists and engineers were establishing a terminology for new things; no one had defined them in the past because such things did not exist—module is an example. As late as 1967, the *Random House Dictionary of the English Language* gave as the fifth definition of module under computer technology, "A readily interchangeable unit containing electronic components, especially one that may be readily plugged in or detached from a computer system." The space world was well ahead of the dictionary because, as every American television viewer knew, a module—command, service, or lunar—was a unit of the spacecraft that went to the Moon.

As real-live astronauts were introduced into the program, their voices from space and the words spoken by their controllers were heard by millions and their slangy speech flew into the larger language. Things aloft did not just go "according to plan" but were, variously, "A-OK," "sitting fat," "tickety-boo," or "copasetic." Astronauts "drove" (rather than "piloted") their spacecraft, and when they did something that was daring and determined, they were given the ultimate accolade: "steely-eyed missile man." When NASA sent the early astronauts on the road to plump for support of the space program, it was known to them as "a week in the barrel," a sly allusion to the political pork barrel.

And while the Space Age brought with it a host of colorful names— names that should resonate for centuries to come from Sputnik through

4. The first use of the word "orbitting" in the press is in 1955 in a description of the Vanguard program in the *Washington Post:* "U.S. to Send up 10 Space Satellites," *Washington Post,* October 7, 1955, p. 1.

Ares and Orion—some of the most memorable naming derived from the old American military tradition of giving your craft a name that is both a call sign or radio "handle" and a nickname. Freedom 7 was a name given by Alan Shepard himself to the first American spacecraft carrying an astronaut. The crew of Apollo 11 picked the name Eagle for their Lunar Module, which was immortalized in the line uttered on July 20, 1969: "Houston, Tranquility base here. The Eagle has landed."

In those early days of human spaceflight, every word seemed to be worthy of attention no matter how seemingly trivial. In the final minutes of Gordon Cooper's 22-orbit Mercury flight in 1963, he was told by his flight controller that he had fired his retro rockets "right on the money." Cooper seems to have replied, "Right on the old gazoo."

Gazoo? Immediately, some insisted that he had really said "bazoo," while others were certain that it was "kazoo." Headlines in some Florida newspapers used "bazoo," and the *New York Times* gave its readers a choice in an article entitled "Right on the Old Bazoo—Or Was It the Old Kazoo? And a Third Version Reports Cooper Said 'Gazoo'—Anyway He Was on It."[5]

Sometimes the impulse to create a new way of speaking went astray and drifted into the wholesale use of abbreviation: acronyms and initialisms. The problem with initialisms, when used in abundance, is that they can be a barrier rather than an aid to comprehension. Consider this verbatim exchange from the log of Apollo 12, Day 5, as the crew prepares for lunar descent:

106:57:29 [Col. Gerald P.] Carr: Intrepid, Houston.

106:57:35 [Charles "Pete"] Conrad: Go ahead, Houston.

106:57:37 Carr: Roger. We're about a minute from LOS. Everything's looking real good. Your computer and everything is fine. The most possible idea we can think for your 1106 alarm was, if you turned your DUA off about when your computer was running, that might have possibly caused it.

106:58:02 [Alan] Bean: What's a DUA?

106:58:06 Carr: Roger. That's your digital uplink assembly.

106:58:12 Bean: Oh! Okay.[6]

Over the decades NASA has created hundreds of thousands of acronyms and initialisms that have allowed simple sets of letter combinations to stand for a multitude of things. Take, for example, the initials LL, which, according to two NASA glossaries, can stand, variously, for Low Level,

5. *New York Times,* May 17, 1963, in *NASA Current News,* May 21, 1963, p. 11.
6. See http://history.nasa.gov/ap12fj/12day5_prep_landing.htm.

Lever Lock, Lower Limit, Lower Left, Launch and Landing, Launch Left, Long Line, and Long Lead. Then there are all the other initialisms that begin with LL:

LLC	Logical Link Control
LLCF	Launch and Landing Computational Facilities
LLI	Limited Life Item
LLIL	Long Lead-time Items List
LLLTV	Low Light Level Television
LLNL	Lawrence Livermore National Laboratory
LLOS	Landmark Line-of-Sight
LLP	Launch and Landing Project
LLPO	Launch and Landing Project Office
LLRF	Lunar Landing Research Facility
LLRV	Lunar Landing Research Vehicle
LLS	Launch and Landing Site
LLT	Long Lead-Time
LLTV	Lunar Landing Training Vehicle[7]

The ultimate three-letter acronym beginning with the letter L may be LOA: List of Acronyms.

The impulse that led to this alphabet soup was, in fact, aimed at getting information recorded quickly and accurately while describing a fast-moving new world. Everything was a shorthand, and a word as simple as "over" could be shortened to OVR in NASA-speak. Little-recalled today was an early attempt by NASA and other space agencies to create its own shorthand version of the proprietary Gregg Shorthand system. Described by the *Houston Chronicle* as a system of "squiggles, fishhooks and gravy ladles," the elements of the NASA system were described in a booklet entitled "Shorthand Symbols for the Glossary of Terms Used in the Exploration of Space."[8]

There are elaborate acronyms that work on paper as clever descriptors but tend to be confusing when spoken or referred to out of context, such as the Space Shuttle experiments known as Project Starshine for Student Tracked Atmospheric Research Satellite Experiment.

---

7. See www.ksc.nasa.gov/inforcenter/acronym.htm and *Space Transportation System and Associated Payloads: Glossary, Acronyms, and Abbreviations*, RP-1059 (Washington, DC: NASA, 1981), p. 166.

8. Stapleton, "Space Age Sign Language." The symbol for zero-gravity, for example, is described as a "squiggle, a curved overline and, underneath, two bat-eyed umlauts lying sideways."

Over time this language became more and more complicated and confusing. In preparing for the retrieval of the Long Duration Exposure Facility (LDEF) on January 12, 1990, the following message was sent to the crew of Shuttle Columbia from the CAPCOM (spacecraft communicator) in the MCC (Mission Control Center) at JSC (the Johnson Space Center in Texas): "We've already had AOS IOS. We expect LOS TDRS East in about four minutes. Probably pick you up on the west side AOS TDRS West at 22:12. However, we will continue with IOS for another six minutes and 40 seconds until LOS."

Martin Metzker, a reporter for Knight-Ridder newspapers, wrote about this in an article entitled "When NASA and Astronauts Communicate, They're Speaking in Tongues." He translated for his readers: "We will continue to hear you through the Indian Ocean station for six minutes and 40 seconds. In four minutes, we'll lose your transmission through the eastern satellite, but will pick you up later through the western satellite." The original message was, in fact, so confusing that the public information officer at the scene told Metzker, "The consensus at mission control is that CAPCOM Frank Culbertson may just have set a record for excessive use of acronyms." Even more to the point, after the message was sent, Columbia co-pilot James Weatherby replied, "Sounds pretty fishy to me."[9]

Fishy? Could anyone confuse this bit of slang with some elaborate initialism or acronym? Although NASA was willing to embrace a dizzying array of shorthand and jargon, for reasons that are not clear it shied away from slang and colloquialisms. NASA's early attempt to define its own terms in the 1965 *Dictionary of Technical Terms for Aerospace Use,* Special Publication 7 (SP-7) stated that "slang would not be included."[10] As a beat reporter covering NASA Headquarters during the mid- to late 1960s, this author can assure the reader that the people who were running the space program may have been writing reports that used the language of SP-7, but they were speaking another that was both slangy and to the point.

This rich and often irreverent slang created a universe parallel to that of the jargon and abbreviations. The slang was part of an older tradition of barnstorming aviators, fighter pilots, and test pilots whose reaction to danger was to make linguistic light of it. The original astronauts who were willing to sit atop a rocket booster in a tiny capsule and be sent into space let it be known that they felt like "spam in a can." Reporters unable to get straight information on the Apollo 1 fire would cynically invoke the

9. Martin Metzker, "When NASA and Astronauts Communicate, They're Speaking in Tongues," *Akron Beacon Journal,* January 13, 1990, p. 4.
10. SP-7, p. vii.

acronym NASA, standing not for the National Aeronautics and Space Administration but for "Never a Straight Answer."[11]

The "common touch" language that the astronauts brought to the space program made it both popular and accessible. The astronauts contributed not just words but also memorable expressions and quotations. At the time of Wally Schirra's death in May 2007, CBS News correspondent Peter King recalled a famous line Schirra had uttered when asked about his thoughts as the clock ticked down to zero before his Mercury flight: "He said, 'Just think of these millions of parts put together by the lowest bidder!' Everybody's used that line since; John Glenn used it, I think they use it on the shuttle today—but it's still a good line!"

Because American space exploration was so public, it took little for NASA-speak and slang to make the jump into common parlance. In an interview with Jessica Weintraub, Jonathan Lighter, editor of the *Historical Dictionary of American Slang,* talked about the phenomenon that occurs when a character uses an expression on a popular television show and millions of people are exposed to it. "The first time this probably happened was in 1961. Alan Shepard's sub-orbital flight was shown live on TV. In Shepard's communication with Shorty Powers, one or both of them said 'A-OK,' a phrase that was probably coined in NASA. Forty years later we're still using it."

The public nature of space has continued into the era of cable television with the advent of *NASA TV,* carried on most U.S. cable systems and many others worldwide, featuring live air-to-ground transmission of Shuttle and Space Station activities. While the broadcasts include narrative and explanation, the chatter is that of an operational vehicle, not necessarily structured for public use but in fact heard by many. In this sense the language of space is an open book.

Yet much of what was heard coming out of the space program had a certain elegance, power, and grace to it befitting the fact that a bold, heroic endeavor had begun. We were learning about black holes (objects whose gravity is so strong that the escape velocity exceeds the speed of light), brown dwarfs (dim objects in space that are too small to be stars

---

11. This may have originated from the writing of William Hines: "NASA's feet of clay were exposed on January 27, 1967—just nineteen years and one day before a bloodier and even more public exposure. A crew of three Astronauts—including one of the original Mercury Seven—was incinerated inside a sealed Apollo spacecraft on a launching pad while rehearsing the countdown for a liftoff scheduled as the maiden Earth orbital test leading to actual moon flights. In the fire's aftermath, the initials NASA acquired new meaning when some said they stood for 'Never a Straight Answer.'" Hines, "The Wrong Stuff," *Progressive* 58 (July 7, 1994): 8.

and too large to be planets), and blue moons (the occasional blue color seen in the Moon due to the Earth's atmosphere).

Some of these words were borrowed from other realms and seemed a bit odd at first in their new contexts: the mundane "housekeeping" used to describe spacecraft maintenance, the romantic "rendezvous" now applied to great hunks of metal meeting in space, and "insertion" becoming the point at which one achieved orbit. "Windows" were now opportunities, and "envelopes" were things to be pushed. "Redundancy" was a virtue, not an excess, and "reentry" meant you were coming home. "Guillotine" was given a reprieve from describing Gallic beheadings and now described a device equipped with explosive blades to cut cables, water lines, and wires during the separation of spacecraft modules from one another or from launch towers (called "umbilical" towers). Old terms were put to uses that had nothing to do with ordinary human activities: "breadboards" that had never seen flour or water and "boilerplate" that had nothing to do with steam.

Still other terms came out of the space-speak and were reapplied to other realms. "Soft landing" became a term used by the likes of former Federal Reserve Chairman Alan Greenspan to describe the national economy, while the "burnout" of rockets and missiles became a descriptor for occupational overload. Such transfers are detailed in the following entries under the heading of "Extended Use."

The language of the Space Age and the Space Race served a broader role in that it enriched the language in general. Its influence ranged from the serious (in the form of entries in the *Oxford English Dictionary*, *Merriam-Webster's Collegiate Dictionary*, *Webster's New World Dictionary*, and other mainstream dictionaries) to the downright frivolous. There was, for example, a moment in 1961 when a major fashion show staged in Washington, D.C., featured creations with "Planet Pinks," "Booster Blues," "Weightless Whites," and "Blast-off Blacks."[12]

Given the reality that the future of space exploration as a publicly funded human endeavor is strongly dependent on public understanding and support, the program should therefore be able to describe itself in language accessible to a larger and larger percentage of an increasingly diverse population. The space program should not be, as one longtime space reporter complained at a 1988 Cape Canaveral launch, "like the priesthood using symbols to keep the outside out."[13]

12. Ruth Wagner, "Fashions Take to Space," *Washington Post,* February 2, 1961, p. C17.

13. "Coping with NASA Jargon 'Like Being Able to Order Your Meals in French," AP dispatch by Deborah Mesce, p. 1. The reporter quoted was Ronald Kotulak of the *Chicago Tribune.*

Scope and Criteria

October 4, 2007, marked the 50th anniversary of the first Earth orbit of Sputnik 1 and the beginning of the Space Age. October 1, 2008, marked the 50th anniversary of the creation of the National Aeronautics and Space Administration (NASA) from portions of the 43-year-old National Advisory Committee for Aeronautics (NACA) and elements from other agencies, largely the Department of Defense. The purpose of this work is to capture the language and terminology of this first half-century of the Space Age.

At a basic level this volume is concerned with the proper names associated directly with NASA, the Communications Satellite Corp., and certain U.S. military and intelligence space programs. At this level, *A Dictionary of the Space Age* is simultaneously a revision, an expansion, and a large-scale updating of *Origins of NASA Names* (SP-4402), which was published in 1976 (directed by Monte D. Wright, Director, NASA History Office, and written by Helen T. Wells, Susan H. Whiteley, and Carrie E. Karegeannes). Because it is now 30 years later, the new work will contain many new entries from areas not covered in the original book, published just before the joint U.S.-Soviet manned Apollo-Soyuz mission. Like *Origins of NASA Names,* this dictionary pays attention to proper names, whether they were defined by a NASA committee or arrived at via a contest, as with the 1999 Terra EOS satellite, which was named by high-school senior Sasha Jones of Brentwood High School in St. Louis.

In addition, this work concerns itself with those proper names associated with the Soviet/Russian space program and the space programs of other nations. In this regard *A Dictionary of the Space Age* serves as a catalog of space programs dating back to the launch of Sputnik.

On a second level, this dictionary attempts to capture a broader foundation for language of the Space Age based on historical principles as employed in the *Oxford English Dictionary* and *Webster's Third New International*—that is, based on considerations of the etymology of a term, its earliest recorded use, and variations in meaning. The definitions adhere to the principles of lexicography and are also written to be understood by a wide audience—academically correct, but accessible. Word histories for major terms are detailed in a conversational tone, and technical terms are deciphered for the interested student or lay reader.

One issue involved in compiling any dictionary is what to include and what to exclude. First and foremost, this is not meant to be a compilation of technical details and terms associated with space but is rather an attempt to capture the culture of the Space Age including slang, colloquialisms, nicknames, and the actual language used by the men and women involved. Some well-established terms from aviation and astronomy that predate our Space Age but are germane to the understanding of the period have been included.

The dictionary is not comprehensive but rather prismatic, in the same sense that a Merriam-Webster's collegiate dictionary is prismatic of an unabridged dictionary: it attempts to cover the most important terms. What will not be addressed and defined in this work are ephemeral or nonce words. (Marsapaloosa, for example, a term created by the Public Affairs Office at the Jet Propulsion Laboratory to create public interest in Mars-bound missions, is not here.) However, simple, seemingly self-explanatory terms that have historical relevance will be included, such as O-ring and gap filler (the fire-retardant strips glued between the Space Shuttle's heat shield tiles that protect the vehicle's exterior).

How to Use This Dictionary

First and foremost, this is a dictionary meant for a multitude of readers, and for that reason it attempts to be both accurate and accessible. But it is also a book meant for browsing, and for that reason there is flexibility in the presentation of entries. If, for example, a good story begs to be told as a digression, it gets told. This is therefore a discursive dictionary. It should also be said that this dictionary is descriptive rather than pre-scriptive. Each definition has been created to reflect the actual use of the term rather than the officially approved one. Case in point: Throughout the period when the United States and the Soviet Union were trying to land a man on the Moon, NASA officially argued that it was not a race. But "race" was in fact the term used by both the press and the public.

Basic Format of Entries

Entries follow this general format:

1. **Term** (in boldface). If there is more than one spelling of a term, the vari-ants are separated by slashes and are listed with the most common one coming first. For example, in the entry "A-OK/A-Okay/A.O.K./AOK," the first spelling is the one used by the *Oxford English Dictionary* and others, the second is the one used by Tom Wolfe in *The Right Stuff,* and the final examples appear at various times in the *New York Times* and other news-papers.[14] Terms that are spelled the same except for style of capitalization—for example, "pogo" (describing the vertical vibration of

---

14. Some of the spelling issues prompted newspapers to address such questions in print. A column on "space semantics" in the *Wall Street Journal* (September 9, 1965, p. 12) said that the paper was having a tough time deciding on the "spelling and capitalization of 'a-okay' (or is it 'A-OK'?)."

a launch vehicle) and "POGO" (acronym for Polar Orbiting Geophysical Observatory)—are defined in separate entries, with the lower-cased term appearing first. Nicknames and terms that are clearly slang or colloquial are so noted.

2. Pronunciation/translation/category/word combination. When appropriate, any or all of these may be given in parentheses.

3. Part of speech. Part of speech is given when there is more than one definition or when there could be confusion.

4. Earliest meaning. If there is more than one definition for the term, the definitions are numbered, with the first being the earliest meaning. This principle is also applied to names. In the entry for Antares, for example, the definition "third stage of a Scout rocket" comes before "call sign for the Apollo 14 Lunar Module."

5. Synonyms and cross-references. A term appearing in italics within or at the end of a definition indicates a synonym or cross-reference. If the cross-reference appears as a term's first definition, the reader should turn to that term's entry for a full definition.

6. Sources. Sources are cited either within parentheses or as a separate paragraph at the end of the entry.

Etymological and Usage Notes

As an aid to the reader, many terms are illustrated in the context of an attributed, dated quotation. I have also included, as part of the definition, the historical background of the concept or object in question. An entry may include commentary under any the following headings:

ETYMOLOGY. When possible, the history of the term is given. If there are several theories about the origin of a term, all are included, often followed by a discussion of their relative merit. Such explanations are not attempted when the term is self-descriptive or suggests its own origin. The etymology of a term like "space vehicle" is self-evident, but that of "space law" is not, because the latter comes out of a set of specific proposals for the creation of a new legal realm that comes into play at a given altitude. The etymology of "space law" thus involves a discussion of the altitude at which space itself begins.

Some terms will have more than one etymology, and these meanings may be in conflict with one another. The principle at work in this dictionary is that all claims should be presented. If there is a bias lurking in this book, it is that words have multiple origins and those origins act cumulatively.

FIRST USE. In the case of selected new terms—those with the greatest impact, in the author's estimation—the entry will supply a dated reference marking the term's debut, often accompanied by a display of a term as it appears in that reference. This dictionary builds on historical principles, and so knowledge of the time and place of the debut of a term is essential to understanding the term as a (for lack of a better expression) "verbal artifact."

In most cases it is, of course, impossible to cite the *very* earliest example that appears in print, so this feature is really meant to give the reader a feel for the relative antiquity and early use of the word or phrase in a public forum. Because of the rigors imposed on the concept of first use, the process is at best problematic. As Dave Wilton points out in his article "Moonshot Terms," Neil Armstrong is credited with two first citations in the *Oxford English Dictionary* for the words "postflight" ("There was some suspicion, lingering in the postflight shock of the first Sputnik, that this was the road the Soviet Union had chosen") and "topo" ("The best we can do on topo features is to advise you to look to the west of the irregularly shaped crater"). Of course, Armstrong is unlikely to have actually coined these words; rather, they were probably common in NASA jargon at the time. Armstrong gets credit because he used them in his 1970 book, *First on the Moon*.[15] "Postflight," for example, was in common newspaper use for years before Armstrong's book.[16]

Lest there be any question, the citation given under "First Use" will be the earliest found in print or in a written transcript. In terms of the daily press, the author's primary newspaper sources are the *Atlanta Constitution, Boston Globe, Chicago Tribune, Christian Science Monitor, Los Angles Times, New York Times, Wall Street Journal,* and *Washington Post,* although others are included.

USAGE. Where appropriate, the entries include commentary on terms that should not be used in certain contexts, or that have come into common use but are objected to by members of the aerospace community. See, for example, the discussion under the entries for "blastoff" or "rocket scientist." This also holds true for the terms "manned" and "unmanned" as

15. Wilton in *A Way with Words: The Weekly Newsletter of Wordorigins* 4, no. 17 (July 22, 2005), www.wordorigins.org. Wilton notes, "67 other words in the OED are given citations from this book by Armstrong, among them A-OK, hypergolic, lift-off, lunar, non-flammable, pitch, playback, preflight, psych, read-out, rocket, rog, selenocentric, slingshot, smack-dab, spaceship, splashdown, transearth, umbilical, undock, and zero-G."
16. An AP dispatch of May 8, 1961, contained this lead sentence: "The 3,000-pound capsule that carried Alan B. Shepard Jr. into space may wind up in the Smithsonian Institution, but for several weeks it will undergo extensive postflight tests." *Washington Post,* May 9, 1961, p. A10.

modifiers to terms like "spacecraft" and "exploration," which are still used widely but are being replaced by the terms "human" and "robotic."

EXTENDED USE. Many terms that originated in the space program have transcended their original meanings. It is through extended meanings that one can see the immense influence of space terminology on the language at large. The more common examples include "window" (for an opportune moment), "destruct," "launch," "countdown," "black box," and "backup."

Sources

Because this is a dictionary rather than a narrative, source citations are embedded in the definition rather than in footnotes or endnotes. This is the style common to all major dictionaries including both the *Oxford English Dictionary* and *Webster's Third International.*

In the case of items carried over from *Origins of NASA Names,* I have compressed the original source notes into one per entry and have eliminated some notes, including the most ephemeral ones—phone conversations—as there is no way of confirming or referring to them 30 years after the fact. Most entries will have light updating and extended references. All entries from *Origins of NASA Names* will be tightened to conform to a terser dictionary-type style.

The principal lexical sources consulted in the compilation of this dictionary are listed in the Bibliography. This aspires to be the most comprehensive list compiled to date on sources pertaining to the language of the Space Age, and it includes ephemeral material. According to a search of the holdings of the National Technical Information Service, there are more than 800 NASA publications that include a glossary—often little more than a list of a dozen or so acronyms replicated elsewhere, but sometimes containing terms of no lexical value: ordinary English words defined in standard dictionaries. For instance, one book about Skylab has a glossary that includes the common words "doffing" and "donning" as terms used for astronauts dressing and undressing.[17]

I have also included narratives that concerned themselves with the spoken language of space—Tom Wolfe's *The Right Stuff,* for example— and a number of books in the NASA Special Publications series that were written as narratives.

Many of the news clippings referred to in this book were found in issues of *NASA Current News,* a Public Affairs publication that served as a regular (daily to weekly) review of NASA press.

17. *Skylab Experiments,* vol. 7, *Living and Working in Space,* EP-116 (Washington, DC: NASA, 1973), p. 37.

Abbreviations

The abbreviations that appear in this dictionary are those currently used in NASA publications as well as those commonly used in lexicography. Books listed in the bibliography are generally cited in the source notes in short form (surname/short title). The following additional abbreviations are used in the source citations.

adj.   adjective

ALSJ Glossary   Garry Kennedy, comp., Glossary appended to NASA's on-line *Apollo Lunar Surface Journal,* http://history.nasa.gov/alsj

ASP Glossary   Astronomical Society of the Pacific, "Introductory Astronomy Glossary," www.astrosociety.org/education/publications/tnl/14/14.html

AP   Associated Press

CP-   NASA Conference Publication

EP-   NASA Educational Publication

FY   fiscal year

"Glossary of Aerospace Age Terms"   "Glossary of Aerospace Age Terms," in "Can You Talk the Language of the Aerospace Age?" (brochure published by the U.S. Air Force Recruiting Service, Wright Patterson Air Force Base, Ohio, 1963)

n.   noun

NASA Names Files   NASA Names Files, in NASA Historical Reference Collection, History Division, NASA Headquarters, Washington, DC

obs.   obsolete

OED   *Oxford English Dictionary*

Paine Report   National Commission on Space, *Pioneering the Space Frontier: Report of the National Commission on Space* (New York: Bantam, 1986)

RP-   NASA Reference Publication

SP-   NASA Special Publication

SP-4402   Helen T. Wells, Susan B. Whiteley, and Carrie Karegeannes,
                Origins of NASA Names, SP-4402 (Washington, DC: GPO, 1976)

SP-7   William H. Allen, *Dictionary of Technical Terms for Aerospace
                Use,* SP-7 (Washington, DC: NASA, 1965)

SP-6001   *Apollo Terminology,* SP-6001 (Washington, DC: NASA,
                August 1963)

TM-   NASA Technical Memorandum

UPI   United Press International

USA   United States Army

USAF   United States Air Force

USN   United States Navy

v.   verb

A Dictionary of the Space Age

A

ablation. The wearing away or burning off of the outer layers of an object *reenter*ing the *atmosphere.* The process cools and protects the outer surfaces of a *spacecraft* or *missile nosecone.* The suborbital *Jupiter* C nosecones were ablative (susceptible to ablation), as was the Apollo Command Module.

Able *(launch vehicle upper stage).* An *upper* stage used in combination with *Thor* or *Atlas* first stages. It was one of several upper stages derived in 1958 from *Vanguard* launch vehicle components by the Department of Defense's *Advanced Research Projects Agency,* Douglas Aircraft Company, and Space Technology Laboratories. The name signified "A" or "first" (from the military practice of having communication code words for each letter of the alphabet). It is sometimes referred to as Project Able. See *Delta.* (SP-4402, p. 5; Milton W. Rosen, Office of Defense Affairs, NASA, telephone interview, February 16, 1965.)

Able and Baker. Names of the two monkeys the United States recovered after *launch* in a *Jupiter* nosecone during a suborbital flight on May 28, 1959. The flight was successful, testing the capability to launch from *Cape Canaveral,* Florida, and to recover *spacecraft* in the Atlantic Ocean, but Able later died.

ETYMOLOGY. Named for the first two vocables in the phonetic alphabet, A and B.

ABMA. Army Ballistic Missile Agency, part of the Army Ordnance Missile Command (AOMC). Prior to the creation of NASA, when it was home to Wernher von Braun and other members of the German *rocket* team, it was a pioneer in early spaceflight responsible for, among other things, the first American *satellite, Explorer* 1.

abort. (1, v.) To cut short or break off an action, especially because of equipment failure; to effect a time-critical termination. (2, n.) An instance of a *rocket, missile,* or *mission* failing to function effectively and not achieving its objective.

USAGE. Although the term was first applied to aviation during World War II for the premature termination of a mission, when applied to rocket launches it had a jarring effect on many Americans because they associated it with what was at the time an illegal medical procedure.

ACE. Advanced Composition Explorer. Major *mission* of the *Explorer* program, launched August 25, 1997. It orbits the L1 *libration point* about 1 million miles (1.6 million km) from Earth and is studying energetic particles originating in the Sun and interstellar space.

acoustic velocity. The speed of propagation of sound waves. "Hopelessly old-fashioned people call it the speed of sound," noted Jack Rice in an article on the language of the *Gemini* program (*St. Louis Post-Dispatch*, September 10, 1965).

acquisition of signal (AOS). When an Earth *tracking station* makes an initial contact with an orbiting *spacecraft* (a process that ends with *loss of signal*). The same could occur in orbit around the Moon or Earth.

active-repeater. A *satellite* that allows a signal transmitted to it to be strengthened and rebroadcast to a ground station. The principle was first tested in 1958 as part of *Project SCORE*.

Adam. See *Project Adam*.

Advanced Composition Explorer. See *ACE*.

Advanced Research Projects Agency (ARPA). Unit within the Department of Defense created by President Eisenhower and Congress immediately following the *launch* of *Sputnik* to assure that the United States would never again be left behind in the area of new technology. It sired, among other things, the first communications relay satellite *(Project SCORE)* and the Internet (in the form of the DARPANET/ARPANET). The name was changed to DARPA (Defense Advanced Research Projects Agency) in 1972.

Advent. See *Project Advent*.

Advisory Committee on the Future of the U.S. Space Program. See *Augustine Report*.

Aerobee/Astrobee *(sounding rocket).* A two-stage sounding rocket designed to carry a 150-pound (68-kg) *payload* to an altitude of 80 miles (130 km). Development of the Aerobee was begun in 1946 by the Aerojet Engineering Corporation (later Aerojet-General Corporation) under contract to the U.S. Navy. The Applied Physics Laboratory (APL) of Johns Hopkins University was assigned technical direction of the project. In 1952, at the request of the Air Force and the Navy, Aerojet undertook design and development of the Aerobee-Hi, a high-performance version of the Aerobee designed expressly for research in the *upper atmosphere.* An improved Aerobee-Hi became the Aerobee 150. The uprated Aerobee 150 was called Astrobee. Aerojet used the prefix Aero to designate its liquid-*propellant* sounding rockets and Astro for its solid-fueled rockets.

ETYMOLOGY. James A. Van Allen, the first director of the project at APL, proposed the name Aerobee. He took the aero- from Aerojet Engineering and the -bee from Bumblebee, the name of the overall project to develop naval rockets that APL was monitoring for the Navy.

SOURCES. Peter T. Eaton, Office of Space Science and Applications, NASA, letter to Historical Staff, NASA, May 2, 1967; James A. Van Allen, Eleanor Pressly, and John W. Townsend Jr., "The Aerobee Rocket," and Townsend et al., "The Aerobee-Hi Rocket," chaps. 4 and 5 in *Sounding Rockets,* ed. Newell; GSFC, *Encyclopedia: Satellites and Sounding Rockets,* p. 321; William R. Corliss, *NASA Sounding Rockets, 1958–1968: A Historical Summary,* SP-4401 (Washington, DC: NASA, 1971), pp. 79–80; Herbert J. Honecker, Advanced Vehicles Section, Flight Performance Branch, Sounding Rocket Division, GSFC, memorandum to John H. Lane, Head, Flight Performance Branch, January 10, 1975; GSFC, *United States Sounding Rocket Program,* p. 38; Edward E. Mayo, Flight Performance Branch, Sounding Rocket Division, GSFC, information sent to Historical Office, NASA, January 30, 1975.

aerobraking. The deliberate use of a planet's *atmospheric drag* to slow a *spacecraft* down and lower its orbital altitude. Among others, the *Mars Global Surveyor* and *Mars Odyssey* spacecraft used this method. See also *atmospheric braking.*

Aeronautics and Space Report of the President. An annual report mandated by the National Aeronautics and Space Act of 1958, to include a "comprehensive description of the programmed activities and the accomplishments of all agencies of the United States in the field of aeronautics and space activities during the preceding calendar year." In recent years the reports have been prepared on a fiscal-year basis. Recent editions are online at www.hq.nasa.gov/office/pao/History/presrep.htm.

aeronomy. The study of the chemical and physical processes taking place in the upper regions of the atmosphere.

Aeros. (1) Name used briefly in the early 1960s by NASA for the Synchronous Meteorological Satellite project. See *SMS.* (2) German research *satellite.* A June 10, 1969, memorandum of understanding between NASA and the German Ministry for Scientific Research (BMwF) initiated a cooperative project that would put into *orbit* a German scientific satellite designed to investigate particle behavior in the Earth's *upper atmosphere.* This was the second U.S.-German research satellite project, following *Azur.* Aeros 1 was launched successfully on December 16, 1972. A second satellite in the series, Aeros 2, was launched on July 16, 1974. (SP-4402, p. 33; NASA News Release 69–91; Lloyd E. Jones Jr., Office of International Affairs, NASA, telephone interview, June 4, 1971; NASA, "Project Approval Document," February 27, 1970; NASA program office.)

ETYMOLOGY (definition 2). BMwF chose the name Aeros (the ancient Greek god of the air) for the proposed *aeronomy* satellite in early 1969. NASA

designated it GRS-A-2 (German Research Satellite A-2) before launch and
gave it the official name Aeros upon launch.

aerospace/aero-space. (1, adj.) Of or pertaining to the Earth's *atmosphere* and the space beyond it. (2, n.) The realm combining the Earth's
atmosphere and space.

ETYMOLOGY / FIRST USE. On October 29, 1957, just three weeks after
Sputnik, the U.S. Air Force sent a memo to all of its commands talking
about "air/space vehicles of the future." This was the beginning of a
campaign to add a new word and concept to the language, making air
and space one environment. The hyphenated word aero-space first
appeared in print in February 1958 in the title of a small glossary entitled
Interim Glossary: Aero-Space Terms, edited by Woodward Agee Heflin and
published by the Air University, Maxwell Air Force Base, Alabama. The
term was also defined there as an adjective: "Of or pertaining to the earth's
envelop [sic] of atmosphere and the space above it, the two considered
as a single realm for activity in the launching, guidance, and control of
ballistic missiles, earth satellites, dirigible space vehicles, and the like" (p. 1).

USAGE. The term was created by the Air Force as a proprietary word
staking out the atmosphere and the space above it as a single realm. If it
could be argued that air and space were one, then the Air Force would be,
logically, America's "Space Force." Writing in *Reporter* magazine in 1963,
David Burnham noted that the Air Force saw the word as its own "secret
weapon in the bitter inter-service battle for the space dollar." The major
rival for those space dollars was the Army, which saw space linked to the
Earth, the platform that allowed one to get into space.

This lexical power play was demonstrated by Gen. Thomas D. White,
then Air Force Chief of Staff, testifying before the House Astronautic and
Space Committee on February 3, 1959. "Aerospace," he told the committee,
"is a term which may be unfamiliar to some of you. Since you will hear it
several times during the course of our presentations, I would like to define
it for the committee. The Air Force has operated throughout its relatively
short history in the sensible atmosphere around the earth. Recent
developments have allowed us to extend our operations further away
from the earth, approaching the environment popularly known as space.
Since there is no dividing line, no natural barrier separating these two
areas, there can be no operational boundary between them. Thus air and
space comprise a single continuous operational field in which the Air Force
must continue to function. This area is aerospace."

After a few minutes of testimony in which White managed to use the
term eight times, Rep. John McCormack (D-Mass.) interrupted to ask
exactly when this new term had been coined. "Within the last year and by
the Air Force, I am willing to add," the General responded proudly.

The term was applied with such abandon that at one point the Air Force actually began calling airplanes "aerospace planes." At the Air Force Academy, cadets puckishly renamed physical education "aerospace dynamics," and chapel attendance became "aerospace theology."

Finally the Army and Navy came to see the Air Force–centric coinage as such a blatant linguistic power play that the Joint Chiefs of Staff were brought in to offer a compromise definition: "The earth's envelope of atmosphere and the space above it, two separate entities considered as a single realm for activity in launching, guiding and controlling of vehicles which will travel in both realms" (NASA Names Files, record no. 17498).

aerospace medicine. The branch of medicine dealing with the effects upon the human body of flight through the atmosphere or in space and with the prevention or cure of physiological or psychological malfunctions arising from these effects.

aerospace power. The entire aeronautical and astronautical capacity of a nation ("Glossary of Aerospace Age Terms," p. 7).

afterbody. Object that trails behind a *satellite* or *spacecraft.*

afterburning. Irregular burning of fuel left in the firing chamber of a *rocket* after fuel cutoff. Also applied to turbojets. (SP-6001, p. 3.)

Agena *(launch vehicle upper stage).* An upper stage used in combination with *Thor* or *Atlas* first stages. Agena was originally developed in the late 1950s for the U.S. Air Force by Lockheed Missiles Systems Division (now Lockheed Martin).

Agena A, the first version of the stage, was followed by the Agena B, which had a larger fuel capacity and engines that could restart in space. The later Agena D was standardized to provide a launch vehicle for a variety of military and NASA payloads. NASA used Atlas-Agena vehicles to *launch* large *Earth satellite*s as well as lunar and inter-planetary *space probe*s. Thor-Agena vehicles launched scientific satellites, such as *OGO* (Orbiting Geophysical Observatory) and *Alouette,* and applications satellites, such as the *Echo* 2 communi-cations satellite and *Nimbus* meteorological satellites. In Project *Gemini* the Agena D, modified to suit the specialized requirements of space *rendezvous* and *docking* maneuvers, became the Gemini Agena Target Vehicle (GATV).

ETYMOLOGY. In 1958 the Department of Defense's *Advanced Research Projects Agency* (ARPA) proposed to name the stage for the star Agena in the constellation Centaurus because it would be a *rocket* "igniting in the sky." The name Agena had first appeared in the *Geography of the Heavens,* published in the 1800s by the "popularizing Connecticut astronomer" Elijah H. Burritt, and was preserved in American dictionaries as the popular name for the star Beta Centauri. Burritt was thought to have coined the name from alpha (A, for first or foremost) and gena (knee) because he had located the star near the "right foreleg" of the constellation. Lockheed approved the choice of the name, since it followed Lockheed's tradition of naming aircraft and missiles after stellar phenomena (e.g., the Constellation aircraft and Polaris intercontinental ballistic *missile*). ARPA formally approved the name in June 1959.

SOURCES. SP-4402, pp. 6–7; W. F. Whitmore, "AGENA: The Spacecraft and the Star," Lockheed Missiles and Space Co. research paper, January 16, 1969; Dick Bissinette, Andrews AFB, letter to Judy Gildea, NASA, March 27, 1963; R. H. Allen, *Star Names: Their Lore and Meaning* (New York: Dover, 1963), p. 154; R. Cargill Hall, Lockheed Missiles and Space Co., letter to Historical Staff, NASA, August 26, 1965; Hall, "The Agena Satellite," unpublished essay, November 1966.

Algol. The first stage of the *Scout rocket,* named for a star in the constellation Perseus. See *Scout.*

all up. Approach to spaceflight testing that did not allow for *learning failures,* or success on the second try. Under this approach, according to Arnold S. Levine, a vehicle "is as complete as practicable for each flight, so that the maximum amount of test information is obtained with a minimum number of flights" (Levine, *Managing NASA in the Apollo Era,* SP-4102 [Washington, DC: NASA, 1982], p. 6). As Levine explains, this meant all three stages of the *Saturn V* were to be flown with the *Apollo Command and Service Module* instead of being flown separately.

ETYMOLOGY. The approach was created by George Mueller, who took over the Office of Manned Space Flight in September 1963. "In Apollo," explained Homer Newell, "the all up philosophy—which called for assembling a complete launcher and attempting to carry out a complete mission even on early test flights—was intended to produce economies as well as to preserve an image of success" (Newell, *Beyond the Atmosphere,* SP-4211, p. 160).

FIRST USE. "More important to the moon program, the flight justified NASA's decision to try the so-called 'all up' concept that was adopted in hopes of meeting the 1970 lunar landing goal. This concept was to assemble and put the entire Saturn 5 rocket into space, even though many of its major components had never been tested in space before" ("Saturn 5's Success Enhances the Odds of US Landing Men on the Moon by 1970," *Wall Street Journal,* November 10, 1957, p. 5).

Alouette. Canadian *satellite* project undertaken in cooperation with NASA. Alouette 1, a topside sounder scientific satellite instrumented to investigate the Earth's ionosphere from beyond the ionospheric layer, was launched into *orbit* by NASA from the *Pacific Missile Range* on

September 28, 1962. It was the first satellite designed and built by a country other than the United States or the Soviet Union and was the first satellite launched by NASA from the West Coast. Alouette 2 was launched later as part of the U.S.-Canadian ISIS project. See *ISIS*.

ETYMOLOGY. The satellite was given its name in May 1961 by the Canadian Defence Research Board. NASA supported the board's choice. The name was selected because, as the French-Canadian word for meadowlark, it suggested flight. Alouette was the title of a popular Canadian song and, in a bilingual country, called attention to Canada's French heritage.

SOURCES. SP-4402, pp. 34–35; Jonathan D. Caspar, "The Alouette (S-27) Program: A Case Study in NASA International Cooperative Activities," HHN-42, 1964 (commented) and 1965 (revised ms.); N. W. Morton, Dept. of National Defence, Canadian Joint Staff, letter to Arnold W. Frutkin, Director of International Programs, NASA, April 27, 1961; Robert C. Seamans Jr., Associate Administrator, NASA, letter to N. W. Morton, May 11, 1961; NASA Announcement 312, May 24, 1961; NASA News Release 64-207; Wallops Station News Release 64–77.

Alpha. Temporary name for the *International Space Station* (ISS).

ALSEP. *Apollo Lunar Surface Experiments Package.*

Altair. The fourth stage of the *Scout rocket,* named after a star in the constellation Aquila. See *Scout.*

America. *Call sign* for the Command Module for *Apollo* 17, the last of the six Apollo missions to the Moon, and the only one to include a scientist-geologist (Harrison Schmitt) as a member of the crew. Ronald Evans piloted the America while Schmitt and Eugene Cernan undertook extended *EVAs* on the lunar surface (22 hours 4 minutes for each) from the *Lunar Module,* named *Challenger.*

ETYMOLOGY. According to Cernan, the name was a way of paying tribute to the American public (quoted in Lattimer, *All We Did Was Fly to the Moon,* p. 94).

American Association for the Advancement of Science (AAAS). A private organization of scientists active in space exploration since its earliest days. Key organization in America's early thrust into space. The abbreviation is sometimes pronounced "triple A, S."

Ames Research Center (ARC). Originally one of NASA's aeronautics centers, now specializing in nanotechnology, information technology, fundamental *space biology, astrobiology,* biotechnology, thermal protection systems, and human factors research. Located at Moffett Field, California, Ames was created by Congress on August 9, 1939, as World War II began, as a second National Advisory Committee for Aeronautics *(NACA)* laboratory for urgent research in aircraft structures.

Ground was broken for the laboratory on September 14, 1939. The NACA facility began operations as the Moffett Field Laboratory in early 1941.

NACA named the facility Ames Aeronautical Laboratory in 1944 in honor of Dr. Joseph S. Ames, a leading aerodynamicist and former president of Johns Hopkins University. He was one of the first members of the NACA and served as NACA chairman from 1927 to 1939. When Dr. Ames retired, he was cited by President Roosevelt for his "inspiring leadership in the development of new research facilities and in the orderly prosecution of comprehensive research programs."

On October 1, 1958, as a facility of the NACA, the laboratory became part of the new National Aeronautics and Space Administration and was renamed Ames Research Center. Mission responsibilities of ARC focused on basic and applied research in the physical and life sciences for aeronautics and space flight. The center managed the *Pioneer* and *Biosatellite* space projects, as well as providing scientific experiments for other missions. It contributed to development of experimental tilt-wing and fan-in-wing aircraft and solutions to high-speed *atmosphere* entry problems, including the blunt-body concept, which is used on every *spacecraft* to prevent burning up upon atmospheric entry.

Wind tunnels are central to Ames' history. Several wind tunnels were opened in the 1940s to test and refine aircraft, guided missiles, satellites, and *reentry* bodies. Of particular note are three tunnels later designated key national resources. The 40- by 80-foot (12- by 24-m) wind tunnel opened in June 1944 to conduct aircraft development work. The 12-foot (3.7 m) *subsonic* wind tunnel opened in July 1946. The Unitary Plan Wind Tunnel enabled Ames to conduct new research; almost all NASA crewed *space vehicle*s, including the *Space Shuttle*, were tested in the Unitary. In the 1970s through the 1990s, all three facilities were renovated. Ames added an 80- by 120-foot (24- by 48-m) tunnel to the 40- by 80-foot tunnel, renamed it the National Full-Scale Aerodynamics Complex (NFAC), and dedicated it in 1987. The 12-foot pressure tunnel was rebuilt in the 1980s and rededicated in 1995. The workhorse Unitary received multiple upgrades in the 1990s. Today, only the Unitary is still in regular use.

When Ames became part of NASA in 1958, its most vital input to NASA's top priority, the lunar program, was the blunt-body concept for reentry *capsule*s, a concept tested and refined in Ames' new arc jet facilities and hypervelocity ranges. The arc jets contributed to thermal protection for all NASA's *crewed* programs, including the Space Shuttle, and also for planetary missions (e.g., the *Galileo* mission's Jupiter *probe*). The complex will continue to be central to the research and development of materials suitable for *heat shield* applications.

In the 1960s Ames emerged as a leading builder of flight simulators. In particular, in 1969 the Flight Simulator for Advanced Aircraft became part of a wide range of simulators, equipment, and facilities developed by Ames to improve pilot workloads, cockpit design, and safety, among other things. Another, the Vertical Motion Simulator, still enables testing of a variety of aircraft and the Space Shuttle.

Ames' longstanding life sciences program conducts research in various centrifuges, two of them unique to the agency, and genome facilities. In addition, Ames took major new strides in supercomputing in 2004 with the Project Columbia facility, which will present researchers with unprecedented computing capabilities. Ames' continuing interest in aviation safety manifests itself in a variety of advanced simulators and facilities, among them Future Flight Central, a sophisticated facility for basic research on movement into and around airports. The center still pursues human factors studies in the Human Performance Research Laboratory (1990), and advanced aerospace technology applications in the Automation Sciences Research Facility (1992).

In the 1990s Ames began to undertake different kinds of research, requiring different kinds of facilities. Its nanotechnology laboratories will help revolutionize space exploration by reducing mass while increasing capability. The center's astrobiology facilities include a world-renowned astrochemistry laboratory to simulate deep space.

For more information about Ames Research Center, see Edwin P. Hartman, *Adventures in Research: A History of Ames Research Center, 1940–1965,* SP-4302 (Washington, DC: NASA, 1970); Elizabeth A. Muenger, *Searching the Horizon: A History of Ames Research Center, 1940–1976,* SP-4304 (Washington, DC: NASA, 1985); Glenn E. Bugos, *Atmosphere of Freedom: Sixty Years at the NASA Ames Research Center,* SP-2000-4314 (Washington, DC: NASA, 2000).

AMR. *Atlantic Missile Range.*

Amundsen. Deep Space 2 instrument probe named for Norwegian explorer Roald Amundsen. See *Mars Polar Lander.*

Angry Alligator. Nickname given to the Augmented Target Docking Adapter on the *Gemini* IX *mission* of June 3–6, 1966, piloted by Thomas P. Stafford and Eugene A. Cernan. When a *rendezvous* with the upper-stage *Agena docking* port was thwarted when its fiberglass shroud opened only partially, the shroud had the appearance of an alligator.
   ETYMOLOGY. As Stafford began slowing the *spacecraft* to rendezvous with the target vehicle, he noted that the shroud was still open: "Then he exclaimed, 'Look at that moose!' As the distance dwindled, he knew that he had been indulging in wishful thinking. 'The shroud is half open on that thing!' Seconds later, Cernan remarked, 'You could almost knock it off!'

When the final braking was completed, the two vehicles were only 30 meters apart and in position for station keeping but it did not seem likely that the spacecraft nose could slip into the mouth of the 'moose' and dock. The crew described the shroud in detail and wondered out loud what could be done to salvage the situation. One of Stafford's remarks—graphic and memorable—became the trademark of the entire mission. His animal analogy switched to reptilian when he said, 'It looks like an angry alligator out here rotating around.' He itched to nudge it with his spacecraft docking bar to open its yawning jaws, but Flight Director Kranz told him to control the urge." (Hacker and Grimwood, *On the Shoulders of Titans*, SP-4203, pp. 333–34.)

Anik (Inuit for brother). Name given to Canadian *Telesat satellite*s when in *orbit*. Anik A-1 was launched from *Cape Canaveral*, Florida, aboard a *Delta rocket* on November 9, 1972. The second Anik was launched on April 20, 1973. The last Shuttle-launched Anik was Anik C1, aboard *STS*-51D in 1985. The latest in the series was launched in April 2007 aboard a proton Breeze M rocket. These satellites, designed to operate for 7 years, were usually retired after about 10 years of service. Four Aniks are still operational, and the duration record is about 14 years. (SP-4402, p. 75; "Canadian Satellite," *Washington Post*, April 16, 1969, p. A17; NASA News Release 71-85.)

anomaly. A deviation from the norm.

   USAGE. Critics of NASA-speak have argued that the term has been used by NASA to take the edge off critical situations. "'Anomaly' turns catastrophes into irregularities," charges Dawn Stover ("Anomaly = Disaster, and Other Handy NASA Euphemisms," *Popular Science*, February 2004). Stover cites a line from a 2003 NASA Press Release ("NASA Mishap Board Identifies Cause of X-43A Failure," July 23, 2003): "Shortly thereafter, the X-43A began to experience a control anomaly characterized by a roll oscillation."

ANS. Astronomical Netherlands Satellite. In June 1970 NASA and the Netherlands Ministries of Economic Affairs and Education and Science reached agreement to *launch* the first Dutch scientific *satellite*. The satellite was designated ANS, and an ANS Program Authority was created by the ministries to direct the cooperative project. NASA provided an experiment and the *Scout launch vehicle*, and the Program Authority designed, built, and tested the *spacecraft* and provided *tracking* and data acquisition. The satellite, launched August 30, 1974, carried an ultraviolet telescope to study selected stellar ultraviolet sources and instruments to investigate both soft and hard x-ray sources. (SP-4402, p. 35; NASA News Release 70-91.)

Ansari X Prize. Competition to *launch* a privately funded *space vehicle*. Created in 1996 for a team to privately build, *launch*, and finance a vehicle capable of carrying a single pilot and added ballast to compensate for the weight of two hypothetical passengers. In order to claim the prize, the same vehicle had to repeat this trip twice within two weeks. According to the rules and guidelines of the competition, as a demonstration of economic reusability, no more than 10 percent of the vehicle's non-*propellant* mass could be replaced between the first and second flights. On October 4, 2004, *SpaceShipOne* won the prize of $10 million by achieving its second suborbital spaceflight. The pilot was Brian Binnie; the first flight was piloted by Mike Melvill. SpaceShipOne was funded solely by Paul G. Allen, designed by Burt Rutan, and built by his company, Scaled Composites. The vehicle achieved the first privately funded human spaceflight in history. The vehicle cost an estimated $30 million to produce. (http://history.nasa.gov/x-prize.htm.)

Antares. (1) The third stage of the Scout *rocket,* named for the brightest star in the constellation Scorpius. See *Scout.* (2) *Call sign* for the *Apollo 14 Lunar Module,* piloted by Alan Shepard and Edgar Mitchell, who went to the lunar surface while Stuart Roosa piloted the Command Module, *Kitty Hawk.* They performed nine hours of Moon walks and brought back 98 pounds (44.45 kg) of lunar material. Apollo 14 was the third U.S. lunar landing *mission,* and the first since the near-disaster of Apollo 13.

ETYMOLOGY. (definition 2). According to Roosa, the name was given to the craft by Edgar Mitchell: "His logic was that Antares was the most visible 'landmark' as they pitched over during the powered descent to the lunar surface." Roosa, who ended up naming the Command Module Kitty Hawk, added, "It is a very difficult task to name a spacecraft. Ed and I spent hours trying to find two names that would be coordinated. Finally I told Ed, 'You name the Lunar Module anything you want, but I'm going with 'Kitty Hawk.''" (Quoted in Lattimer, *All We Did Was Fly to the Moon,* p. 82.)

anti-g. Designed to counteract the effects of high acceleration.

anti-g suit. A garment to protect *astronaut*s from certain physiological effects of acceleration.

anti-matter. A form of matter of which the atoms are composed of anti-particles, as protons, electrons, etc. assumed to carry charges opposite to those associated with ordinary matter. Particles having such properties have been produced in particle accelerators.

anti-satellite (ASAT). A *satellite* or other device whose purpose is to destroy or otherwise negate the *mission* of an enemy's operational satellite. Methods employed could be the physical destruction of a satellite (exploding ballistic satellite) or interference with the satellite's communications or power systems (laser beam or particle beam). The need for such a device was seen within days of the *Sputnik* launching. The first use of such technology was the Chinese ASAT test in January 2007, in which China destroyed one of its own satellites.

A-OK/A-Okay/A.O.K./AOK. Shorthand signifying that everything is in perfect working order. The term has a strong association with the earliest days of *crewed* spaceflight.

ETYMOLOGY / FIRST USE. "In reporting the Freedom 7 flight, the press attributed the term to astronaut [Alan B.] Shepard, and indeed a NASA News Release of May 5, 1961, has Shepard report 'A-OK' shortly after impact. A replay of the flight voice communications tape disclosed that Shepard himself did not use the term. It was Col. John A. 'Shorty' Powers who reported Shepard's condition as 'A-OK' in a description of the flight." (Swenson, Grimwood, and Alexander, *This New Ocean*, SP-4201, p. 375.)

During the second *Mercury* flight, Powers used the phrase at least four times, according to a *Sunday Oregonian* article about the phrase: "'Gus reports he is in very good condition; his trajectory is A-OK.' 'Reports here in the Mercury Control Center are A-OK all the way.' 'Our instruments here indicate A-OK.' 'The flight surgeon reports [Grissom] came through the high 'G' of re-entry in A-OK condition." ("Father of 'A-OK,'" *Sunday Oregonian*, December 18, 1961, in *NASA Current News*, December 22, 1961, p. 6.) The writer then adds, "But the phrase is absent from the official transcript of Grissom's radio communications released by NASA after his flight." The article goes on to bestow on Powers the title of "the father of 'A-OK.'"

Powers got the term from engineers building the voice networks for human flight. It was attributed to Paul Lein, of the Western Electric Company, who about four months earlier when confronted with static and background noise used the letter A with its "brilliant sound" in order to be heard clearly. In *The Right Stuff*, Tom Wolfe holds that the term was emblematic "shorthand for Shepard's triumph over the odds and for astronaut coolness under stress" (p. 270). (NASA Names Files, record no. 17497.)

AOS. *Acquisition of signal.*

Apache *(sounding rocket upper stage).* Solid-*propellant rocket* stage developed by Thiokol Chemical Corporation (later Thiokol Corporation) and used with the *Nike* first stage. Identical in appearance to the Nike-*Cajun*, the Nike-Apache could reach higher altitudes because the Apache propellant burning time was longer (6.4 seconds versus Cajun's 4 seconds). It could carry 75-pound (34-kg) payloads to an operating altitude of 130 miles (210 km), or 200 pounds (100 kg) to 78 miles (125 km).

ETYMOLOGY. The Apache, named for the American Indian tribe, followed the Thiokol tradition of ethnic and Indian-related names, a tradition that had begun with *Cajun* and also included the *Hawk, Malemute,* and *Tomahawk.*

SOURCES. Rosenthal, *Venture into Space,* SP-4301, pp. 127–29; *Space: The New Frontier,* EP-6, p. 38; GFSC, Vehicles Section, Spacecraft Integration and Sounding Rocket Division, telephone interview, March 19, 1970; GSFC, *United States Sounding Rocket Program,* pp. 38, 47.

aphelion. Point at which a body (planet, *comet,* etc.) in solar *orbit* is furthest from the Sun. See also *perihelion.*

apogee. (1, n.) Point at which a body in *orbit* around the Earth or another celestial body is furthest from that body. See also *perigee.* Strictly speaking, apogee and perigee refer only to Earth orbit, but the terms are often applied to orbits around other celestial bodies. (2, v.) For a rocket or satellite to reach its highest point.

Apollo. American *space program* of 19 missions, which included six piloted lunar landings between 1969 and 1972, four *Skylab* missions in 1973, and the *Apollo-Soyuz Test Project* in 1975. It has been termed "the most famous name in space exploration."

Project Apollo took on more urgency when the goal of a *human* lunar landing was proposed to the Congress by President John F. Kennedy on May 25, 1961, and was subsequently approved by the Congress. It was a program of three-man flights, leading to the landing of men on the Moon. *Rendezvous* and *docking* in lunar *orbit* of Apollo *spacecraft* components were vital techniques for the intricate flight to and return from the Moon.

The Apollo spacecraft consisted of the Command Module, serving as the crew's quarters and flight control section; the Service Module, containing propulsion and *spacecraft* support systems; and the *Lunar Module,* carrying two crewmen to the lunar surface, supporting them on the Moon, and returning them to the *Command and Service Module* in lunar orbit. Module designations came into use in 1962, when NASA made basic decisions on the flight mode (lunar orbit rendezvous), the boosters, and the spacecraft for Project Apollo. From that time until June 1966, the Lunar Module was called the *Lunar Excursion Module* (*LEM*), or sometimes the *bug.* It was renamed Lunar Module by the NASA Project Designation Committee because the word excursion implied mobility on the Moon, and this vehicle did not have that capability. The later Apollo flights, beginning with Apollo 15, carried the *Lunar Roving Vehicle,* or *rover,* to provide greater mobility for the *astronaut*s while on the surface of the Moon.

Beginning with the flight of Apollo 9, *call sign*s for both the Command and Service Module (CSM, sometimes just CM) and the Lunar Module (LM) were chosen by the astronauts who were to fly on each *mission.* The call signs for the *craft* were: Apollo 9, *Gumdrop* (CM), *Spider* (LM); Apollo 10, *Charlie Brown* (CM), *Snoopy* (LM); Apollo 11,

Columbia (CM), *Eagle* (LM); Apollo 12, *Yankee Clipper* (CM), *Intrepid* (LM); Apollo 13, *Odyssey* (CM), *Aquarius* (LM); Apollo 14, *Kitty Hawk* (CM), *Antares* (LM); Apollo 15, *Endeavour* (CM), *Falcon* (LM); Apollo 16, *Casper* (CM), *Orion* (LM); Apollo 17, *America* (CM), *Challenger* (LM).

The formula for numbering Apollo missions was altered when the three astronauts scheduled for the first human flight lost their lives in a flash fire during *launch* rehearsal on January 27, 1967. In honor of astronauts Virgil I. Grissom, Edward H. White II, and Roger B. Chaffee, the planned mission was given the name Apollo 1 even though it had not been launched. Carrying the prelaunch designation AS-204 (for the fourth launch in the Apollo *Saturn IB* series), the mission was officially recorded as "first manned Apollo Saturn flight, failed on ground test." Manned Spacecraft Center Deputy Director George M. Low had urged consideration of the request from the astronauts' widows that the designation Apollo 1, used by the astronauts publicly and included on their insignia, be retained. NASA Headquarters, Office of Manned Space Flight, therefore recommended the new numbering, and the NASA Project Designation Committee announced its approval on April 3, 1967.

The earlier, uncrewed Apollo Saturn IB missions AS-201, AS-202, and AS-203 were not given "Apollo" flight numbers, and no missions were named Apollo 2 or Apollo 3. The next mission flown, the first *Saturn V* flight (AS-501, for the first launch in the Apollo Saturn V series), became Apollo 4 after launch into orbit on November 9, 1967. Subsequent flights continued the sequence through 17.

The Apollo program carried the first men beyond the Earth's field of gravity and around the Moon on Apollo 8 in December 1968 and landed the first men on the Moon in Apollo 11 on July 20, 1969. The program concluded with Apollo 17 in December 1972 after having put 27 men into lunar orbit and 12 of them on the surface of the Moon. Data, photos, and lunar samples brought to Earth by the astronauts and data from experiments they left on the Moon began to give a picture of the Moon's origin and nature, contributing to the understanding of how the Earth had evolved.

ETYMOLOGY. In July 1960 NASA was preparing to implement its long-range plan beyond Project *Mercury* and to introduce a human circumlunar mission project, then unnamed, at the NASA–Industry Program Plans Conference in Washington. Abe Silverstein, Director of Space Flight Development, proposed the name Apollo because it was a name from ancient Greek mythology with attractive connotations, and the precedent for naming piloted spaceflight projects for mythological gods and heroes had been set with Mercury (which had also been named by Silverstein).

In a 1969 interview with the *Cleveland Plain Dealer,* Silverstein said, "I thought the image of the god Apollo riding his chariot across the sun gave the best representation of the grand scale of the proposed program. So, I chose it." Apollo was the god of archery, prophecy, poetry, and music, and most significantly, he was the charioteer of the Sun, pulling the Sun across the sky each day in his horse-drawn golden chariot. NASA approved the name and publicly announced Project Apollo at the July 28–29, 1960, conference.

FIRST USE. "The National Aeronautics and Space Administration yesterday disclosed plans for 260 major launchings in the next decade, including a manned spacecraft called Apollo that will follow Project Mercury" ("US Plans 260 Major Space Launchings in Next Decade," *Los Angeles Times,* July 29, 1960, p. B1).

SOURCES. NASA Names Files, record no. 17512; Merle G. Waugh, Office of Manned Space Flight, NASA, letter to James M. Grimwood, Historian, MSC, November 5, 1963; NASA Ad Hoc Committee to Name Space Projects and Objects, minutes of meeting, May 16, 1960; Bulfinch, *Mythology,* pp. 17, 40ff; Abe Silverstein, Director, Office of Space Flight Programs, NASA, memorandum to Harry J. Goett, Director, GSFC, July 25, 1960; Julian W. Scheer, Assistant Administrator for Public Affairs, NASA, memorandum from Project Designation Committee, June 9, 1966; George E. Mueller, Associate Administrator for Manned Space Flight, NASA, memorandum to Robert C. Seamans Jr., Deputy Administrator, NASA, February 9, 1967; Scheer, memorandum to Seamans, February 17, 1967; Mueller, memorandum to Scheer, March 28, 1967; George M. Low, Deputy Director, MSC, letter to Mueller, March 30, 1967; Scheer, memorandum to distribution, April 3, 1967; Mueller, TWX to KSC, MSFC, and MSC, Apollo and AAP Mission Designation, March 24 and April 24, 1967.

Apollo Lunar Surface Experiments Package (ALSEP). The collection of experiments flown to the lunar surface by *Apollo* 12, 14, 15, 16, and 17.

Apollo on steroids. Name used by NASA Administrator Michael Griffin in 2005 to typify a second lunar landing program that would return humans to the Moon as early as 2018 with a new *spacecraft* that would replace the *Space Shuttle.* "Think of it as Apollo on steroids," he told reporters at NASA headquarters in Washington at a September 18, 2005, news conference.

The term contains its own implied criticism: that a 21st-century spacecraft/program is based on 1960s technology. "This is not good," wrote Abram Katz in response to the term. "Thirty-six years after Apollo 11 landed on the Sea of Tranquility, we're simply planning to pursue an expanded version of 1960s technology" (*New Haven Register,* October 9, 2005, p. F6). Intended or not, the term so firmly attached itself to the new lunar initiative that many news accounts mentioned the nickname when the announcement was made on August 31, 2006, that the Martin-Marietta Company had been selected to build the *Orion* spacecraft: "In picking Lockheed Martin for Orion, described by NASA's chief as 'Apollo on steroids,' NASA bypassed Apollo throwbacks Northrop Grumman of Los Angeles and its chief subcontractor Boeing of Chicago" (Seth Borernstein, "Lockheed Martin Wins NASA Contract," *Houston Chronicle,* September 1, 2006, p. A1).

FIRST USE. "Mott described the Boeing OSP [Orbital Space Plane] concept as 'Apollo on steroids,' with a crew capsule attached to different modules to meet different mission requirements" (Mike Mott, General Manager, Boeing NASA Systems, quoted in *Aviation Week & Space Technology,* January 25, 2004).

USAGE. When NASA officials dabble in slang, the results can be difficult to control. A case in point is the use of "on steroids" to mean a more powerful version of something, like an athlete who has used anabolic steroids to enhance his strength. For reasons unclear, in the wake of "Apollo on steroids" the phrase was attached to a number of NASA projects and phenomena, including a "flying wing on steroids," "solar flares on steroids," "x-rays on steroids" (describing gamma rays), and "the Adirondacks on steroids" (a mapping enhancer).

Apollo-Saturn (AS). The program that would use Saturn *rocket*s to put *Apollo spacecraft* in *orbit.*

Apollo-Soyuz Test Project (ASTP). The first international *human* space project, carried out in July 1975. The joint U.S.-Soviet *rendezvous* and *docking mission* took its name from the *spacecraft* to be used, the American *Apollo* and the Soviet *Soyuz.* On September 15, 1969, two months after the Apollo 11 lunar landing *mission,* the President's *Space Task Group* made its recommendations on the future U.S. *space program.* One objective was broad international participation, and President Nixon included this goal in his March 1970 Space Policy Statement. The President earlier had approved NASA plans for increasing international cooperation in an informal meeting with Secretary of State William P. Rogers, Presidential Assistant for National Security Affairs Henry A. Kissinger, and NASA Administrator Thomas O. Paine aboard Air Force One while flying to the July Apollo 11 *splashdown.*

The United States had invited the Soviet Union to participate in experiments and information exchange in previous years. Now Dr. Paine sent Mstislav V. Keldysh, President of the Soviet Academy of Sciences, a copy of the U.S. post-Apollo plans and suggested exploration of cooperative programs. In April 1970 Dr. Paine suggested, in an informal meeting with academician Anatoly A. Blagonravov in New York, that the two nations cooperate on *astronaut* safety, including compatible docking equipment on *space station*s and shuttles to permit rescue operations in space emergencies. Further discussions led to a October 28, 1970, agreement on joint efforts to design compatible docking arrangements. Three working groups were set up. Agreements on further details were reached in Houston, Texas, June 21–25, 1971, and in Moscow November 16–December 2, 1971. NASA Deputy Administrator George M. Low and a delegation met with a Soviet delegation in Moscow April 4–6, 1972, to draw up a plan for docking a U.S. Apollo spacecraft with a Russian Soyuz in Earth *orbit* in 1975.

Final official approval came in Moscow on May 24, 1972. President Nixon and Soviet Premier Aleksey N. Kosygin signed the Agreement Concerning Cooperation in the Exploration and Use of Outer Space for Peaceful Purposes, including development of compatible spacecraft docking systems to improve the safety of human spaceflight and to make joint scientific experiments possible. The first flight to test the systems was to be in 1975, with modified Apollo and Soyuz spacecraft. Beyond this mission, future human spacecraft of the two nations would be able to dock with each other.

During the work that followed, engineers at the Manned Spacecraft Center (renamed *Johnson Space Center* in 1973) shortened the lengthy phrase "joint rendezvous and docking mission" to *rendock,* as a handy project name, but in June 1972 the NASA Project Designation Committee approved the official designation Apollo Soyuz Test Project (ASTP), incorporating the names of the U.S. and Soviet spacecraft. NASA and the Soviet Academy of Sciences announced the official ASTP emblem in March 1974. The circular emblem displayed the English word Apollo and the Russian word Soyuz on either side of a center globe with a superimposed silhouette of the docked spacecraft.

Apollo 18 and Soyuz 19 were launched on July 15, 1975, with a "meeting in space" on July 17. Soyuz 19 landed July 21, and Apollo 18 splashed down three days later. This, the first international *crewed* space mission, carried out experiments with astronauts and *cosmonaut*s working together, in addition to testing the new docking systems and procedures.

SOURCES. *The Post-Apollo Space Program: Directions for the Future,* Space Task Group report to the President, General Services Administration, National Archives and Records Service, Office of the Federal Register, Weekly Compilation of Presidential Documents 5 (September 22, 1969), p. 1291; *Public Papers of the Presidents of the United States: Richard Nixon*

(Washington, DC: GPO, 1971), pp. 250–53; Thomas O. Paine, "Man's Future in Space," 1972 Tizard Memorial Lecture, Westminster School, London, March 14, 1972; NASA News Release, "Text of US/USSR Space Agreement," May 24, 1972; NASA News Release 72-109; "Washington Roundup," *Aviation Week & Space Technology* 96 (May 15, 1972): 13; Richard D. Lyons, "Chief Astronaut Foresees Further Cuts in the Corps," *New York Times,* May 28, 1972, p. 1; John P. Donnelly, Assistant Administrator for Public Affairs, NASA, memorandum to Dale D. Myers, Associate Administrator for Space Flight, NASA, June 30, 1972; Apollo-Soyuz Test Project, Project Approval Document, Office of Manned Space Flight, NASA, December 19, 1972, and October 6, 1973. For more information about the Apollo program, see Ezell and Ezell, *The Partnership,* SP-4209, online at www.hq.nasa.gov/office/pao/History/SP-4209/cover.htm.

Aqua. EOS PM-1. A multinational NASA scientific research *satellite* in *orbit* around the Earth, studying the precipitation, evaporation, and cycling of water. Additional variables being measured by Aqua include radioactive energy fluxes, aerosols, vegetation cover on the land, phytoplankton and dissolved organic matter in the oceans, and air, land, and water temperatures. It is the second major component of the Earth Observing System *(EOS),* following on *Terra* (launched 1999) and followed by *Aura* (launched 2004).

ETYMOLOGY. Aqua (Latin for water) was named to be complementary to the satellite Terra (Latin for Earth). (http://aqua.nasa.gov/.)

Aquarius. *Call sign* for the *Apollo* 13 *Lunar Module,* the third lunar landing attempt. The *mission* was aborted after the rupture of one of the Service Module oxygen tanks, which damaged several of the power, electrical, and life support systems, leaving the module unable to supply sufficient air, water, and electricity to return the three crew members—James A. Lovell Jr., John L. Swigert Jr., and Fred W. Haise Jr.—to Earth. The Aquarius LM, a self-contained *spacecraft* unaffected by the accident, was used as a "lifeboat" to provide austere life support for the return trip. The mission was termed a "successful failure" because of the experience gained in rescuing crew.

ETYMOLOGY. According to Apollo 13 Commander Jim Lovell, "Contrary to popular belief, [the LM] was not named after the song in the play *Hair,* but after the Egyptian goddess Aquarius. She was symbolized as a water carrier who brought fertility and therefore life and knowledge to the Nile Valley, and we hoped our Lunar Module, Aquarius, would bring life back from the Moon." (Quoted in Lattimer, *All We Did Was Fly to the Moon,* p. 77.)

ARC. *Ames Research Center.*

Arcas. All-purpose Rocket for Collecting Atmospheric Soundings *(sounding rocket).* A small solid-*propellant sounding rocket* named in

1959 by its producer, Atlantic Research Corporation. It was no accident that the first three letters of the acronym were also the producer's initials. An inexpensive vehicle designed specifically for meteorological research, Arcas could carry an 11-pound (5-kg) *payload* to an altitude of 40 miles (64 km). Later versions were the Boosted Arcas, Boosted Arcas II, and Super Arcas, all of which NASA used.

Two other sounding rockets developed by Atlantic Research were used briefly by NASA. The Arcon was named by Atlantic Research, and the Iris was named by Eleanor Pressly of *Goddard Space Flight Center,* which managed the rockets.

SOURCES. Atlantic Research Corp., announcement released by U.S. Army Missile Support Agency, January 26, 1959; Peter T. Eaton, Office of Space Science and Applications, NASA, letter to Historical Staff, NASA, May 2, 1967; W. C. Roberts Jr. and R. C. Webster, Atlantic Research Corp., "Arcas Rocketsonde System Development," September 3, 1959; GSFC, *Encyclopedia: Satellites and Sounding Rockets,* p. 321.

ARES. Advanced Rover Engineering and Science Tool. The software scientists and engineers used to "drive" the Mars Exploration Rovers *(MERs)*.

Ares (pronounced air-eez or ah-rays). Name for the next generation of *launch vehicle*s that will return humans to the Moon, announced officially on June 30, 2006. It includes a crew launch vehicle, referred to as Ares I, and a cargo launch vehicle, which will be called Ares V.

Ares I will use a single five-segment solid *rocket booster,* a derivative of the *Space Shuttle*'s solid rocket booster, for the first stage. A liquid oxygen/liquid hydrogen J-2X *engine* derived from the J-2 engine used on *Apollo*'s second stage will power the *Crew Exploration Vehicle*'s second stage. The Ares I can lift more than 55,000 pounds (24,947 kg) to *low Earth orbit.* Ares V, a heavy lift launch vehicle, will use five RS-68 liquid oxygen/liquid hydrogen engines mounted below a larger version of the Space Shuttle's external tank, and two five-segment solid-*propellant rocket* boosters for the first stage. The *upper stage* will use the same J-2X engine as the Ares I. The Ares V can lift more than 286,000 pounds (129,727 kg) to low Earth orbit and stands approximately 360 feet (110 m) tall. This versatile system will be used to carry cargo and the components into orbit needed to go to the Moon and later to Mars. ("Ares: NASA's New Rockets Get Name," NASA News Feature, June 20, 2006, www.nasa.gov/mission_pages/exploration/spacecraft/ares_naming.html.)

ETYMOLOGY. Greek god of war. Ares was the son of Zeus and Hera. He was often accompanied in battle by his sister Eris, the goddess of discord, and by Hades, the lord of the dead. Deimos and Phobos were two of Ares' sons. "It's appropriate that we named these vehicles Ares, which is a pseudonym for Mars," said Scott Horowitz, Associate Administrator for NASA's Exploration Systems Mission Directorate, Washington, at the Cape Canaveral press conference announcing the naming. "We honor the past with the number designations and salute the future with a name that resonates with NASA's exploration mission." (The I and V designations pay homage to the Apollo program's *Saturn I* and *Saturn V* rockets, the first large U.S. *space vehicle*s conceived and developed specifically for human spaceflight.) According to the *Palm Beach Post,* Horowitz said that hundreds of names were rejected before Ares was chosen: "I couldn't possibly go through all the names. All the constellations in the sky, all the Greek and Roman gods, all their children, their cousins—I mean it went on and on and on" ("Ares New God of Space, NASA Officials Say," *Palm Beach Post,* July 1, 2006, p. A2). "Ares is a name that is used to refer to Mars, and it connects to our vision to go to the moon and on to Mars," Horowitz explained at the news conference ("New NASA Craft's Name Is Ares, Greek for Mars," *Deseret Morning News,* July 2, 2006, p. A17). ("Ares: NASA's New Rockets Get Name," NASA News Feature, June 20, 2006, www.nasa.gov/mission_pages/exploration/spacecraft/ares_naming.html.)

USAGE. "A-R-I-E-S is the constellation Aries," explained Horowitz at the time of the announcement, "A-R-E-S is the Roman name, the synonym for Mars." According to standard sources on mythology, the two were in fact not synonymous. "Ares," to quote the *World Book Encyclopedia* (vol. 1, 1989, p. 645), "represented the most brutal and violent aspects of War. The Greeks, who placed little value on these qualities, did not respect Ares highly." Ares is also not very smart. In the *Iliad,* Diogenes (a mortal) is able to wound Ares with an assist from the goddess Athena.

Argo *(sounding rocket).* The first *sounding rocket* in this series, developed by the Aerolab Company (later a division of Atlantic Research Corporation), was called Jason. Subsequent vehicles in the series were given names also beginning with the letter J: Argo D-4 and D-8 were named Javelin and Journeyman. The D-4 and D-8 designations referred to the number of stages (D for four) and to the design revision (fourth and eighth).

Javelin was designed to carry *payload*s of 88–154 pounds (40–70 kg) to altitudes of 497–621 miles (800–1,100 km). Journeyman could carry 44–154 pounds (20–70 kg) to 932–1,305 miles (1,500–2,100 km). The last Javelin was used by NASA in 1976, but Journeyman was discontinued in 1965. Javelin was also mated to the *Nike* first stage for heavier payloads. (Rosenthal, *Venture into Space,* SP-4301, pp. 127–29; Wallops Station News Release 71-12.)

ETYMOLOGY. From the ancient Greek myth of Jason's travels aboard the

ship Argo (named for its builder, Argus) in search of the Golden Fleece
(Bulfinch, *Mythology*, p. 108).

Ariane. Name for a series of European civilian *launch vehicles* developed
under the auspices of the *European Space Agency* (ESA). It was originally proposed by the French Space Agency after the failure of the
Europa launch vehicle, was approved in 1973, and first flew in 1979.
There have been five vehicles in the Ariane family.

ETYMOLOGY. Ariane is the French spelling of the name of the mythological character Ariadne, beautiful daughter of Minos and Pasiphae. She
fell in love with Theseus and gave him the thread with which he found his
way out of the Minotaur's labyrinth. It is also the name of a type of
hummingbird.

Arianespace. The commercial entity created in 1980 to handle the
production, operations, and marketing of the Ariane vehicles. This
organization paved the way for Europe's independent access to space.
It became the world leader in commercial launch services and introduced the concept of dual-payload launches for telecommunications
satellites.

Ariel (ionospheric research *satellite*). The world's first international
satellite, a cooperative project between the United Kingdom and
NASA. Other satellites followed in the program.

Ariel 1 (UK-1 before *orbit*), launched from *Cape Canaveral* on April
26, 1962, was built by NASA's *Goddard Space Flight Center* and instrumented with six British experiments to make integrated measurements
in the ionosphere. Ariel 2, containing three U.K.-built experiments, was
placed in orbit March 27, 1964. Ariel 3, designed and built in the
United Kingdom, was launched May 5, 1967, with five experiments.
The U.K.-built Ariel 4 carried four U.K. experiments and one U.S.
experiment into orbit December 11, 1971, to investigate plasma,
charged particles, and electromagnetic waves in the ionosphere.
Ariel 5 (UK-5), also British-built, was launched October 15, 1974, to
study x-ray sources. Ariel 6 (UK-6) was launched June 3, 1979, and
carried six experiments.

The UK-X4 satellite was in a different series from the Ariels. The
letter X added to the prelaunch designation indicated that it was
experimental, and when launched in March 1974 to test *spacecraft*
systems and sensors, it was christened Miranda. It was a U.K. satellite
launched by NASA under a contract for reimbursable services rather
than a joint research *mission*.

The United Kingdom's Skynet satellites belonged to still another
series. The Skynet I and II series of U.K. Ministry of Defence communications satellites were launched by NASA, beginning in 1969,

under agreement with the U.S. Air Force, which reimbursed NASA for *launch vehicles* and services. The Skynet satellites continued with a 4 series and a 5 series. The latest Skynet (Skynet 5A) was launched in March 2007. The last U.S.-launched Skynet was Skynet 4D in January 1998.

ETYMOLOGY. The Ariel satellite was named in February 1962 for the spirit of the air who was released by Prospero in Shakespeare's play *The Tempest*. The name, a traditional one in British aeronautics, was chosen by the U.K. Minister of Science and endorsed by NASA.

SOURCES. SP-4402, p. 35–36; D. J. Gerhard, Office of the Scientific Attaché, U.K. Scientific Mission, Washington, DC, letter to Arnold W. Frutkin, Director of International Programs, NASA, December 14, 1961; Frutkin, memorandum to Robert C. Seamans Jr., Associate Administrator, NASA, December 15, 1961; Boyd C. Myers II (Chairman, NASA Project Designation Committee), Director, Program Review and Resources Management, NASA, memorandum to Seamans, February 2, 1962, with approval signature of Dr. Seamans; NASA News Release 74-36; NASA program office; U.N. Document A/AC.105/INF.289, April 18, 1974.

Aries *(sounding rocket).* In 1974 NASA was working with the Naval Research Laboratory, Sandia Laboratories, and West Germany to develop a new sounding rocket, the Aries, using surplus second stages from the Department of Defense Minuteman intercontinental ballistic *missile.* The rocket, which had flown three test flights by December 1974, would lift larger payloads for longer flight times than other rockets in astronomy, physics, and space-processing research projects. Aries was flown by NASA until the early 1990s, and the military used Aries as a target launch vehicle for ABM missiles through the end of 2004.

The Aries would have greater volume for carrying experiment instruments than provided by the *Aerobee* 350 sounding rocket and would carry scientific *payloads* of 397–1,984 pounds (180–900 kg) to altitudes that would permit 11–7 minutes of viewing time above 57 miles (92 km), appreciably longer than the viewing time of the Aerobee 350 and the *Black Brant* VC; the first test flights had carried 1,801 pounds (817 kg) to 168 and 186 miles (270 and 299 km). It also was expected to give 11–8 minutes in weightless conditions for materials-processing experiment payloads of 99–1,000 pounds (45–454 kg).

ETYMOLOGY. When the project was first conceived, the new vehicle was called Fat Albert after the television cartoon character, because its short, fat shape contrasted with that of other rockets. The Naval Research Laboratory asked Robert D. Arritt of its Space Science Division to choose a more dignified name. Arritt and a group of his colleagues chose Aries. Tt was the name of a constellation (appropriate because the *rocket* would be used for astronomy projects), and it was "a name that was available." It also was Arritt's zodiac sign.

SOURCES. U.S. Congress, House of Representatives, Committee on Science and Astronautics, Subcommittee on Space Science and Applications, *Hearings: 1975 NASA Authorization, Pt. 3, February and March 1974* (Washington, DC: GPO, 1974), pp. 117, 456, 560–61.

arming tower. Steel structure used to load *propellant*s, liquid hydrogen and liquid oxygen. The term is borrowed from similar structures used to load ordnance into missiles. (SP-6001, p. 6.)

FIRST USE. An AP story tells of building an arming tower at the new NASA facility at *Cape Canaveral:* "Here explosives necessary for flight can be attached to the Saturn" ("Moon Rocket Plans Told by NASA," *Chicago Tribune,* July 22, 1962, p. 10).

Army Ballistic Missile Agency. See *ABMA.*

ARPA. *Advanced Research Projects Agency.*

artificial gravity. A simulation of gravity in *outer space* or free-fall. It is desirable for long-term space travel for ease of mobility and to avoid the adverse health effects of *weightlessness.*

artificial moon. The popular name for the early series of Russian and American Earth-circling satellites.

FIRST USE. The term was first applied in the late 1940s for nuclear-powered orbiting vehicles. A *New York Times* article takes the position that the only way to voyage into space "within our lifetime" is with the *fission* of uranium: "The present type of rocket cannot reach the Moon or any other planet because it cannot carry enough fuel" (W. K., "Artificial Moon May Circle Earth," *New York Times,* July 6, 1947, p. 81).

artificial satellite. Manufactured equipment that orbits around the Earth or the Moon. The term could also refer to a *spacecraft* orbiting another planet, but in general these are simply called spacecraft.

ETYMOLOGY / FIRST USE. "The astronauts even suggest that artificial satellites be created to revolve around the Earth and Venus at predetermined distances (Waldemar Kaempffert, "Rocket Ships and a Visit to Mars," *Forum and Century,* October 1930, p. 235). The term appears in its modern sense in a 1945 proposal from Arthur C. Clarke: "An 'artificial satellite' at the correct distance from the earth would make one revolution every 24-hours; i.e., it would be within optical range of nearly half the earth's surface. Three repeater stations, 120 degrees apart in the correct orbit, could give television and microwave coverage to the entire planet." (Clarke, "Peacetime Uses for V2," *Wireless World,* February 1945, p. 58.)

AS- . *Apollo-Saturn* (followed by *mission* number).

ASAT. See *anti-satellite*.

Asp *(sounding rocket)*. Acronym for Atmospheric Sounding Projectile. Designed to carry up to 79 pounds (36 kg) of *payload,* the solid-*propellant* sounding rocket Asp was developed by Cooper Development Corporation for the Navy's Bureau of Ships. The first prototype was launched December 27, 1955. NASA used Asp as an *upper stage* in the *Nike*-Asp briefly. It was test flown several times in 1960, but a need for the vehicle did not develop. (Robert B. Cox, "Asp," in *Sounding Rockets,* ed. Newell, p. 105.)

asteroid. Any of the thousands of small rocky objects that *orbit* the Sun, most of them between the orbits of Mars and Jupiter (although some pass closer to the Sun than Earth does and others have orbits that take them well beyond Jupiter). Asteroids are larger than *meteoroids*. The largest asteroid is named Ceres; it is about as wide as the state of Texas. See also *near-Earth objects*. (ASP Glossary.)

ASTP. *Apollo-Soyuz Test Project.*

astrionics. Electronics used in *astronautics.*

astro-. Ubiquitous *Space Age* prefix denoting outer space.

　　ETYMOLOGY. Latin *astrum* (star, constellation).

　　USAGE. If a monkey went aloft, it became an astromonk; if an *astronaut* picked up a camera, he became an astrophotographer. Although scores of astro-compounds were created, few survived. For instance, the term astrogation (for astral navigation, interstellar space travel) was created in 1961, existed for a moment, and then was never heard again.

Astrobee. See *Aerobee*.

astrobiology. The branch of biology concerned with the discovery or study of life beyond Earth. Also known as *exobiology* since the early 1960s. The term was first mentioned in a document in the NASA strategic plan for 1996, where it was defined as "the study of the living universe." (SP-6001, p. 7; Dick and Strick, *Living Universe.)*

astronaut. (1) A human who travels in space. (2) Beginning in 1959, any American who had been selected to fly in space. Edward Givens Jr., who was killed in a car accident in 1967, was identified as an astronaut even though he had never been in space. (3, obs.) One who advocates space exploration, as in, "The astronauts even suggest that artificial satellites be created to revolve around the Earth and Venus at pre-determined distances" (Waldemar Kaempffert, "Rocket Ships and a Visit to Mars," *Forum and Century,* October 1930, p. 235).

　　ETYMOLOGY. Astro- + naut (sailor). Literally, space sailor. The word first appeared in 1880 in *Across the Zodiac,* a science fiction novel by Percy Greg. The *OED* notes a number of uses prior to its adoption by NASA, the earliest nonfiction citation being a 1929 allusion to "the first obstacle encountered by the would-be 'Astronaut'" (*Journal of the British Astronomical Association,* June 1929, p. 331). The term was used liberally during the infancy of the jet plane (at that time commonly called the "rocket plane"), as in a *New York Times* article that opens with this sentence: "Evidently the astronauts who dreamed of kicking themselves from the earth to Mars were not mad" ("The Rocket Plane Is Here," *New York Times,* January 8, 1944, p. 12).

According to journalist James Schefter, the name was picked by Bob Gilruth (who would later head the *Mercury* program), who had first seen it in *Time* magazine and felt that it had a nice ring to it. Schefter wrote, "Dr. Hugh Dryden, deputy administrator of NASA wanted cosmonaut. 'Astro' referred to stars, and Dryden argued that they were just at the beginning, exploring the cosmos, not the stars. So they passed the question around and Dryden lost. 'Everyone we talked to seemed to prefer astronaut,' Gilruth wrote in his memoir, 'and this was the name that stuck. That is fortunate, I believe, because now when we say astronaut, we know we mean Americans, and when we say cosmonaut, we know we mean Russians.'" (Schefter, *The Race: The Uncensored Story of How America Beat Russia to the Moon* [New York: Doubleday, 1999]. The memoir Schefter quotes is Robert R. Gilruth, "From Wallops Island to Project Mercury," in *Essays on the History of Rocketry and Astronautics: Proceedings of the Third through the Sixth History Symposia of the International Academy of Astronautics,* ed. R. Cargill Hall, TM-X, vol. 2 [Washington, DC: NASA, 1977].)

On April 19, 1959, NASA announced the selection of the first seven U.S. space travelers, or astronauts. At the time, it was stated that the term followed the semantic tradition that began with Argonauts (the legendary Greeks who traveled far and wide in search of the Golden Fleece) and continued with Aeronauts (pioneers of balloon flight). The Soviets used the term *cosmonaut,* and the Chinese used *taikonaut.*

One early variant of the term was "astronette," for a "woman engaged in space-flight exploration," according to C. H. McLaughlin's *Space Age Dictionary* (1962) and several other early glossaries. Along with "astronautess" and "cosmonette," it was among the shortest-lived of the new terms. When American women finally did go into space, they were called astronauts, just as the first Soviet woman in space was called a cosmonaut.

One of the suggested names for the first U.S. piloted *space program* was Project Astronaut, but the name lost out in favor of Project *Mercury.*

FIRST USE. "The NASA said that the portion of the earth shown is what an astronaut pilot would see from an altitude of 120 miles above Cuba" ("Dummy Space Capsules Given Air-Ocean Tests," *Los Angeles Times,* March 27, 1959, p. 7).

USAGE. Scott Carpenter notes that Gus Grissom detested the word, "sensing the phony P.R. aspect, and announced one day to everyone's amusement, 'I'm not *ass* anything. I'm a *pilot*. Isn't that good enough, fer chrissake?'" (Carpenter, *For Spacious Skies,* p. 218). An article on Israeli astronaut Ilan Ramon stated that in Israel the term astronaut "has roughly been equivalent to 'space cadet' in American slang—that is someone out of touch with the real world" (*Los Angeles Times,* December 24, 2002, p. A2).

astronaut altitude. The distance aloft at which a human is deemed to be in space. The distance is approximately 62 miles (100 km). The term came to prominence with the announcement of the *Ansari X Prize* of $10 million to be awarded to the first private team that successfully sends a vehicle, a single pilot, and added ballast equivalent to two more passengers to this altitude and back twice in two weeks ("Ambitious Entrepreneurs Planning to Send Tourists into Astronaut Altitude," *New York Times,* February 17, 1998, p. A16).

astronautics. The art, skill, or activity of designing, building, and navigating *space vehicles.*

ETYMOLOGY / FIRST USE. Introduced in 1927 by R. Esnault-Pelterie, a French pioneer in the field of aircraft design, to denote the science of space travel, borrowed, by his own admission, from the work of novelist J. H. Rosny to describe the "art" of voyaging from star to star. Esnault-Pelterie used the term as the subject of an essay contest ("Astronautics," *New York Times* editorial, March 8, 1928, p. 24). It was put to quick use, but as a hypothetical concept: "Astronautics, a coined word meaning 'methods of navigating space,' is still among the unproved sciences, but because its stirring implications make the transatlantic flights of our boldest aviators seem puerile and inconsequential in comparison, it will also be a tempting field for the engineer, physicist and the dreamer" (Waldemar Kaempffert, "Rocket Ships and a Visit to Mars," *Forum and Century,* October 1930, p. 233).

Astronomical Netherlands Satellite. See *ANS*. Also known as Netherlands Astronomical Satellite.

astronomical unit. A unit of measurement equal to the average distance between the Earth and the Sun. Usually abbreviated A.U., an astronomical unit is equal to about 93 million miles (150 million km), a distance that light takes about 8 minutes to cover. It is a handy unit for expressing distances in the solar system. For example, the diameter of the *orbit* of the most distant planet, Pluto (designated a dwarf planet in 2006 by the International Astronomical Union), is about 80 A.U. (ASP Glossary.)

astronym (astro- + acronym). A short-lived term used to describe the proliferation of aerospace acronyms and initialisms in the early days of the *Space Age.* The letter A alone produced such marvelous constructs as AESOP (Artificial Earth Satellite Observation Program), ALARM (Alert,

Locate, and Report Missiles), ALERT (Attack and Launch Early Reporting to Theater), and ARISTOTLE (Annual Review and Information
Symposium on the Technology of Training and Learning, an Air Force
formulation). "Like all government agencies since 1950, NASA made
extensive use of acronyms. In February 1971, the Documents Department of the Kennedy Space Center Library compiled a selective list of
acronyms and abbreviations. It contained more than 9,500 entries."
(Benson and Faherty, *Moonport,* SP-4204.)

astrophysics. That part of the astronomy dealing with the physics of
astronomical objects and phenomena.

Atlantic. Series of Intelsat communications *satellite*s. See *Intelsat.*

Atlantic Missile Range (AMR). Formerly the USAF Long Range Proving
Ground, then the Eastern Test Range, and now the *Eastern Space and
Missile Center.* See *Kennedy Space Center.*

Atlantis. *Space Shuttle* orbiter *OV*-104. It was the fourth orbiter to
become operational.

ETYMOLOGY. Named after the primary research vessel for the Woods Hole
Oceanographic Institute in Massachusetts, 1930–66. The two-masted 460-
ton ketch was the first U.S. vessel to be used for oceanographic research.
Such research was considered to be one of the last bastions of the sailing
vessel as steam- and diesel-powered vessels came to dominate the
waterways. The steel-hulled ocean research ship was approximately 140
feet (43 m) long and 29 feet (9 m) wide to enhance her stability. She
featured a crew of 17 and room for 5 scientists. The research personnel
worked in two onboard laboratories, examining water samples and marine
life brought to the surface by two large winches from thousands of feet
below the surface. The water samples taken at different depths varied in
temperature, providing clues to the flow of ocean currents. The crew also
used the first electronic sounding devices to map the ocean floor.

Atlas *(launch vehicle).* The U.S. Air Force intercontinental ballistic *missile*
(ICBM) family, which played an important role as an early space launch
vehicle and continues in use today as the Atlas V version. It was an
Atlas *missile* that went into *orbit* at the end of 1958 carrying *Project
SCORE.* A modified Atlas took four *Mercury* astronauts into orbit. The
Atlas launch vehicle was an adaptation of the USAF Atlas ICMB. After
two *human* suborbital flights using the *Redstone* rocket, the modified
Atlas launched the four human orbital flights in Project Mercury, and
NASA used it with the *Agena* or *Centaur* upper stages for a variety of
uncrewed space missions.

ETYMOLOGY. Early in 1951 Karel J. Bossart, head of the design team at
Convair (Consolidated Vultee Aircraft Corporation), which was working on
the missile project for the Air Force, decided that the project (officially

listed as MX-1593) should have a popular name. He asked some of his staff for ideas. and they considered several possibilities before agreeing upon Atlas, Bossart's own suggestion. The missile they were designing would be the biggest and most powerful yet devised. Bossart recalled that Atlas was the mighty god of ancient Greek mythology who supported the world on his powerful shoulders. The appropriateness of the name was confirmed by the fact that the parent company of Convair was the Atlas Corporation. The suggestion was submitted to the Air Force and was approved by the Department of Defense Research and Development Board's Committee on Guided Missiles in August 1951.

SOURCES. SP-4402, p. 9; John L. Chapman, *Atlas: The Story of a Missile* (New York: Harper and Brothers, 1960, p. 62; R. T. Blair Jr., Convair Division, General Dynamics Corp., letter to Historical Staff, NASA, September 10, 1965; Robert F. Piper, Historical Office, Air Force Space Systems Division, letter to Historical Staff, NASA, August 31, 1965; Brig. Gen. D. N. Yates, Director of Research and Development, Office of the Deputy Chief of Staff for Development, Hq. USAF, memorandum to the Chairman, Committee on Guided Missiles, Research and Development Board, DOD, July 30, 1951; S. D. Cornell, Acting Executive Secretary, Committee on Guided Missiles, memorandum to Brig. Gen. D. N. Yates, August 6, 1951; Yates, memorandum to Commanding General, Air Research and Development Command, August 27, 1951; NASA News Release 75-19. For more information about Atlas, see Launius and Jenkins, eds., *To Reach the High Frontier*.

atmosphere. All the gases that surround a star, like our Sun, or a planet, like our Earth.

atmospheric braking. The action of *atmospheric drag* in slowing down an object that is approaching a planet or some other body with an *atmosphere*. This effect can be deliberately used, where enough atmosphere exists, to alter the *orbit* of a *spacecraft* or to decrease a vehicle's velocity prior to landing. See also *aerobraking*.

atmospheric drag. The resistance offered by the Earth's *atmosphere* to a body moving through it. Sometimes also applied to other planetary atmospheres.

Atmospheric Sounding Projectile *(sounding rocket)*. See *Asp.*

ATS. Applications Technology Satellite. The name referred to the *mission* to test technological experiments and techniques for new practical applications of *Earth satellites*. The name evolved through several transitions, beginning with the project's study phase. In 1962–63, at NASA's request, Hughes Aircraft Company conducted feasibility and preliminary design studies for an Advanced *Syncom* satellite. The concept was of a communications *spacecraft* in synchronous *orbit* with a new stabilization system and a multiple-access communication capability. Other names in use were Advanced Synchronous Orbit Satellite, Advanced Synchronous Satellite, and Advanced Synchronous Communications Satellite.

By March 1964 NASA had decided that Advanced Syncom should not only test communications technology but also support development of "meteorological sensing elements, measurements of the space environment in various orbits such as the synchronous orbit, and the conduct of experiments on general stabilization systems which apply not only to communications systems but to other systems." As the concept of the satellite changed, so did its name, becoming Advanced Technological Satellite. Hughes was selected to build five ATS spacecraft.

The name change to Applications Technology Satellite came in October 1964. Dr. Homer Newell and Dr. John F. Clark, Director of Space Sciences, had concluded that the adjectives "Advanced" and "Technological" were undesirable because they seemed to overlap with responsibilities of NASA's Office of Advanced Research and Technology. On October 2, Dr. Newell formally proposed, and Associate Administrator Robert C. Seamans Jr. approved, the change to Applications Technology Satellite, bringing "the name of the project more into line with its purpose, applications technology, while retaining the initials ATS by which it is commonly known."

Launched December 6, 1966, ATS 1 took the first high-quality U.S. photographs of the Earth from synchronous orbit, showing the changing cloud-cover patterns. In addition to weather data, the satellite relayed color television across the United States and voice signals from the ground to aircraft in flight. ATS 3, launched November 5, 1967, carried advanced communications, meteorology, and navigation experiments and made high-resolution color photographs of one complete side of the Earth. ATS 6 was launched May 30, 1974, to support public health and education experiments in the United States and India. It was the first communications satellite with the power to broadcast TV photos to small local receivers.

SOURCES. SP-4402, pp. 36–37; U.S. Congress, Senate, Committee on Aeronautical and Space Sciences, *Hearings: NASA Authorization for Fiscal Year 1964, Pt. 1, April 1963* (Washington, DC: GPO, 1963), pp. 8, 143, 433–34; Homer E. Newell, Associate Administrator for Space Science and Applications, NASA, in U.S. Congress, Senate, Committee on Aeronautical and Space Sciences, *Hearings: NASA Authorization for Fiscal Year 1965, Pt. 2, March 1964* (Washington, DC: GPO, 1964), p. 559; NASA News Release 64–50; Robert F. Garbarini, Director of Engineering, Office of Space Science and Applications, NASA, memorandum to Director, Communication and

Navigation Programs Division, NASA, September 11, 1964; Newell, memorandum to Robert C. Seamans Jr., Associate Administrator, NASA, October 2, 1964, with approval signature of Dr. Seamans.

ATV. Automated Transfer Vehicle. The *European Space Agency* ATV is an autonomous logistical resupply vehicle designed to dock to the *International Space Station* and provide the crew with dry cargo, atmospheric gas, water, and *propellant*. After the cargo is unloaded, the ATV is reloaded with trash and waste products, undocks, and is incinerated during *reentry*. (*Reference Guide to the International Space Station*, SP-2006-557.)

Augustine Report. Common name for the 1990 report of the Advisory Committee on the Future of the U.S. Space Program. In 1990 President George H. W. Bush chartered the committee under the leadership of Norman Augustine, Chief Executive Officer of Martin Marietta. On December 17, 1990, Augustine submitted his commission's report, delineating the chief objectives of NASA and recommending several key actions. All of these related to the need to create a balanced *space program*—one that included human spaceflight, robotic *probes*, *space science*, applications, and exploration—within a tightly constrained budget.

aunty. Colloquialism for *anti-satellite* or an anti-satellite device.

Aura. Earth Observing System *(EOS)* mission launched July 15, 2004. Aura studies the composition, chemistry, and dynamics of the Earth's atmosphere, complementing its sister EOS systems *Aqua* and *Terra*.

Aurora 7. *Mercury capsule* flown by M. Scott Carpenter, which made three Earth orbits on May 24, 1962. This was the first *mission* in which an *astronaut* ate food in space.

ETYMOLOGY. According to Carpenter, "I considered naming my capsule Rampart 7 after the Colorado mountain chain. That probably would have been a better name than Aurora. It's shorter and more positive. It would have come through the static better. But I liked Aurora 7. It has a celestial significance and it had a sentimental meaning to me because my address as a child back in Colorado was on the corner of Aurora and 7th Streets in Boulder." (Quoted in Lattimer, *All We Did Was Fly to the Moon*, p. 13.)

Aurorae. ESRO IA *satellite*. See *ESRO*.

Automated Transfer Vehicle. See *ATV*.

avionics (aviation + electronics). The application of electronics to systems and equipment used in aircraft and *spacecraft*.

ETYMOLOGY. "Avionics is a new word coined by *Aviation Week* as a simple and much needed term to describe generically all the applications of electricity to the field of aeronautics" (*Aviation Week & Space Technology*, October 28, 1949, p. 17). Philip Klass, longtime senior editor at *Aviation*

Week & Space Technology, is credited with having created the term (Obituary, "Philip Klass, 85, Dies: Aviation Journalist, UFO Debunker," *Washington Post,* August 11, 2005, p. B5).

AXAF. Advanced X-ray Astrophysics Facility. Former name of the *Chandra X-ray Observatory.*

Azur. German research *satellite.* A July 17, 1965, memorandum of understanding between NASA and the German Ministry for Scientific Research (BMwF) initiated a cooperative project that would put into *orbit* a German scientific satellite designed to investigate the innermost of the Earth's *radiation belts.* The agreement provided for the *launch* of the satellite after a successful series of *sounding rocket* tests to check out the proposed satellite instrumentation. NASA would provide the *Scout launch vehicle,* conduct launch operations, provide *tracking* and data acquisition, and train BMwF personnel. In June 1966 NASA designated the satellite GRS-A, for German Research Satellite A. GRS-A became GRS-A-1 when an agreement was reached to launch a second research satellite, *Aeros,* designated GRS-A-2. (SP-4402, pp. 38, 39; Memorandum of Understanding between the German Ministry for Scientific Research and the United States National Aeronautics and Space Administration, attachment to NASA News Release 69-146; NASA, "Project Approval Document," June 15, 1966; Charles F. Rice Jr., GSFC [former AZUR Project Coordinator at GSFC], telephone interview, June 2, 1971.)

ETYMOLOGY. BMwF chose the name Azur (German for sky blue) for the satellite in early 1968, and GRS-A was officially designated Azur by NASA after launch on November 8, 1969.

B

babble. The aggregate crosstalk from a large number of communications channels.

backout. (1, n.). A reversal of the *countdown* sequence because of a failure in the *rocket* or a *hold* of unacceptable duration. (2, v.) To undo things already done during a countdown, usually in reverse order. (SP-7.)

backup. (1) A substitute *rocket* used to save time in an emergency. (2) Substitute crew who train alongside prime crew and are available to replace prime crew in the event of illness or accident.

ETYMOLOGY. The use of the term in the sense of a substitute term dates to the early 1950s. It achieved wide application when the Army *satellite* later to be called *Explorer* was the backup for *Vanguard*.

Bag Restraint Assembly. See *BRA*.

Baikonur Cosmodrome. The world's oldest and largest working space *launch* facility. It was originally built by the Soviets and is now under Russian regulation, although located in Kazakhstan. It was the *launch complex* where *Sputnik* 1, Earth's first *artificial satellite*, was launched. The *rocket* that lifted Yuri Gagarin (the first human in space) into *orbit* was also launched from Baikonur. In fact, all Russian *crewed* missions are launched from Baikonur, as well as all *geostationary*, lunar, planetary, and ocean surveillance missions. Baikonur is the only Russian site that has launched satellites into retrograde orbits.

USAGE. The name Baikonur is misleading. The former Soviet Union used the name and coordinates of a small mining town, Baikonur, to describe the location of its rocket complex. In fact, the complex is about 200 miles (322 km) southwest of the mining town. This misrepresentation was done intentionally to hide the actual location of the launch complex. Although the true location is now known, the launch complex is still referred to as Baikonur. (http://liftoff.msfc.nasa.gov/rsa/pads.html.)

ballistics. The science that deals with the motion, behavior, appearance, or modification of *missile*s acted upon by *propellant*s, rifling, wind, gravity, temperature, or other modifying conditions of force.

ballute (balloon + parachute). A parachute braking device used to land certain *spacecraft*.

barbecue mode. Popular name for what is officially termed passive thermal control. It is the process during which a *spacecraft* is put into a slow, gentle rotation so that solar heating and cooling are relatively uniform. The term came into play during the *Apollo* program but has been used in the context of other programs including the *Space Shuttle*.

ETYMOLOGY. From the rotating spit used to cook meat, especially the electric spit of the home barbecue unit.

FIRST USE. The term first came to prominence during the Apollo 8 *mission*. From the spacecraft: "Jim [Lovell] is going to take a shot of us from the lower equipment bay, and then we have to get back to our passive thermal control in the barbecue mode so that we don't get one side of the spacecraft too hot for too long a time" (transcript of December 22, 1968, transmission in "Excerpts from Conversation with the Spacemen," *New York Times,* December 23, 1968, p. 26). By the time of Apollo 13, the term is appearing in the NASA Press Kit: "During coast periods between mid-course corrections, the spacecraft will be in the passive thermal control (PTC) or 'barbecue' mode in which the spacecraft will rotate slowly about its roll axis to stabilize spacecraft thermal response to the continuous solar exposure" (Apollo 13 Press Kit, April 2, 1970, Release 70-50).

USAGE. The term has headline appeal: "'Barbecue Mode' Keeps Apollo Spacecraft Warm" (*Washington Post,* July 18, 1969, p. A13); "Shuttle Slips into Last 'Barbecue Mode'" (*Chicago Tribune,* July 4, 1982, p. 8).

barber chair. Adjustable seat that can go from upright to supine, in the manner of a barbershop chair, to increase tolerance for high acceleration. ("Space Age Slang," *Time* magazine, August 10, 1962, p. 12.)

base. A permanently occupied center for people on the Moon, on Mars, or in space that provides life support and work facilities. Bases would evolve from *outposts*. See *settlement*. (Paine Report, p. 198.)

Baykonur Cosmodrome. See *Baikonur Cosmodrome.*

Beagle 2. Unsuccessful *Mars Express lander* (the Mars Express *orbiter* itself was a success). Named after the ship in which Charles Darwin sailed in 1831 to explore uncharted areas of the Earth. Beagle 2's aim was to settle the question of whether life exists, or ever existed, on Mars.

beast. Early *Space Age* colloquialism for a large *rocket*. See *Webb's Giant*. (SP-6001, p. 9; "Space Age Slang," *Time* magazine, August 10, 1962, p. 12.)

Beidou. Chinese *satellite* navigation system.

bent-pipe communication. Use of relay stations to achieve non-line-of-sight transmission links.

Big Bang. The primeval explosion that most astronomers think gave rise to the *universe* as we see it today, in which clusters of galaxies are moving apart from one another. By "running the film backward"— projecting the galaxies' motion backward in time—astronomers calculate that the Big Bang happened about 13.7 billion years ago. (ASP Glossary.)

Big Crunch. Hypothesis that the *universe* will stop expanding and start to collapse upon itself. A counterpart to the *Big Bang*.

big dumb booster. Name given in the 1960s to an unbuilt *rocket* that could serve as America's workhorse cargo carrier in the post-*Apollo* era. Defined by journalist Victor Cohn in the context of the *Space Shuttle*: "It might be launched by what spacemen call a 'big dumb booster,' a simpler, cheaper kind of rocket than today's complex Saturn" (*Washington Post,* July 13, 1969, p. 108).

ETYMOLOGY. "It derives its name from the fact that it is a very simple rocket, based partly on solid-fuel technology, which lacks the sophisticated plumbing and electronics of the Saturn series" ("Moon-Race Activities Resume Next Month," *Sunday Star,* September 15, 1968, p. C2).

Big Joe *(launch vehicle)*. The name of a single *Atlas booster* and its test flight. Part of Project *Mercury*, Big Joe tested a full-scale Mercury *capsule* at full operational speed for the critical *reentry* into the Earth's *atmosphere.* It was a key test of the *heat shield,* in preparation for Mercury's *crewed* orbital space flights. The name, conferred in 1958, was attributed to Maxime A. Faget, then at *Langley Research Center.* The project was a logical progression from the previously named Little Joe, a smaller test booster for demonstration flight tests in Project Mercury. See *Little Joe.* (SP-4402, p. 10; Swenson, Grimwood, and Alexander, *This New Ocean,* SP-4201, p. 125; Paul E. Purser, MSC, handwritten note to James M. Grimwood, Historian, MSC, October 1963.)

bioastronautics. *Astronautics* considered for its effect on plant and animal life; the study of the biological, behavioral, and medical problems pertaining to astronautics. This includes systems functioning in the environments expected to be found in space, vehicles designed to travel in space, and conditions on celestial bodies other than Earth. (Pitts, *The Human Factor,* SP-4213, p. 255; SP-6001, p. 10.)

ETYMOLOGY / FIRST USE. Term created by the Air Force Systems Command in consolidation of all "bioastronautics research and development" under a single command, a move announced on September 30, 1961. "The consolidation included aerospace medicine, life sciences, biosciences, biomedicine, human factors, behavioral sciences, space medicine, biotechnology, human engineering, human resources, aviation medicine, and space biology" (*New York Times,* October 1, 1961, p. 23).

bionics. The study of systems, particularly electronic systems, that function after the manner of, or in a manner characteristic of, or resembling, living systems.

biosatellite (biological +satellite). Generic term for an artificial *satellite* specifically designed to contain and support humans, animals, or other living organisms in a reasonably normal manner, and that, particularly for humans and animals, possesses the means for safe return to Earth. (Pitts, *The Human Factor,* SP-4213, p. 256.)

Biosatellite. Series of *satellites* launched during the 1960s to conduct space experiments with living organisms, both plant and animal, in order to study the physiological effects of exposure to prolonged *weightlessness, radiation,* and other conditions of the space environment.

In June 1962, the Director of Bioscience Programs at NASA asked the Project Designation Committee to consider an official name for such a project should it be initiated. The committee devised the name Biosatellite, a contraction of the phrase biological satellite. The shorter form Bios formed the basis for the new name and occasionally appeared as a synonym. But the Biosatellite program should not be confused with BIOS (Biological Investigation of Space), the name of a separate project undertaken in 1961.

In March 1963 NASA contracted for *spacecraft* feasibility studies for a "bio-satellite program." After evaluating the results of these studies and obtaining funding for the project, NASA selected the General Electric Company to build the spacecraft and later chose the biological experiments to be flown on them. By early 1964 the project was well under way, and the name Biosatellite had been adopted.

Biosatellite 1 was launched December 14, 1966. It functioned normally in orbital spaceflight but failed to reenter as it should have three days later. Biosatellite 2, launched September 7, 1967, obtained information on the effects of radiation and weightlessness on plant and low-order animal life forms. The program ended with the flight of Biosatellite 3, launched June 28, 1969, which was prematurely terminated after eight and a half days. Analysis of the death of the pigtailed monkey that was carried aboard that flight provided additional information on the effects of prolonged weightlessness during *crewed* flights.

SOURCES. SP-4402, p. 38–39; NASA News Release 66-312; O. E. Reynolds, Director of Bioscience Programs, NASA, memorandum to Harold L. Goodwin (Member, NASA Project Designation Committee), Director, Office of Program Development, NASA, June 4, 1962; Jack Posner, Office of Space Science and Applications, NASA, telephone interview, August 10, 1965; NASA Project Designation Committee, minutes of meeting January 9, 1963.

biosphere. (1) The total environment of the Earth that supports self-sustaining and self-regulating human, plant, and animal life. (2) An artificial closed-ecology system in which biological systems provide mutual support and recycling of air, water, and food. (Paine Report, p. 198.)

biotelemetry. The remote measuring and evaluation of life functions, as in *spacecraft* and *artificial satellites*.

bird. Colloquial term for a *rocket, satellite,* or *spacecraft.* Any inanimate flying object. (SP-6001, p. 10.)

birdwatcher. Name given to those who lined the beaches around *Cape Canaveral / Cape Kennedy* to watch space launches. Sometimes also applied to the press.

black box. Any electronic component or piece of equipment such as an amplifier or radio that can be mounted or removed as a single package, such as a *satellite* experiment (SP-6001, p. 10).
ETYMOLOGY. Originally applied to, among other things, the generic early-20th-century box camera.

FIRST USE. The term is first applied to electronics in conjunction with strategic missiles and is used in expressions like "black box brain" or "black box intelligence." For example: "Its automatic celestial navigation, an intricate electronics wizardry developed by Northrop, replaces the human navigator, his sextant, charts and astronomical tables with a black box brain for star guidance" ("Strategic Missiles Being Built Here," *Los Angeles Times,* April 17, 1955, p. 18). Early association with NASA dates to the *Vanguard* program, when a mock black box was kept on hand during the early launches to simulate blowing up the control center and the range safety officer if anything went wrong with the launch. This story came out when the first Vanguard satellite went into Earth *orbit* and Milton Bracker of the *New York Times* told of the humor employed by the Vanguard crew during early failures ("A Rocket of Vanguard 1 Also Lifts Load from Crew's Shoulders," *New York Times,* March 18, 1958, p. 14).

black box man. Generic name for engineers with experience in digital component design, circuit design, and other specialties required to design new electronic components. Help- wanted ads beginning in the late 1950s were sometimes headed "Black Box Men" (ad for a Santa Ana company, *Los Angeles Times*, January 5, 1958, p. H13).

Black Brant *(sounding rocket).* A series of sounding rockets developed by Bristol Aerospace Ltd. of Canada with the Canadian government. The first rocket was launched in 1959. By the end of 1974 close to 300 Black Brants had been launched, and vehicles were in inventories of research agencies in Canada, Europe, and the United States, including the Navy, the Air Force, and NASA.

NASA took Black Brants into its sounding rocket inventory in 1970 and was using the Black Brant IVA and VC in 1974. The Black Brant IVA used a modified *upper stage* and a more powerful *engine* than previous models, to boost it to 560 miles (900 km). The Black Brant V series consisted of three 17-inch (43-cm) diameter sounding rockets with all components interchangeable.

The Black Brant VA (or BBVA) used stabilizer components with the Black Brant II's engine and carried 300-pound (136-kg) *payload*s to 100 miles (160 km), to fill a need for that altitude range. The BBVB, using an engine giving rocket performance double that of the Black Brant II, was designed to meet requirements for scientific investigations above 200 miles (320 km) altitude.

The Black Brant VC was used by NASA to support the 1973–74 *Skylab* Orbital Workshop missions by evaluating and calibrating Workshop instruments. The three-fin solid-fueled Black Brant VB was converted to a four-fin model suitable for launching from White Sands Missile Range and permitting *recovery* of the rocket payloads. The changes decreased performance somewhat but increased stability and allowed greater variations in payload length and weight on the VC. NASA launched the Black Brant VC on two flights during each of the three *crewed* missions to the Skylab Workshop.

NASA still uses Black Brant VB and VC, as well as Black Brant IX–XII.

ETYMOLOGY. The Canadian Armament Research and Development Establishment (CARDE) selected the name Black Brant, a small, dark, fast-flying goose common to the northwest coast and Arctic regions of Canada. The Canadian government kept the name with the addition of numbers (currently up to XII) for different members of the series, rather than giving a code name to each version, to emphasize that they were sounding rockets rather than weapons.

SOURCES. SP-4402, pp. 128–29; A. W. Fia, Vice President, Rocket and Space Division, Bristol Aerospace Ltd., "Canadian Sounding Rockets: Their History and Future Prospects," *Canadian Aeronautics and Space Journal* 20, no. 8 (October 1974): 396–406.

black hole. An invisible object in *outer space* formed when a massive star collapses from its own gravity. A black hole has such a strong pull of gravity that within a certain distance of it nothing can escape, not even light. Black holes are thought to result from the collapse of certain very massive stars at the end of their lives, but other kinds have been postulated as well: "mini black holes," for example, which might have been formed in turbulence shortly after the *Big Bang*; and "super massive black holes" (with masses millions of times the Sun's), which may exist in the cores of large galaxies, including our own Milky Way galaxy. (ASP Glossary.)

ETYMOLOGY. The term was coined by American scientist John Wheeler in 1969.

blackout. (1) Physiological: a temporary loss of vision and/or consciousness when a person is subjected to high accelerations. (2) Radio: a temporary loss of radio and *telemetry* contact between a *spacecraft* reentering the *atmosphere* and ground stations that occurs because of an ionized sheath of plasma that develops around the vehicle. Also known as fadeout. (SP-6001, p. 10.)

blast-off/blastoff. Officially recognized primarily as a colloquial term for *launch* or *takeoff* (SP-6001, p. 10).

ETYMOLOGY. Defined in *Travelers of Space,* ed. Martin Greenberg, as "the initial expenditure of energy by a spaceship leaving a planet, or in emergency takeoffs" (New York: Gnome, 1951, p. 20). Prior uses of the term as a slang verb meaning "to depart; take one's leave" are noted in Mary Reifer, *Dictionary of New Words* (New York: Philosophical Library, 1955). All the early citations of the term come from science fiction writers, among them Arthur C. Clarke, who applied it as science fact, as in this line: "The first rocketeers, crushed in their acceleration couches under the strain of blast-off" (*Holiday* magazine, November 1953, p. 78). There are a number of science fiction uses beginning in 1952: "The Sol-ship blasted-off during the night" (*Science Fiction*, June 1952, p. 121); "I am always nervous at blast off" (*Boy's Life*, September 1952, p. 30). The term was quickly adopted by the space shows for kids such as *Captain Video* as a term of departure. (To "hit space" was to leave in a hurry in this jargon, and to be confused was to be "riding the wrong orbit." Contained in NASA Names Files, record no. 15137. This file contains Tom Corbett comic strips, ads for space shows for kids, and an unattributed "Galaxy Glossary" that included the language of such early television shows as *Rod Brown of the Rocket Rangers*.)

FIRST USE. The first public use of the term by an American associated with the *space program* occurred on the eve of the ill-fated *Vanguard launch* of December 6, 1957, when a high-level Naval Research Laboratory official was quoted as saying, "There were a few bugs in the rocket, but I think we've ironed them out. It looks like we will be ready for blast off tomorrow." (*Springfield-Union*, December 4, 1957, p. 1.)

USAGE. The term is frowned on by some space purists, who consider it spurious and prefer *liftoff*. An irate letter to *Spaceflight* magazine in 1982 offered many alternatives ("rise-off," "committed to launch," etc.) but insisted, "never for God's sake blast off," which was "like describing a birth by saying 'The baby popped out'" (letter from Deane Davis, *Spaceflight*, September–October, 1982, p. 384). A 1966 letter to *Science* magazine termed it "spurious, having been coined by newspapermen (along with A-OK and spin off)" (letter from David McNeill, *Science*, May 13, 1966, p. 878). That letter occasioned a reply from William Hines, a reporter who covered space for the *Washington Star*: "Blast off seems to have been carried over into science (fact) writing from science (fiction) writing, rather than having been invented *de novo* by newsmen. Whether blast off is 'spurious,' as McNeill suggests, is a matter of opinion; it is as precise in meaning as the engineer-approved word liftoff and a good deal more descriptive" (*Science*, August 12, 1966, p. 694). *New York Times* word maven William Safire has considered the controversy and confesses that he gets "more of a lift out of blastoff" (*New York Times*, December 24, 1989).

blockhouse. A heavily reinforced protective shelter near a *launch pad*. During the *Apollo* program the term referred to the Launch Control Center (LCC). (SP-6001, p. 11.) In earlier times it denoted "a reinforced concrete structure, often built underground or half underground, and sometimes dome shaped, to provide protection against blast, heat, or explosion during rocket launchings or related activities; specifically, such a structure at a launch site that houses electronic control instruments used in launching a rocket." (SP-7.)

blowoff. The term used to describe the *separation* of a stage of a *rocket* or part of a *missile* by explosive force (SP-6001, p. 11).

Blue Scout *(launch vehicle)*. The version of the *Scout* that was used by the U.S. Air Force to *launch* Department of Defense spacecraft.

boattail. The rear portion of an elongated body, such as a *rocket*, having decreasing cross-sectional area toward the rear (SP-7; SP-6001, p. 11).

boattail section. The rear portion of a body having decreasing cross-sectional area toward the rear, applied especially to *boosters* and *missiles* (NASA, *Glossary/Congressional Budget Submission*).

boilerplate. A replica of the flight model (e.g. of a *spacecraft*), but usually heavier and cruder, built for test purposes (SP-6001, p. 11).

ETYMOLOGY. An interesting transfer of an old newspaper term to a new realm. Its first meaning was syndicated material supplied to newspapers in plate form.

FIRST USE. "The overall contract [for Apollo spacecraft with North American Aviation] is now extended to February 15, 1966. [North American] will provide NASA's Manned Spacecraft Center at Houston with 14 spacecrafts, 16 boilerplate spacecrafts, 10 full-scale mockups, five engineering simulators and evaluators and two mission simulators." (*Los Angeles Times,* September 5, 1964, p. 10.)

booster. Short for booster *engine* or booster *rocket*. A first-stage rocket used to give an initial propulsion to another rocket or rockets. The term was created to distinguish a *spacecraft* from the rocket that is used to send (or boost) it aloft.

FIRST USE. First applied in the popular press immediately after World War II in describing the propulsion system for the Corporal guided missile: "The missile leaps skyward along a 100-foot steel-railed tower, impelled by a 'Tiny Tim' booster rocket similar to those which dropped during the war from Navy bomber planes" ("Caltech Lifts Secrecy on Five Years of Experiments with Jet Propulsion," *Los Angeles Times,* June 23, 1946, p. A1).

booster engine. An *engine,* especially a *booster rocket,* that adds to the *thrust* of the sustainer engine; a *launch vehicle.*

bootstrap. Adjective describing a self-generating or self-sustaining process. When the operation of a *rocket engine* is no longer dependent on outside power, it is said to be in bootstrap operation. (SP-6001, p. 11.)

ETYMOLOGY. From the notion of independence conveyed by the phrase, "by one's own bootstraps."

bootstrapping. The process that enables any device, such as a turbo pump, to feed back part of its output so as to create more energy and thus function independently.

Boreas. ESRO IB *satellite.* See *ESRO.*

bow shock. A sharp front formed in the *solar wind* ahead of the *magnetosphere,* marked by a sudden slowing down of the flow near Earth. It is quite similar to the shock forming ahead of the wing of a supersonic airplane.

BRA. Bag Restraint Assembly. A two-compartment mesh bag used by *Apollo* astronauts to keep their helmets out of the way during in-cabin operations.

ETYMOLOGY. "'Bra' is probably an after-the-fact, made-up acronym. The name undoubtedly comes from its resemblance to a brassiere" (*ALSJ* Glossary).

breadboard. Experimental setup of an electronic system or circuit, frequently wired on a board, used to test its feasibility (NASA, *Glossary/Congressional Budget Submission,* p. 5).

Brilliant Eyes. Generic name for a family of small, space-based *missile* early-warning satellites.

Brilliant Pebbles. Generic name for a small, space-based *missile* interceptor with a nonexplosive warhead.

brown dwarf. An object in space that is too small to be a star but too large to be a planet.

BSLSS. *Buddy Secondary Life-Support System.*

bubble colony. Colony of persons placed on the Moon or other spatial body provided with individual or group environmental *capsule*s (SP-6001, p. 12).

Buddy Secondary Life-Support System (BSLSS). Set of hoses and connectors that allowed the *Apollo* astronauts to share cooling water in the event that one of the Portable Life Support Systems failed (*ALSJ* Glossary).

bug. Early nickname for the *Lunar Excursion Module* (*LEM*) before it was renamed the *Lunar Module* (*LM*). The term is used in early NASA budget discussions to describe the module that would land on the Moon (NASA, *Glossary/Congressional Budget Submission,* p. 23).

Bumblebee. USN *rocket* project. See *Aerobee/Astrobee.*

Bumper. Early rocket program inaugurated in 1947, which used the V-2 rocket (see *V-1, V-2*) together with the *Wac Corporal.* The goals were to refine operations of two-stage rockets and reach record altitudes. Early Bumpers were launched from White Sands, New Mexico; numerous others were launched later from *Cape Canaveral,* Florida.

Buran (Russian for snowstorm or blizzard). Soviet reusable *spacecraft* program that began in 1976 as a response to the U.S. *Space Shuttle* program. Soviet politicians were convinced that the U.S. Shuttle could be used for military purposes, posing a potential threat to the balance of power during the Cold War. The Buran project is thought to be the largest and the most expensive in the history of Soviet space exploration. It was canceled after one test flight in November 1988.

burn (n.). Combustion and consumption of *propellant* by a *rocket*. Most propulsion in space is achieved through a sequence of burns.

burnout. An act or instance of fuel or oxidizer depletion or, ideally, the simultaneous depletion of both; the point at which a *rocket* shuts down.

FIRST USE. First employed to describe the early three-stage *Atlas missile*. The *Los Angeles Times* of April 17, 1955, p. 18, contains a diagram with three stages of burnout labeled as such: first burnout, second burnout, and final burnout.

EXTENDED USE. The use of this term to describe debilitating emotional exhaustion can traced to Herbert J. Freudenberger, a psychologist, who debuted it in his 1974 book, *Burnout: The High Cost of High Achievement*.

burnout velocity. The speed of a *rocket* after its *engine* has stopped.

Cajun *(sounding rocket upper stage)*. A solid-*propellant rocket* stage designed and developed under the Pilotless Aircraft Research Division of the *NACA*'s Langley Aeronautical Laboratory (later NASA's *Langley Research Center*).

Design of the Cajun motor was based on the Deacon motor, begun during World War II by Allegany Ballistics Laboratory for the National Defense Research Council. NACA purchased Deacon propellant grains from Allegany to propel its aerodynamic research models. Deacon was used with the *Nike* first stage. In 1956 Langley contracted with Thiokol to develop the improved Deacon, named Cajun.

The *Nike*-Cajun, lifting 77-pound (35-kg) instrumented payloads to a 100-mile (160-km) altitude, was one of NASA's most frequently used sounding rockets prior to the 1980s.

ETYMOLOGY. The project's manager, Joseph G. Thibodaux Jr., formerly of Louisiana, suggested the new motor be called Cajun because of the term's association with southern Louisiana French culture (developed from the blending of Acadian settlers from Nova Scotia in the late 1700s with other immigrants coming from France, Haiti, Spain, Britain, and Germany in the late 1800s). Allen E. Williams, Director of Engineering in Thiokol Chemical Corporation's Elkton (Md.) Division, agreed to the name, and later the

Elkton Division decided to continue giving its rocket motors ethnic, regional, and Indian-related names.

SOURCES. SP-4402, pp. 130–31; William J. O'Sullivan Jr., "Deacon and Cajun," chap. 6 in *Sounding Rockets,* ed. Newell, pp. 96–97, 100–101; Peter T. Eaton, Office of Space Science and Applications, NASA, letter to Historical Staff, NASA, May 2, 1967; Rosenthal, *Venture into Space,* SP-4301, pp. 127–29; GSFC, *United States Sounding Rocket Program,* p. 38.

CALIPSO. Cloud-Aerosol Lidar and Infrared Pathfinder Satellite Observations. Satellite launched in 2006 to provide the next generation of climate observations, drastically improving our ability to predict climate change and to study the air we breathe. CALIPSO is a collaboration between NASA and the French *space agency CNES* (Centre National d'Études Spatiales*)*. Langley is leading the CALIPSO *mission* and providing overall project management, systems engineering, and *payload* mission operations. NASA's *Goddard Space Flight Center* provides support for system engineering and project and program management. (NASA News Release 06–190, "NASA Launches Satellites for Weather, Climate, Air-Quality Studies," April 26, 2006.)

call sign. Name used to identify either end of a radio communication. In the famous transmission of July 20, 1969, "Houston, Tranquility base here. The Eagle has landed." Houston is the *call sign* for the *Johnson Space Center,* and *Eagle* the call sign for the *Lunar Module* operating from the surface of the Moon.

CAMEO. See *Project CAMEO.*

Canadian Space Agency (CSA). As a federal organization established in 1989, the Canadian Space Agency contributes to the development of the Canadian civil *space industry.*

Canary Bird. Nickname for Intelsat II-C (Atlantic 2) communications *satellite.* See *Intelsat.*

the Cape. *Cape Canaveral.*

Cape Canaveral. Island tip on Florida's east coast with NASA's primary *rocket launch* site, the *Kennedy Space Center.* Defense units began using Cape Canaveral for *missile* testing after World War II, when rockets became more powerful and missiles developed wide ranges that made sites such as White Sands in New Mexico too dangerous. Cape Canaveral Missile Test Annex (now Cape Canaveral Air Force Station) must be differentiated from the Kennedy Space Center, and from the Cape Canaveral of the surrounding township.

Cape Kennedy. The name given to *Cape Canaveral* after the death of President John F. Kennedy. The name was in place from 1963 into 1973.

According to Benson and Faherty, *Moonport* (SP-4204), "On November 16, 1963, President Kennedy made a whirlwind visit to

Canaveral and Merritt Island, his third visit in 21 months. Administrator Webb, Dr. Debus, and General Davis greeted the President as his Boeing 707 landed. At *Launch Complex* 37 he was briefed on the Saturn program." The next week the President died by an assassin's bullet in Dallas. The new president, Lyndon B. Johnson, announced that he was renaming the Cape Canaveral Auxiliary Air Force Base and NASA Launch Operations Center the John F. *Kennedy Space Center.* With the support of Governor Farris Bryant of Florida, the President also changed the name of Cape Canaveral to Cape Kennedy. The next day he followed up his statement with Executive Order no. 11129. In this he did not mention a new name for the *Cape,* but did join the civilian and military installations under one name, thus causing some confusion. To clarify the matter, Administrator Webb issued a NASA directive changing the name of the Launch Operations Center to the John F. Kennedy Space Center, NASA, and an Air Force general order changed the name of the air base to the Cape Kennedy Air Force Station. The United States Board of Geographic Names of the Department of the Interior officially accepted the name Cape Kennedy for Cape Canaveral the following year.

Space employees seemed to approve the naming of the *spaceport* as a memorial to President Kennedy. Up to that time the Launch Operations Center had had only the descriptive name. Debus wrote a little later, "The renaming of our facilities to the John F. Kennedy Space Center, NASA, is the result of an Executive Order, but to me it is also fitting recognition to [President Kennedy's] personal and intense involvement in the National space program." Many in the Brevard area, however, felt that changing the name of Cape Canaveral—one of the oldest place-names in the country, dating back to the earliest days of Spanish exploration—was a misguided gesture. After a stirring debate in the town council, the city of Cape Canaveral declined to change its name.

Efforts to have Congress restore the name Canaveral to the Cape failed, but Governor Reuben Askew signed a bill on May 29, 1973, returning the name on Florida state maps and documents. On October 9, 1973, the Board of Geographic Names did likewise for federal usage.

capsule. (1) A small, sealed, pressurized cabin with an internal environment that will support life in space. (2) A fancy term for the cockpit in high-speed aircraft and *spaceships.*

FIRST USE. Applied to *Sputnik* 2 and the housing for the dog *Laika.* Capsules are often employed as an escape mechanism. Early glossaries treated it with an element of disdain. ("Glossary of Space Talk," *Popular Mechanics,* March 1959, p. 72.)

Carl Sagan Memorial Station. Name bestowed on the *Mars Pathfinder lander* in tribute to the astronomer Carl Sagan (1934–96), who did so much to advance the cause of Mars exploration both professionally and in the popular arena.

Casper. *Call sign* for the *Apollo* 16 Command Module for the *mission* of April 16–27, 1972, commanded by Thomas K. Mattingly II while John W. Young and Charles M. Duke Jr. explored the lunar surface with the *Lunar Roving Vehicle* and gathered 213 pounds (95.8 kg) of lunar material.

ETYMOLOGY. Mattingly said, "You may recall that the Teflon flight clothing was white and shapeless. I overheard some kids who were watching one of the onboard T.V. shots of an earlier crew, remark that they looked like Casper the friendly ghost. He was a comic strip character that had been around for a number of years. I liked the idea of something that was not so serious and which kids could identify with." (Quoted in Lattimer, *All We Did Was Fly to the Moon,* p. 90.)

Cassini. Orbiter portion of the NASA–*European Space Agency Cassini-Huygens mission* to Saturn, 1997. Named after astronomer Gian Domenico Cassini (1625–1712), born in Italy, later a naturalized French citizen. He discovered four of Saturn's satellites and observed a dark division in the planet's ring (the Cassini Division).

Cassini-Huygens. An uncrewed space *mission* to study Saturn and its moons undertaken jointly by NASA, the *European Space Agency,* and the space communications company Alenia Spazio (Italy). The *spacecraft* consists of two main elements: the Cassini *orbiter* and the Huygens *probe.* It was launched on October 15, 1997, and entered Saturn's *orbit* on July 1, 2004. The Cassini spacecraft is the first to explore Saturn's system of rings and moons from orbit. Cassini immediately began sending back intriguing images and data. The Huygens probe dove into Titan's thick *atmosphere* in January 2005. The sophisticated instruments on both spacecraft are providing scientists with vital data and the best views ever of this mysterious, vast region of our solar system. (http://saturn.jpl.nasa.gov/home/index.cfm.)

Castor. The second stage of the Scout *rocket,* named for the "tamer of the horses" in the constellation Gemini. See *Scout.*

catastrophic failure. A sudden failure without warning, as opposed to a degradation failure (SP-6001, p. 14).

CELSS. *Closed-ecology life support system.*

Centaur *(launch vehicle upper stage).* A stage known from 1956 to 1958 simply as the "high-energy upper stage" because it proposed to make first use of the theoretically powerful but troublesome liquid hydrogen as fuel. The Department of Defense's *Advanced Research Projects*

Agency (ARPA) awarded the initial contract for six research and development flight-test vehicles to Convair/Astronautics Division of General Dynamics Corporation in November 1958. The Centaur stage was required to increase the *payload* capability of the *Atlas* and to provide a versatile second stage for use in complex space missions.

NASA, which received management responsibility for the Atlas-Centaur, used the launch vehicle in the *Intelsat* IV series of *comsat*s and the *Surveyor* series of *space probes*. Centaur was also used to *launch* some of the larger *satellites* and space probes such as *OAO* 2 and 3, *ATS* 5, and the heavier *Mariner* and *Pioneer* space probes, and was mated with the Air Force *Titan* III for the heavier payloads flown in the mid-1970s. NASA launched the U.S.-German *Helios* 1 into *orbit* around the Sun on a Titan IIIE–Centaur on December 10, 1974. Centaur continues to be used in combination with the Atlas V and may be used with other rockets in the future. See *Atlas* and *Titan*.

ETYMOLOGY. The name derived from the legendary Centaur, half man and half horse. The horse portion represented the "workhorse" Atlas, the "brawn" of the launch vehicle; the man portion represented the Centaur stage which, containing the *payload* and guidance, was in effect the "brain" of the Atlas-Centaur combination. Krafft Ehricke of General Dynamics, who conceived the vehicle and directed its development, proposed the name, and ARPA approved it. Eugene C. Keefer of Convair is credited with proposing the name to Ehricke.

SOURCES. SP-4402, p. 11–12; Frank Kerr, Astronautics Division, General Dynamics Corp., teletype message to Lynn Manley, LeRC, December 10, 1963; Lynn Manley, letter to Historical Staff, NASA, December 11, 1963.

Centre Spatial Guyanais (CSG). The official name for *Europe's Spaceport*, located in *Kourou*, French Guiana.

CETI. Communication with Extraterrestrial Intelligence, as distinguished from Search for Extraterrestrial Intelligence *(SETI)*. The former term implies actual communication and was used by Soviet astronomers beginning in the 1960s, while the U.S. program adopted the less controversial SETI terminology beginning in 1976.

CEV. *Crew Exploration Vehicle*. Renamed *Orion* in August 2007.

Challenger. (1) *Call sign* for the *Apollo* 17 *Lunar Module*, the final artificial object left on the lunar surface, where it rests on a mountaintop overlooking the valley of Taurus-Littrow. Apollo 17 was the last of the six Apollo missions to the Moon. Harrison Schmitt and Eugene Cernan crewed the Challenger while Ronald Evans piloted the Command Module.

ETYMOLOGY. "The name of the Lunar Module came down in the final analysis between the names Heritage and Challenger," said Cernan.

"Challenger just seemed to describe more of what the future for America really held, and that was a challenge." (Quoted in Lattimer, *All We Did Was Fly to the Moon,* p. 94.)

(2) The second *STS* or *Space Shuttle orbiter* to become operational at *Kennedy Space Center*. Like their predecessors, Space Shuttle Challenger and its crews made significant contributions to America's scientific growth. Astronauts Robert L. Crippen and Frederick H. Hauck piloted Challenger (STS-7) on a mission to *launch* two communications satellites and the reusable Shuttle Pallet Satellite (SPAS 01). Sally K. Ride, one of three mission specialists on the first Shuttle flight with five crew members, became the first woman *astronaut*. Challenger joined the NASA fleet of reusable winged *spaceship*s in July 1982. It flew nine successful Space Shuttle missions.

On January 28, 1986, the Challenger's seven-member crew—Francis R. "Dick" Scobee, Michael J. Smith, Judith A. Resnik, Ronald E. McNair, Ellison S. Onizuka, Gregory B. Jarvis, and Christa McAuliffe—were all killed during *launch* from the Kennedy Space Center about 11:40 A.M. The explosion occurred 73 seconds into the flight as a result of a failure of an O-ring seal between two *solid rocket booster* segments that ignited the main liquid fuel tank. The explosion became one of the most significant events of the 1980s, as millions around the world saw the accident on television and empathized with the seven crew members killed.

ETYMOLOGY. Space Shuttle Challenger was named after the HMS Challenger, an English research vessel operating from 1872 to 1876. The voyage of HMS Challenger lasted 1,000 days and covered more than 68,000 nautical miles (125,000 km). Many consider it to have been the first true oceanographic expedition because it yielded a wealth of information about the marine environment.

Challenger Commission. Popular name for the Presidential Commission on the Space Shuttle Challenger Accident issued on June 6, 1986. The White House–appointed commission, chaired by former Secretary of State William P. Rogers and therefore also called the Rogers Commission, was deliberate and thorough, and its findings gave as much emphasis to the accident's managerial as to its technical origins.

Chandra X-ray Observatory (CXO). One of NASA's Great Observatories in Earth *orbit,* launched in July 1999 and named after S. Chandrasekhar. It was previously called the Advanced X-ray Astrophysics Facility *(AXAF)*.

ETYMOLOGY. Subrahmanyan Chandrasekhar (1910–95) was an Indian astrophysicist renowned for creating theoretical models of *white dwarf* stars, among other achievements. His equations explained the underlying physics behind the creation of *white dwarfs, neutron star*s, and other compact objects.

Charlie Brown. Call sign for the *Apollo* 10 Command Module, crewed by

Eugene A. Cernan, John W. Young, and Thomas P. Stafford. This *mission* was the "dress rehearsal" for the Moon landing.

ETYMOLOGY. Named after the prime character in Charles M. Schulz's *Peanuts* comic strip. As in the comic, Charlie Brown would be the guardian of *Snoopy* (the *Lunar Module*).

cherry picker. Jointed crane with a small cupola at the end for lifting a person up to conduct repairs to structures such as power lines. Used in the early days of the *human spaceflight program* as a means of swift egress from the gantry in case of a rocket failure before launch.

chief designer. Name used by the Soviet Union to hide the identity of Sergei Korolev. He became the mysterious, elusive, faceless "chief designer" hailed after each early space accomplishment. The world would soon become obsessed with trying to determine his identity. At the time of Korolev's death in 1966, Stuart H. Loory reported from Moscow for the *New York Herald Tribune* that his identity was as closely guarded behind the Iron Curtain as it was in the West. The first rumor that Korolev was the "chief designer" came after Nikita Khrushchev mentioned him by name in a toast proposed at the marriage of Cosmonaut Valentina Tereshkova, the first woman in space, to Cosmonaut Andrian Nikoleyev. In the West the veil of secrecy worked so well that NASA Administrator James E. Webb could not bring himself to send a note of condolence to his Soviet counterpart because, in part, "Korolev has never been known." With the collapse of the Soviet Union, details of Korolev's life have become much better known. Several biographies have been written, including one by Korolev's daughter.

CHIPS. Cosmic Hot Interstellar Plasma Spectrometer, a NASA astronomy *spacecraft* launched by a *Delta* 2 *rocket* from Vandenberg Air Force Base on January 13, 2003, with a spectrograph scanning the entire sky for hot and diffuse nebulae.

chuffing. (1) Sporadic or intermittent burning of fuel *(OED)*. (2) A noise resulting from a combustion instability, especially in a liquid-*propellant rocket engine,* characterized by a pulsing operation at low frequency (SP-6001, p. 15).

ETYMOLOGY. From chuff, a regularly repeated sharp puffing sound such as that produced by a locomotive or one-cylinder engine.

circumterrestrial space. Inner space. That region abutting the Earth's *atmosphere* extending from about 60 miles (97 km) to about 50,000 miles (80,500 km).

cislunar. Pertaining to the area between the Earth and Moon or between the Earth's *orbit* and that of the Moon (NASA, *Glossary/Congressional Budget Submission,* p. 7).

ETYMOLOGY. Latin *cis* (on this side) + *luna* (Moon).

Clarke orbit. *Geostationary orbit* at 22,300 miles (36,000 km), so called because of writer Arthur C. Clarke's early suggestion that such an orbit would be useful for communications and relay (Clarke, "Extra Terrestrial Relays: Can Rocket Stations Give World Wide Radio Coverage?" *Wireless World,* October 1945, 305–8).

clean room. A space that is virtually free of dust or bacteria, used in laboratory work and in assembly or repair of precision equipment, including satellites and planetary *probes.*

Clementine. Department of Defense–NASA lunar mapping mission, 1994. The principal objective of Clementine was to space-qualify lightweight imaging sensors and component technologies for the next generation of Department of Defense *spacecraft.* Intended targets for its sensors included the Moon, Geographos (a near-Earth asteroid), and the spacecraft's own interstage adapter. After entering lunar *orbit* in 1994, Clementine provided more than 1.8 million images of the lunar surface. Because of a malfunction, it never rendezvoused with Geographos.

closed-ecology life support system (CELSS). A mechanical or biological system that recycles the air, water, and food needed to sustain human life on a *space station* or *base* (Paine Report, p. 197).

closed-loop system. Control system with a feedback loop that is active, thus achieving some level of self-correction. Opposite of an *open-loop system.*

Cloud-Aerosol Lidar and Infrared Pathfinder Satellite Observations. See *CALIPSO.*

CM. Command Module.

CNES (pronounced see-ness). Centre National d'Études Spatiales. French *space agency.*

colonize. To establish a *settlement* or settlements.

USAGE. The term was used in science fiction of the 1950s for settlements on the Moon, notably in Arthur C. Clarke's *Earthlight,* published in 1955. It took on new overtones in the wake of *Sputnik:* "I believe the Soviets are planning to 'take over' outer space—to colonize it with their sputniks and space platforms" (quotation attributed to "a leading research expert in the American earth satellite program," *Christian Science Monitor,* October 22, 1957, p. 1).

Columbia. (1) *Call sign* for the *Apollo* 11 Command Module. This was the first lunar landing *mission.* On July 20, 1969, the *Lunar Module*—piloted by Neil A. Armstrong and Edwin E. Aldrin—landed on the lunar surface while Michael Collins orbited overhead in the Command Module. ETYMOLOGY (definition 1). The module was named after the Columbiad in Jules Verne's 1867 novel *From the Earth to the Moon* (the name referred not to the spaceship itself but to the cannon used to launch it). The name was also chosen for its association with America's origins. Mike Collins explained: "Columbia had almost become the name of our country. Finally, the lyrics 'Columbia, Gem of the Ocean' kept popping into my mind and they argued well for the recovery of a spacecraft which hopefully would float on the ocean." (Quoted in Lattimer, *All We Did Was Fly to the Moon*, p. 66.)

(2) Shuttle *orbiter OV-102*. *Space Shuttle* Columbia had enjoyed a long and distinguished service life when it broke up on February 1, 2003, about 15 minutes before the scheduled landing of the *STS-107* mission. All seven astronauts aboard lost their lives. It was the first flight in recent years that had not been related to activities of the *International Space Station*. The seven crew members—Rick Husband, William H. McCool, Michael Anderson, David Brown, Kalpana Chawla, Laurel Clark, and Ilan Ramon—had helped oversee 80 *microgravity* experiments ranging from projects of interest to schoolchildren to experiments having significant commercial and scientific potential. After a 16-day *mission,* begun on January 16, 2003, the crew was lost during *reentry,* when communications failed and the orbiter disintegrated over western Texas on its path toward *Cape Canaveral.* Disintegration was caused by a hole in the leading edge of the wing caused by a piece of foam from the external tank striking the wing during launch. It was the 28th mission for the Columbia, then the oldest orbiter in the Shuttle fleet.

ETYMOLOGY (definition 2). The orbiter was named after the Boston-based sloop captained by American explorer Robert Gray. On May 11, 1792, Gray and his crew maneuvered the Columbia past a dangerous sandbar at the mouth of what proved to be largest river in the Pacific Northwest of North America. The river was later named after the ship. Gray also led Columbia and its crew on the first American circumnavigation of the globe, carrying a cargo of otter skins to China and then returning to Boston. Columbia is considered to be the feminine personification of the United States. The name is derived from that of another famous explorer, Christopher Columbus. (http://science.ksc.nasa.gov/shuttle/resources/orbiters/columbia.html.)

Columbus Research Laboratory. Europe's largest contribution to the *International Space Station,* it was successfully launched on February 7, 2008, to support scientific and technological research in a *microgravity* environment. (*Reference Guide to the International Space Station,* SP-2006–557.)

comet. A small chunk of ice, dust, and rocky material only a few miles across that, when it comes close enough to the Sun, can develop a tenuous "tail." The tails of comets are composed of gas and dust driven off the comet's surface by the Sun's energy, and they always point away from the Sun (no matter what direction the comet is moving). Comets spend most of their time very far from the Sun and are active only for a short period of time (a few months at most) as they move quickly around the Sun on their elongated orbits. See also *dirty snowball*. (ASP Glossary.)

Comet Nucleus Tour. See *CONTOUR*.

command and control. System that directs the course of a rocket or missile.

Command and Service Module (CSM). Name for the portion of the Apollo spacecraft that remained conjoined from *launch* until the beginning of *reentry*. The Command Module contained the living space, computers, and control units. The Service Module contained the thrust and attitude control engines, the communications antennas, fuel cells, and hydrogen and oxygen gas containers. The terms Command and Service Module and Command Module (CM) are sometimes used interchangeably, even though the CM is only one part of the CSM.

Command Module (CM). See *Command and Service Module*.

Commercial Orbital Transportation Services. See *COTS*.

comsat. Communications satellite.

COMSAT/Comsat/ComSat. The Communications Satellite Corporation. Created in 1962 with the passage of the Communications Satellite Act. The act authorized the formation of a private corporation to administer *satellite* communications for the United States. It served as the U.S. signatory to *INTELSAT*. In 2000, COMSAT became part of the Lockheed Martin Corporation.

Constellation. See *Project Constellation*.

consumables. NASA jargon for food, air, water, and power on human missions.

CONTOUR. Comet Nucleus Tour. Failed heliospheric *spacecraft* that was launched by a *Delta* 2 *rocket* from *Cape Canaveral* on July 3, 2002. It was designed to meet at least two *comets*, Comet Encke on November 12, 2003, and Comet Schwassmann-Wachmann 3 (SW3) on June 19, 2006. Ground-based telescope images indicated that CONTOUR had broken up near the scheduled end of its rocket *burn*, and after numerous attempts at contact, the spacecraft was presumed lost. (http://nssdc.gsfc.nasa.gov/database/MasterCatalog?sc=2002-034A.)

copacetic/copasetic. Slang term meaning completely satisfactory, as it should be. *A-OK* and *tickety-boo* mean more or less the same thing, but copasetic implies a more beatific mood.

USAGE. Although the term was in use before the *Space Age*, it became associated with the *space program* when, according to an article on words being accepted for inclusion in *Webster's New World Dictionary:* "An astronaut used the word while in outer space and millions of people heard it on TV and radio. Copacetic was widely reported in newspapers and magazines and was so commonly used thereafter that it made it into the 1982 edition of Webster's." ("Word Sleuths," *Cleveland Plain Dealer,* June 16, 1996, p. 8.)

Copernicus. See *OAO* (Orbiting Astronomical Observatory).

CORONA. America's first *satellite* intelligence program, a series of satellites with increasingly accurate cameras, which provided coverage of the Soviet Union, China, and other areas from the Middle East to Southeast Asia, from February 1958 until the satellite's retirement in 1972. The first successful recovery of film from a CORONA drop-bucket satellite occurred in August 1960. Satellite imagery from CORONA was used for a variety of analytical purposes, from assessing military strength to estimating the size of grain production. See also *Discoverer Project.*

USAGE. Though not an acronym, the name is traditionally spelled with capital letters.

COS-B. Cosmic Ray Observation Satellite. See *ESRO.*

cosmodrome. A launching site for *spacecraft* in Russia.

cosmology. Branch of astronomy that addresses the origin, large-scale properties, and evolution of the observable *universe.*

cosmonaut. (1) Traveler in the cosmos. (2) *Astronaut* of the Soviet Union / Russia.

ETYMOLOGY. Cosmos + naut (sailor). Literally, a cosmic sailor.

FIRST USE. The term first came to the attention of the West in the October 1959 issue of the Soviet magazine *Ogonek* (*New York Times* News Service piece in *Los Angeles Times,* "US Believes Russians Have Space Man Plans," November 27, 1959, p. 6).

COTS. Commercial Orbital Transportation Services. NASA initiative that seeks commercial providers of launch and return logistics services to support the *International Space Station* after the *Space Shuttle* is retired (*Reference Guide to the International Space Station,* SP-2006-557).

countdown. (1) A step-by-step process that culminates in a climactic event, each step being performed in accordance with a schedule marked by a count in inverse numerical order; specifically, this process is used in leading up to the *launch* of a large or complicated *rocket* vehicle, or in leading up to a captive test, a readiness firing, a mock firing, or other firing test. (SP-7.) (2) The act of counting inversely during this process. Should something go wrong, the countdown is halted until the problem is corrected. (SP-6001, p. 20.)

USAGE. In definition 2, the countdown ends with *T-time*; thus, T minus 60 minutes indicates there are 60 minutes to go, excepting for holds and recycling. The countdown may be measured in days, hours, minutes, or seconds. At the end, it narrows down to seconds: 4—3—2—1—0.

ETYMOLOGY. The actual practice of the countdown was first rendered graphically in the German science fiction movie *Frau im Mond* (1929), where it is used for dramatic effect. British science fiction writer George W. Griffith (1857–1907) used a 10–1 countdown in his novelette *The Crellin Comet* (which appeared in *Pearson's Weekly,* Extra Christmas Number, November 1897) and a 20–1 countdown in his novel *World Peril* (published by F. V. White, July 1910). (NASA Names Files, record no. 17494.)

FIRST USE. The word in its modern sense—excluding its extensive use in 20th-century sports—is first used in connection with the testing of atomic weapons. "Just before the drop he announced that the time of the fall would be forty-two seconds. The B-36 was invisible in the darkness, but Mr. Felt reported 'bombs away' and then began the 'countdown'—30 seconds—15 seconds—5—4—3—2—." ("Mightiest of Atom Blast of Tests Unleashed on Nevada Desert," *New York Times,* June 5, 1953, p. 1.) A countdown was used even earlier during the first atomic bomb test (Trinity) in July 1945. In *Space Age* terms it comes into use with the first American attempt to launch a *Vanguard satellite* in late October 1957 ("Fueling a Final Hazard in Launching Vanguard," *Chicago Tribune,* October 25, 1957, p. 8).

CRAF. Comet Rendezvous / Asteroid Flyby. This *space probe* was intended to gather information about the early solar system by examining a *comet* (Kopff) and an asteroid (449 Hamburga) at close range. The project was eventually canceled when it went over budget, and its scientific objectives were folded into the *Rosetta Mission.*

craft. Any machine designed to fly through space. Short for *spacecraft.*

crater. An impression left in a celestial body when a smaller incoming object strikes it. All rocky planets, and some satellites and asteroids, have craters.

Crawler-Transporter/Crawler Transporter/Crawler (CT). Immense tracked transport vehicle able to carry a *launch vehicle* and *spacecraft* to its *launch pad,* used in both the *Apollo* and *Space Shuttle* programs (SP-6001, p. 20). Dating back to the Apollo program, there have been two CTs (CT 1 and CT 2), which underwent a major upgrade beginning in 2003. Moving at a mere 6 miles per hour (10 kph), these are true workhorses of the *space program.* They were originally built by the Marion Power Shovel Company.

By 2004 it was determined that the two CTs at *Kennedy Space Center* had logged 3,400 miles (5,470 km), approximately the distance between Miami and Seattle.

FIRST USE. "Artist's drawing of Apollo spacecraft and Saturn V launch vehicle atop crawler-transporter, which is about the size of a baseball diamond" ("Crawler Must Move 12 Million Lb. Mooncraft without a Jiggle," *Chicago Tribune*, February 12, 1963, p. 1).

USAGE. During the Shuttle era the hyphen in the original term was dropped.

crawlerway. Road connecting *Vehicle Assembly Building* (VAB) and the *launch* pad at *Kennedy Space Center* (SP-6001, p. 21).

the creeps. *Mercury-Gemini*-era *astronaut* slang for itchy skin caused by low pressure in a *capsule* ("Space Age Slang," *Time* magazine, August 10, 1962, p. 12).

crewed spaceflight. Synonym for *human spaceflight* or *piloted space-flight*. Gender-neutral term suggested as an alternative to *manned* when discussing projects like the *International Space Station*. The term was front and center at the 1992 International Colloquium on Manned Space Flight in Cologne, Germany, where some delegates questioned the name of the conference ("Sexless Talk: Crewed Is Good, Manned Is Crude," *Space News,* June 1–7, 1992, p. 6). See *human spaceflight*.

Crew Exploration Vehicle. Name for the *spacecraft* that will return American crews to the Moon and then to Mars. It was formally renamed *Orion* in August 2006. The first flights are planned for 2012–14 and will support the *International Space Station*.

crust. The outer layer of the Earth and other celestial bodies.

cryogenic. Requiring or involving the use of very low temperature. Describing a *rocket* fuel or oxidizer that is liquid only at very low temperatures, e.g., liquid hydrogen.

CSM. *Command and Service Module.*

CTS-A. Communications Technology Satellite. Formerly ISIS-C and CAS-C. See *ISIS*.

cyberspace. Term coined by science fiction writer William Gibson in 1984; now used to describe the space of interconnected communication networks on the World Wide Web.

cycling spaceship. A *space station* designed for human habitation that permanently cycles back and forth between the orbits of Earth and Mars (Paine Report, p. 198).

Cyclops. See *Project Cyclops*.

Cytherean. Pertaining to the planet Venus. Glossary *Apollo Terminology* contains a definition for *Cytherean atmosphere* (SP-6001, p. 22).

ETYMOLOGY. The term came to prominence when plans were being made for the first *probes* of Venus in the early 1960s, when planners and policy makers shied away from the logical choice of adjective, venereal, which was too closely associated with human sexual activity. A very early appearance is in the NASA fiscal year 1964 budget (NASA, *Glossary/Congressional*

Budget Submission, p. 9). The next most logical choice was the Greek Aphrodisian, which had similar sexual overtones. The final choice of Cytherean derives from the island of Cythera, from which Aphrodite, the Greek Venus, emerged. *Merriam-Webster's Collegiate Dictionary,* 10 ed., dates the adjective from 1855.

D

dark matter. Name given to the amount of mass whose existence, until now, has escaped all detection, but is deduced from the analysis of gravitational effects on visible matter, and in particular *galaxy* rotation curves. The ability to understand dark matter has been advanced by the *Hubble Space Telescope.*

DARPA. Defense Advanced Research Projects Agency. Current name for ARPA. See *Advanced Research Projects Agency.*

Daughter *(satellite).* International Sun-Earth Explorer ISEE-B. See *Explorer.*

dead man controls. Devices for shutting off or rendering machines safe in case of accident or illness of the operator (SP-6001, p. 23).

dead time. Inoperative or nonproductive time (NASA, *Glossary/ Congressional Budget Submission,* p. 10).

Deal Project. Interim name for the pioneering *satellite* program called *Explorer* after the name and the project called Orbiter were turned down.

> ETYMOLOGY. When Orbiter was turned down, John Small, a section chief at the *Jet Propulsion Laboratory* (JPL), made a pronouncement that is familiar to poker players: "The winners laugh and joke and the losers yell, 'Deal.'" The name stuck. On the launching of Explorer on January 31, 1958, Secretary of the Army Wilbur Bruckner called officials at the *Cape* to tell them that he and Maxwell Taylor, Army Chief of Staff, had selected the name Explorer for the Army satellite. All sorts of names were suggested: John Medaris had wanted to call it Highball, Bruckner favored *Top Kick,* and a contingent from JPL wanted to hold on to the name Deal.

Death-O-Meter. Early *telemetry* display that would monitor vital signs of humans at high altitude or in space. Recalled in Scott Carpenter's *For Spacious Skies,* p. 151.

deboost. To lower the *orbit* of a *spacecraft.*

> FIRST USE. The term (identified as space slang) came into play during the *orbit* of *Lunar Orbiter* 1 when it was dropped 6 miles (9.6 km) in an attempt to open the eye of a malfunctioning camera ("Lunar Orbiter Descends 6

Miles in Attempt to Open Camera Eye," *Washington Post,* August 26, 1966, p. A3).

debugging. The process of detecting and eliminating mistakes in equipment and computer programs or other software. Also called troubleshooting.

decay. Degeneration of a *satellite's orbit* from an ellipse as it is pulled back into the Earth's *atmosphere.*

Deep Impact. NASA spacecraft that *rendezvous*ed and impacted *Comet Tempel 1* in 2005 in order to analyze the composition of the comet. Launched on January 12, 2005, Deep Impact was composed of two parts, a *flyby spacecraft* and a smaller impactor. (See *impactor object.*) The impactor was released into the comet's path for a collision on July 4, 2005. The *crater* produced by the impactor was football field sized and 2 to 14 stories deep.

ETYMOLOGY. Speculation that the *mission* was named for the 1999 film of the same name about a comet heading toward Earth was disputed by the National Optical Astronomy Observatory: "Deep Impact, a space mission to study the structure of a comet's deep interior . . . whose name was chosen well before the motion picture of the same name was announced, will be led by PI Michael A. Hearn (Maryland). The Deputy PI will be Michael J. S. Belton (NOAO)." (*NOAO Newsletter,* September 1999, p. 1.)

deep space. By general consensus, any region beyond the *Earth-Moon system; outer space.*

Deep Space I. NASA *flyby mission* to asteroid 1992 KD, 1998. It also rendezvoused with *Comet* Borrelly in 2001.

Deep Space II. NASA penetration *mission* to Mars, 1999.

Deep Space Climate Observatory (DSCOVR). Formerly *Triana.*

Deep Space Network (DSN). Communications and *tracking* system maintained by the *Jet Propulsion Laboratory* for all *spacecraft* navigating the solar system beyond Earth orbit. It consists of very large diameter radio antennas located in California, Spain, and Australia.

Defense Satellite Communications System (DSCS). Since the launch of its prototype in 1967, the "workhorse" of long-haul military satellite communications. DSCS satellites are placed in *geosynchronous orbit* to provide high-volume, secure voice and data communications. Phases II and III were successors to the IDSCS (Initial Defense Satellite Communications System) program, which began in 1967 with the launch of the first eight satellites of this constellation. The DSCS system has been an extremely valuable asset for supporting military and government communications over the past several decades.

Defense Support Program (DSP). Also known as Satellite Early Warning System. An Air Force *geosynchronous satellite* system, first established in 1970, that uses infrared detectors to sense heat from missile plumes against the Earth background and detect and report missile launches, space launches, and nuclear detonations. Its effectiveness was proven during the Persian Gulf conflict, which took place from August 1990 through February 1991. During Operation Desert Storm, DSP detected the launch of Iraqi Scud missiles and provided timely warning to civilian populations and coalition forces in Israel and Saudi Arabia. See http://www.af.mil/factsheets/factsheet.asp?id=96.

Delta *(launch vehicle).* When NASA was formed in 1958, it inherited from the Department of Defense's *Advanced Research Projects Agency* the *booster* programs using combinations of *Thor* or *Atlas* boosters with *Vanguard upper stage*s. The first of these upper-stage configurations was designated *Able.* The Delta was similar to the previous Thor-based combinations and was a fourth or D version.

Over the years the Thor-Delta was repeatedly uprated by additions and modifications. The *liftoff thrust* of the Thor first stage was increased in 1964 by adding three strapped-on solid-*propellant rocket* motors. With the Delta second stage, the launch vehicle was called Thrust-Augmented Delta (TAD). In 1964 NASA undertook upgrading the Delta capability by enlarging the second-stage fuel tanks. When this more powerful version, introduced in 1965 and designated Improved Delta, was used with the Thrust-Augmented Thor first stage, the vehicle was called Thrust-Augmented Improved Delta (TAID). In 1968 NASA incorporated an elongated Thor first stage with added fuel capacity for heavier payloads, and the three strapped-on motors were uprated. This version, with the improved Delta second stage, was called Long-Tank Thrust-Augmented Thor-Delta (LTTAT-Delta), or Thrust-Augmented Long-Tank Delta.

The "Super Six" version, with six strap-on *Castor* rockets for extra thrust, was first used in 1970, and nine strap-ons went into use in 1972. A more powerful third stage, TE-364-4, was also introduced in 1972, as was the "Straight Eight" Thor-Delta, with an 8-foot (2.4-m) diameter for all three stages including the fairing. The wider fairing could accommodate larger *spacecraft.*

In 1960 the Thor-Delta placed 132 pounds (60 kg) into a 995-mile (1,600-km) *orbit.* By the end of 1974 the vehicle could *launch* a 1,500-pound (700-kg) spacecraft into orbit for transfer to a 22,000-mile (35,500-km) *geosynchronous orbit,* a 4,000-pound (1,800-kg) *payload* into a 115-mile (185-km) orbit, or 850 pounds (386 kg) on a *trajectory* to Mars or Venus.

The economical, reliable Thor-Delta was a workhorse vehicle used for a wide range of medium satellites and small *space probe*s in two-stage or three-stage combinations, with three, six, or nine strap-on thrust-augmenter rockets. Among its many credits were meteorological satellites (Tiros, TOS), communications satellites *(Echo, Telstar, Relay, Syncom, Intelsat)*, scientific satellites *(Ariel, Explorer, OSO)*, and the Earth Resources Satellite ERTS 1. The vehicle's first three-satellite launch put NOAA 4, *OSCAR* 7, and *INTASAT* into orbit on November 15, 1974.

Thor-Delta was discontinued in the early 1980s when the *Space Shuttle* was to become the sole launcher for satellites. After the *Challenger* accident this policy was reevaluated, and the Delta II was brought into service in 1989. The Delta III followed in 1998 and the Delta IV heavy lift vehicle in 2002. Delta II and IV are still active launch vehicles.

ETYMOLOGY. Milton W. Rosen of NASA was responsible for the name. He had been referring to the combination as Delta, "which became the firm choice in January 1959 when a name was required because NASA was signing a contract for the booster." The vehicle was variously called Delta and Thor-Delta.

SOURCES. SP-4402, pp. 12-13; Robert L. Perry, "The Atlas, Thor, Titan, and Minuteman," in *The History of Rocket Technology,* ed. Eugene M. Emme (Detroit: Wayne State University Press, 1964), p. 160; Milton W. Rosen, Office of Defense Affairs, NASA, telephone interview, February 6, 1965; NASA News Releases 60-237, 60-242, 64-133, 67-306, 70-2, 72-206, 74-77K, 75-19; NASA program office; Communications Satellite Corp., Intelsat III Press Kit, 1968; Robert J. Goss, "Delta Vehicle Improvements," in *Significant Accomplishments in Technology, Proceedings of Symposium at Goddard Space Flight Center, November 7–8, 1972,* SP-326 (Washington, DC: NASA, 1973), 11–13. For more information about Delta, see Launius and Jenkins, eds., *To Reach the High Frontier.*

deorbit/de-orbit. (1, n.) A *maneuver* that brings a *spacecraft* or *rocket* stage out of *orbit* and into descent. (2, v.) To move a "dead" or weakened *satellite* out of *geosynchronous orbit* by raising or lowering it to a level where it will not pose a hazard to active satellites. ("Definitions of the Term De-orbit," *Spaceflight,* July 2002, p. 307.)

de-scrub. To return a *mission* to flight status. See *scrub.*

Destiny. U.S. laboratory module on the *International Space Station* (ISS). Destiny, launched February 7, 2001, on *STS*-98, provides the ISS with research and habitation accommodations. The cylindrically shaped module has a length of 28 feet (8.8 m) and is 14 feet (4.3 m) in diameter. (*Reference Guide to the International Space Station,* SP-2006-557.)

destruct. (1, n.) A *rocket* or other *space vehicle* that has been deliberately destroyed. (2, v.) To deliberately explode a *rocket* after it has been launched; to destroy.

ETYMOLOGY / FIRST USE. As a term pertaining to rockets and satellites, the word came into the language in 1958 as a probable back-formation from "destruction." It is listed as a new term in *Interim Glossary Aero-Space Terms* (1958): "Destructs are executed when the missile gets off its plotted course or functions in a way as to become a hazard" (Heflin, *Interim Glossary Aero-Space Terms,* p. 9). It quickly entered the wider language, especially as *self-destruct.*

USAGE. "This is a word needed in the most serious emergencies and in split-second urgency. It had, therefore, to be clear and unequivocal. It had to be a word not in common meaning lest its common meaning lead to some confusion. But it had to be sufficiently close to a common word to convey its meaning instantaneously. 'Destruct' is a good coinage—though it has agitated many who feel that it is 'bad English.'" (Bergen Evans, "New World, New Words," *New York Times,* April 9, 1961.)

DFRC. *Dryden Flight Research Center.*

dirty snowball. Characterization of a *comet* after the NASA *ICE* (International Cometary Explorer) probe passed through the 14,000-mile (23,000-km) thick tail of the Giacobini-Zinner comet in 1985.

ETYMOLOGY / FIRST USE. The term originates with astronomer Fred Whipple in 1950. Against the prevailing theory of sandy composition, Whipple proposed that comets consist of ice with some rocks mixed in. The term has often been used since that time. For example, Dr. John Bryant, chief of the Space Physics Laboratory at *Goddard Space Flight Center,* stated on September 13, 1985, that the early results from ICE fit in "nicely with our picture of a comet as a large dirty snowball." (The comment was repeated in reports of September 14, 1985, such as the *Washington Post* piece "Probe of Comet Supports Theory of Composition–Scientist Likens It to 'Dirty Snowball,'" p. A3.)

Discoverer Project. Air Force project to test techniques for future military space systems. Objectives include precise *orbit* stabilization and *recovery* of an instrumented *capsule* ejected upon command. Initial *launch* was in February 1959. Discoverer was the cover name for the *satellite* series dubbed *CORONA* by the CIA, also known as Key Hole or Keyhole and designated KH-1, etc.

Discovery. (1) The third orbiter, Orbiter Vehicle (OV) 103, to become operational at *Kennedy Space Center.*

ETYMOLOGY (definition 1). The orbiter was named after one of the two ships used by the British explorer James Cook during the 1770s for voyages in the South Pacific that led to the discovery of the Hawaiian Islands. Cook also used Discovery to explore the coasts of southern Alaska and northwestern

Canada. During the American Revolutionary War, Benjamin Franklin made a safe-conduct request for the British vessel because of the scientific importance of its research.

(2) A continuing line in NASA's budget dedicated to small planetary missions characterized by a three-year development schedule and a budget cap of $150 million (FY 1992), now $299 million (FY 2007). *Mars Pathfinder* and *Near-Earth Asteroid Rendezvous* (NEAR), the first two Discovery missions, were granted new starts in NASA's fiscal year 1994 budget. Two additional missions, *Lunar Prospector* and Stardust, were competitively selected for new starts in NASA's fiscal year 1995 and 1996 budgets, respectively. Additional Discovery missions are *Deep Impact,* Genesis (partially successful), *CONTOUR* (failed), *Messenger,* Aspera-3 and Dawn (in progress), and Kepler, Moon Mineralogy Mapper, and GRAIL (future missions).

Dixie Cup. Nickname for the cup-shaped individual sample bags that were used on *Apollo* 12, 14, and 17.

ETYMOLOGY. The registered trademark of a waxed-paper drinking cup. "The first disposable, individual drinking cups were developed by Lawrence Luellen in 1907–8 and, in the next decade, gained market acceptance due to increasing concerns about disease spread by use of common-use dippers and glasses. The company was later headed by Hugh Moore and, in 1919, the cup, which had been known as the Health Cup, acquired the Dixie Cup brand-name and national prominence as a result of the Influenza Epidemic that struck after World War I." (*ALSJ* Glossary.)

docking. (1) The act of joining a *space vehicle* to another in space. (2) The act of becoming so joined. The term achieved wide popular use during the *Apollo* 10 dress rehearsal for the Moon landing when there was a docking and crew transfer operation between the *Command and Service Module* and the *Lunar Module.*

FIRST USE. The *OED* cites this as the first use of the term in the context of space: "The idea of 'docking' a spaceship inside a space-station is suicidal lunacy" (*Journal of the British Interplanetary Society,* 1951).

DODGE. See *Project DODGE.*

Donatello. One of the Italian-built Multi-Purpose Logistics Modules *(MPLMs)* that are carried in the cargo bay of the *Space Shuttle* and ferried to the *International Space Station.* They contain experiments and supplies. (*Reference Guide to the International Space Station,* SP-2006-557.)

Doppler effect. The change in frequency with which energy reaches a receiver when the receiver and the energy source are in motion relative to each other; also called Doppler shift.

ETYMOLOGY. Eponym for 18th-century physicist Christian J. Doppler.

downlink. A connection wherein a *spacecraft* can communicate to Earth at radio wavelengths.

downrange. Direction away from *launch* site along intended flight line (SP-6001, p. 27).

down the slot. Describing a successful *shot* of a *rocket* or *missile* down the test range.

downtime. Period during which equipment is not operating correctly because of machine failure (SP-6001, p. 27).

Dryden Flight Research Center (DFRC). Formerly the *Flight Research Center.* The center is one of NASA's sites for aeronautical research, in addition to *Langley Research Center, Ames Research Center,* and *Glenn Research Center.*

In October 1946, five engineers arrived at Muroc Army Air Field from the *NACA* Langley Research Center. By December there were thirteen NACA employees at the base, detailed to the remote location for the X-1 research project. Muroc's main allure was the dry lakebed, the largest such geological feature in the world, and the seemingly endless sunny days with cloudless skies. The Army provided the NACA with space in one of two main hangars on South Base. Meanwhile, the Army conducted its most sensitive operations at the more remote North Base facility several miles further north on the lakeshore. In addition to the NACA, elements of the Army Air Forces and aviation contractors directly involved in flight-testing and research were at South Base. Given the remoteness of the Army field, many of the employees found rudimentary housing on the base, in barracks dubbed "kerosene flats" for the permeating odor of the heating and cooking fuel. All NACA activity between 1946 and 1954 was conducted out of the work area provided on South Base as well as several adobe revetments where rocket engines were test-run.

In 1951 Congress approved funding for a NACA facility at what was now called Edwards Air Force Base. The Air Force (the new name under which the Army Air Forces were reorganized in 1947) leased to the NACA just under a square mile of land located north of the Air Force's own new facilities, also on the western shore of the lakebed. Construction began on a new facility in early 1953, and by mid-1954 NACA employees had moved into the complex, a move coincident with the unit achieving independent status within the NACA. Located at the edge of the lakebed, the new structure consisted of a three-story main building flanked by two hangars and a ramp that led from the flight line directly to the lakebed. Initially, NACA aircraft wanting to use the Air Force's new concrete runway had to taxi on the lakebed to gain access to it. Only later was a long taxiway built linking the center with the main Air Force runway and complex. In any case, most operations in the 1950s benefited from the 44-square-mile (114-sq-km) lakebed expanse, and flights left and returned to the center directly from the concrete-like surface of the dry lake. The main structure of the High Speed Flight Research Station building, at 4800 Lilly Drive, housed all engineers, administrators, and control rooms—even the credit union. The adjoining hangars made for good coordination among engineers, mechanics, and pilots. As the ranks of employees grew, space became scarce and the two open courtyards where employees ate lunch were enclosed to create office space.

Since 1954, new structures have been added to the center, including a Flight Loads Laboratory used for structural analysis, a Fabrication Shop to make all manner of custom parts on demand, and an additional structure known as the Research Aircraft Integration Facility (RAIF) with three separate hangar bays and offices. The RAIF also houses an array of flight simulators used in research projects. Set back from the flight line stands the Data Analysis Facility, which receives telemetry from test aircraft for analysis and distribution. Clustered at the northern end of the center are a variety of buildings established in support of *Space Shuttle* operations. Most visible among these are a hangar capable of housing the Shuttle, and the Mate-Demate Device (MDD). The MDD facilitates loading of the Shuttle onto the Shuttle carrier aircraft.

ETYMOLOGY. Named in honor of the late Dr. Hugh L. Dryden, one of America's most prominent aeronautical engineers, on March 26, 1976. At the time of his death in 1965, he was NASA's Deputy Administrator. For more information about Dryden Flight Research Center, see Richard P. Hallion, *On the Frontier: Flight Research at Dryden, 1946–1981*, SP-4303 (Washington, DC: GPO, 1984); Lane E. Wallace, *Flights of Discovery: An Illustrated History of the Dryden Flight Research Center*, SP-4309 (Washington, DC: GPO, 1996).

DSCOVR. *Deep Space Climate Observatory.* Formerly *Triana.*

DSCS. *Defense Satellite Communications System.*

DSN. *Deep Space Network.*

DUA. Digital *uplink* assembly.

Dyna-Soar/Dynasoar (dynamic + soaring). Hypersonic boost-glide vehicle, 1958–63, also known as the X-20. Dyna-Soar was a *Sputnik*-inspired program that began in 1958 to develop a single-piloted Earth-orbiting vehicle capable of maneuverable *reentry* through the *atmosphere.* It was canceled on December 10, 1963, after $400 million had been spent on the program. It was beset by numerous confusing incarnations and changes in purpose. Was it to be a piloted space bomber, a reconnaissance platform, or a high-speed test vehicle? (Robert Redding and Bill Yenne, *Boeing: Planemaker to the World* [New York: Crescent, 1989], p. 229.)

FIRST USE. In testimony before the House Armed Services Committee on February 25, 1958, Lt. Gen. D. L. Putt referred to the "'Dynasoar Project' as a winged vehicle employing the principles of both centrifugal and aerodynamic flight and a vehicle of rather unique configuration and characteristics" (reported in "AF Asks 'Go' on Moon Shot," *Washington Post,* February 26, 1958, p. A10).

USAGE. If any name of the *Space Age* sent the wrong message, it was this one. "Dinosaur" had long been used in the slang sense to indicate something or someone out of touch with the modern world. For example: "Dyna-Soar or Dinosaur? Another Military Space Vehicle Due to Become Extinct in Conflict of Ideas" (*New York Times,* March 16, 1963, p. 4).

Eagle. *Call sign* for the *Apollo* 11 *Lunar Module,* crewed by Neil A. Armstrong and Edwin E. Aldrin Jr., which accomplished the first lunar landing while Michael Collins piloted the Command Module *Columbia.* Armstrong set foot on the surface, telling the millions of listeners that it was "one small step for [a] man—one giant leap for mankind." Aldrin soon followed him.

ETYMOLOGY. The name Eagle was selected after the eagle had been chosen for the *mission*'s insignia, which according to Collins had been suggested by Jim Lovell. Collins added an olive branch to the eagle's beak, but the branch was later moved to the talons because NASA felt that the bare talons looked too hostile. (Lattimer, *All We Did Was Fly to the Moon,* p. 65.)

Early Bird. Intelsat I communications satellite. See *Intelsat.*

Earthlight. Illumination of the Moon from reflected light from Earth.

Earth-Moon system. The space and all of its contents extending about 480,000 miles (772,000 km) in every direction from the Earth's center.

Earth Observatory Satellite (EOS). See *ERTS, EOS,* and *SEOS.*

Earth Observing System (EOS). See *EOS.*

Earth probes. A series of small Earth observation satellites including TRMM *(Tropical Rainfall Measuring Mission).*

Earth Resources Experiment Package (EREP). See *ERTS, EOS,* and *SEOS.*

Earth Resources Observation Satellite (EROS). See *ERTS, EOS,* and *SEOS.*

Earth Resources Survey Satellite (ERS). See *ERTS, EOS,* and *SEOS.*

Earth Resources Technology Satellite (ERTS). Renamed *Landsat.* See *ERTS, EOS,* and *SEOS.*

Earthrise. Name given to the image presented to the world on Christmas Eve 1968, when an *Apollo* 8 *orbit*—crewed by Frank Borman, James A. Lovell, and William A. Anders—came around the Moon providing the first full color look at ourselves as one: the Earth seeming to rise over the lunar surface. The first black-and-white Earthrise picture had been taken by Lunar Orbiter 1 in 1966, but is far less known. Mythology scholar Joseph Campbell commented on the Apollo missions in his book *Myths to Live By* (1972): "The earth is a heavenly body, most beautiful of all, and all poetry now is archaic that fails to match the wonder of this view. Now there is a telling image: this earth, the one oasis in all space, an extraordinary kind of sacred grove, as it were, set apart for the rituals of life; and not simply one part or section of this earth, but the entire globe now a sanctuary, a set-apart Blessed Place."

The Apollo 8 photo is said to have helped inspire Earth Day, the event and the movement. It also served as a reminder that there was a vast gulf between the peaceful image of a planet in space and the reality of life on Earth. The same month Apollo 8 orbited the Moon, the U.S. death toll in Vietnam exceeded 30,000.

FIRST USE. The term first appears in print as the title line in a caption for an AP wire photo of the image ("Earthrise on the Moon," *Chicago Tribune,* December 30, 1968, p. 6).

Earth satellite. A body that orbits about the Earth. Specifically, an artificial satellite placed in *orbit* by human beings (SP-7).

ETYMOLOGY. This term replaced "artificial moon" as the common popular name for satellites in the early 1960s. Although the term is modern, the concept is not. In his book *Philosophiae Naturalis Principia Mathematica* (1687), Sir Isaac Newton pointed out that a cannonball shot at a sufficient velocity from atop a mountain in a direction parallel to the horizon would go all the way around the Earth before falling. The force of gravity would put it in a stable orbit.

Earth station. A radio station located either on the Earth's surface or in its lower *atmosphere,* used to contact one or more *space station*s of similar kind by means of reflecting *satellite*s or other celestial objects.

Earth System Science Pathfinder. A line of small Earth observation satellites characterized by a 36-month development schedule and capped life-cycle costs. The first two such missions, the Vegetation Canopy Lidar and the Gravity Recovery and Climate Experiment, were launched in 2000 and 2001, respectively. *CALIPSO* and CloudSat are also included in this program.

Eastern Space and Missile Center (ESMC). Formerly the USAF Long Range Proving Ground, then the Atlantic Missile Range, and after that the Eastern Test Range. This primary U.S. *launch* site is located at *Cape Canaveral* on the east coast of Florida at 28.5 deg N, 80.6 deg W. This site is primarily used for low-inclination orbits (including *geostationary*) and for all crewed missions. See *Kennedy Space Center.*

Eastern Test Range (ETR). Formerly the USAF Long Range Proving Ground, then the Atlantic Missile Range, and now the *Eastern Space and Missile Center.* See *Kennedy Space Center.*

eccentric. Describing a noncircular, highly elliptical *orbit.*

eccentricity. The amount of separation between the two foci of an ellipse and, hence, the degree to which an elliptical *orbit* deviates from a circular shape.

Echo. Passive communications *satellite.* The idea of an inflatable, spherical space satellite was conceived in January 1956 by William J. O'Sullivan Jr., aeronautical engineer at the *NACA's* Langley Aeronautical Laboratory (later NASA's *Langley Research Center*), and proposed as an air-density experiment for the *International Geophysical Year* (IGY), July 1, 1957–December 31, 1958. The balloon satellite was similar to one described by John R. Pierce of Bell Telephone Laboratories in his 1955 article, "Orbital Radio Relays." Pierce was interested in the orbiting inflated sphere for use as a reflector for radio signals, and he proposed a cooperative communication experiment using O'Sullivan's balloon satellite. By early 1959 O'Sullivan's original proposal for IGY air-density studies had become NASA's passive communications satellite project.

O'Sullivan's design was tested in a series of *Shotput* launches, and the Echo project proved that an aluminized-*Mylar* sphere could be carried aloft by a *rocket,* be inflated in space, and remain in *orbit* to provide a means of measuring atmospheric density as well as a surface for reflecting radio communications between distant points on the Earth.

Echo I, a passive communications satellite, launched by NASA on August 12, 1960, was the fruition of O'Sullivan's labors. His inflatable-sphere concept was also employed in three air-density *Explorer* satellites, in Echo 2, and in *PAGEOS* 1.

ETYMOLOGY. The word echo was often used in the radio and radar sense to describe the reflection of ground-transmitted signals from the surface of an orbiting balloon. The name Project Echo, derived through informal use, was given to the 98-foot (30-m) inflatable-structure satellite.

SOURCES. SP-4402, pp. 40–41; William J. O'Sullivan, "Notes on Project Echo" (ms.), n.d.; NASA News Release 61-252; John R. Pierce, "Orbital Radio Relays," *Jet Propulsion* 25 (April 1955): 153–57; E. W. Morse, "Preliminary

History of the Origins of Project Syncom" (ms.), 1964; Robert W. Mulac, LaRC, letter to Historical Staff, NASA, December 10, 1963.

EGADS button. Button used by the range safety officer to *destruct* a *missile* in flight. This concept was popularized in the term "panic button." The acronym EGADS stands for Electronic Ground Automatic Destruct Sequencer. (SP-6001, p. 30.)

FIRST USE. A 1957 article about *Cape Canaveral* described the destruction of missiles that have gone off course and the cry of EGADS that goes up when that happens: "There is gloom on Earthstrip No. 1 when a missile ends with EGADS" ("Visit to Earthstrip No. 1," *New York Times Sunday Magazine,* September 8, 1957, pp. 13-16). Earthstrip No. 1 was a short-lived nickname for the missile facility located at Patrick Air Force Base in Florida that would eventually become *Kennedy Space Center.*

Einstein. The second of NASA's High Energy Astrophysical Observatories, *HEAO*-2, renamed Einstein after *launch* on November 12, 1978. It was the first fully imaging x-ray telescope put into space. It was named for Albert Einstein (1879–1955), the German-born theoretical physicist widely regarded as one of the greatest scientists of all time. (http://heasarc.gsfc.nasa.gov/docs/einstein/heao2.html.)

electronic intelligence (ELINT). The interception and analysis of other countries' electromagnetic signals, including radar, radio, telephone, and microwave transmissions. Done effectively utilizing satellites.

Electronics Research Center (ERC). NASA component located in Cambridge, Massachusetts, that served to develop the agency's *in-house* expertise in electronics during the *Apollo* era. ERC opened in September 1964, taking over the administration of contracts, grants, and other NASA business in New England from the antecedent North Eastern Operations Office, created in July 1962 and closed in June 1970 because of budget reductions. The facility was transferred to the Department of Transportation for use in research and development efforts and was renamed the Transportation Development Center. (SP-4402, pp. 139–40; NASA News Releases 64-219, 69-171; NASA Announcement 64-189; NASA Circulars 320, 32; General Services Administration, National Archives and Records Service, Office of the Federal Register, *Weekly Compilation of Presidential Documents* 6, no. 13 [March 30, 1970]: 446; Department of Transportation Release 6870.)

EMU. *Extravehicular Mobility Unit.*

Endeavour. (1) *Call sign* for the *Apollo* 15 Command Module. This was the first *mission* to employ the *Lunar Roving Vehicle*. The mission—crewed by David R. Scott, commander; Alfred J. Worden, Command Module pilot; and James B. Irwin, *Lunar Module* (call sign *Falcon*) pilot—was launched on schedule from the NASA *Kennedy Space Center* on July 26,

1971. (2) The fifth *orbiter* in the *STS* program, *OV* 105, the "replacement orbiter." Congress authorized the construction of Endeavour in 1987 to replace *Challenger,* which was lost in 1986. Structural spares from the construction of other shuttles were used

in its assembly. The decision to build Endeavour was favored over refitting *Enterprise,* in part because it was cheaper. The shuttle was delivered by Rockwell International in May 1991 and launched a year later. On its first *mission* it captured and redeployed the stranded *Intelsat* VI communications *satellite,* and in 1993 it made the first servicing mission to the *Hubble Space Telescope.*

ETYMOLOGY. Both *spacecraft* were named after the first ship commanded by James Cook, the 18th-century British explorer, navigator, and astronomer. On Endeavour's maiden voyage in August 1768, Cook sailed to the South Pacific, in order to observe and record the infrequent event of the planet Venus passing between the Earth and the Sun. Determining the transit of Venus enabled early astronomers to find the distance of the Sun from the Earth, which could then be used as a unit of measurement in calculating the parameters of the *universe.* In 1769, Cook was the first person to fully chart New Zealand (previously visited in 1642 by the Dutchman Abel Tasman from the Dutch province of Zeeland). Cook also surveyed the eastern coast of Australia , navigated the Great Barrier Reef, and traveled to Hawaii.

The decision to give this name to the orbiter came out of a nationwide competition involving over 71,000 schoolchildren. On March 10, 1986, Congressman Tom Lewis (R., Florida) first called for the name to be selected from suggestions submitted by students in elementary and secondary schools. In October 1987, Congress authorized this move in House Joint Resolution 559. (NASA Press Release 89-70, "President Bush Names Replacement Orbiter Endeavour," May 10, 1989.)

USAGE. The name of both the Command Module and the orbiter is spelled in the British manner.

engine. In *spacecraft,* a *rocket* or thruster that burns liquid *propellant*s and can be throttled to adjust *thrust.*

Enterprise. The first *Space Shuttle orbiter.* Designated *OV*-101, the vehicle was rolled out of Rockwell's Air Force Plant 42, Site 1 assembly facility (Palmdale, Calif.) on September 17, 1976. On January 31, 1977, it was transported 36 miles (58 km) overland from the assembly facility to NASA's Dryden Flight Research Facility at Edwards Air Force Base for the approach and landing test. The Enterprise also made its first free-flight test at Dryden on August 12, 1977.

ETYMOLOGY. OV-101 was originally to have been named Constitution (in honor of the United States Constitution), and NASA had planned to unveil the first orbiter on September 17, 1976—Constitution Day. However, viewers of the popular television science fiction show *Star Trek* initiated a write-in campaign urging the White House (under President Gerald Ford) to adopt the name Enterprise (the starship of the TV series). Ford agreed to the change. ("And Now, a Real Space Shuttle Named 'Enterprise,'" *Washington Post,* September 9, 1976, p. E9.)

   This decision did not disrupt NASA's plan to name all of the orbiters after famous seagoing vessels because Enterprise was also the name of a sailing ship that took part in an important Arctic expedition between 1851 and 1854 as well as the name of the first nuclear-powered aircraft carrier.

Environmental Research Satellite (ERS). See *ERTS, EOS,* and SEOS.

Environmental Science Services Administration. See *ESSA.*

Eole. French meteorological *satellite* CAS-A. NASA and the French *space agency CNES* (Centre National d'Études Spatiales) signed a memorandum of understanding on May 27, 1966, providing for development of a cooperative satellite-and-instrumented-balloon network to collect meteorological data for long-range weather forecasts. Known as FR-2 (see *FR-1*) and also as simply French Satellite until December 1968, the project was redesignated by NASA as CAS-A, an acronym for the first in a series of international Cooperative Applications Satellite(s). The satellite was given its permanent name Eole after successful *launch* into *orbit* on August 16, 1971. (SP-4402, pp. 52–53; NASA News Releases 66-156, 70-222; NASA, "Project Approval Document," December 7, 1966; NASA, "Research and Development: Cooperative Effort/Flight," FY 1969 Project Approval Document, December 2, 1968.)

   ETYMOLOGY. Eole, the French name for Aeolus, ancient Greek god of the winds, was chosen by CNES as the name for the satellite project.

EOS. (1) Earth Observatory Satellite. A early generic name for *satellites* that made observations of the Earth. The term is applied in names like SEOS (Synchronous Earth Observatory Satellite). (2) Earth Observing System. A series of Earth-observing satellites at the core of NASA's Earth Science Program. Key areas of study include clouds; water and energy cycles; oceans; chemistry of the atmosphere; land surface changes; water and ecosystem processes; glaciers and polar ice sheets; and the solid Earth. Phase I of EOS consisted of focused, free-flying satellites, *Space Shuttle* missions, and various airborne and ground-based studies. Phase II began in December 1999 with the launch of the first EOS satellite, *Terra,* and later *Aqua* and *Aura.*

equatorial orbit. An *orbit* in the plane of the equator.

ERC. *Electronics Research Center.*

EROS. Earth Resources Observation Satellite. See *ERTS, EOS,* and *SEOS.*

ERS. Earth Resources Survey Satellite. See *ERTS, EOS,* and *SEOS.*

ERTS, EOS. Between 1964 and 1966, studies of remote-sensing applications were conducted jointly by NASA and the Departments of Interior and Agriculture, and NASA initiated a program of aircraft flights to define sensor systems for remote-sensing technology. The studies indicated that an automated remote-sensing *satellite* appeared feasible and that a program should be initiated for the development of an experimental satellite.

In early 1967 NASA began definition studies for the proposed satellite, by then designated ERTS (for Earth Resources Technology Satellite), and by early 1969 the project was approved. Two satellites, ERTS-A and ERTS-B, were subsequently planned for *launch*. ERTS-A became ERTS 1 upon launch in July 1972; it was still transmitting data on Earth resources, pollution, and environment at the end of 1974 for users worldwide and was turned off in January 1978. ERTS-B was launched on January 2, 1975.

NASA announced on January 14, 1975, that ERTS 1 had been renamed *Landsat* 1, and ERTS-B became Landsat 2 when launched. Associate Administrator for Applications Charles W. Mathews said that since NASA planned a *SEASAT* satellite to study the oceans, Landsat was an appropriate name for a satellite that studied the land. Dr. George M. Low, Deputy Administrator of NASA, had suggested that a new name with more public appeal be found for ERTS. John P. Donnelly, Assistant Administrator for Public Affairs, had therefore asked NASA office heads and centers to submit ideas for new names by the end of December 1974. The NASA Project Designation Committee made its recommendation from among a number of the replies received, and Landsat was approved.

The early nomenclature for both the program and the proposed satellites was confusing. The Earth Resources Program was variously known as the Natural Resources Program, the Earth Resources Survey Program, and the Earth Resources Observation Program.

The designation Earth Resources Survey Program was eventually used to include ERTS and remote-sensing aircraft programs, as well as the Earth Resources Experiment Package (EREP) flown on *Skylab* missions in 1973–74. These programs formed a part of NASA's overall Earth Observations Programs, which also included the meteorology and Earth physics program.

Before 1967 several names were in use for the proposed Earth resources satellite, including the designation ERS, an acronym for Earth Resources Survey satellite, which was in conflict with an identical designation for an Air Force satellite project known as the Environmental Research Satellite.

Further confusion arose when the Department of the Interior, which in cooperation with NASA had been studying the application of remote-sensing techniques, announced the name EROS (an acronym for Earth Resources Observation Satellite) for the satellite project.

In early 1967, when NASA initiated the definition studies of the experimental satellite, the name ERTS came into use.

In early 1970 the NASA Project Designation Committee met to choose a new name for the ERTS satellites, and several names were suggested, including Earth, Survey, and Ceres (the ancient Greek goddess of the harvest). The committee favored Earth, but after submitting the name to the other government agencies in the program and receiving unfavorable responses from some, it dropped the name Earth, and ERTS was used up through the end of 1974.

ETYMOLOGY. The name ERTS as an acronym for Earth Resources Technology Satellite was a functional designation; it was derived from early concepts of an "Earth resources" satellite system to provide information on the environment by using remote-sensing techniques.

SOURCES. SP-4402, pp. 42–45; John Hanessian Jr., "International Aspects of Earth Resources Survey Satellite Programs," *Journal of the British Interplanetary Society* 23 (Spring 1970): 535, 541; NASA News Release 75-15; Charles W. Mathews, ERTS-B Mission Briefing, NASA Hq., January 14, 1975; Howard G. Allaway, Public Affairs Officer, NASA, telephone interview, February 13, 1975; Bernice Taylor, Administrative Assistant to Assistant Administrator for Public Affairs, NASA, telephone interview, February 12, 1975; Eldon D. Taylor, Director of Program Review and Resources Management, NASA, memorandum to Chief, Management Issuances Section, NASA, July 7, 1966; W. Fred Boone, Assistant Administrator for Defense Affairs, NASA, memorandum to Leonard Jaffe, Director of Space Applications Programs, NASA, July 22, 1966; John E. Naugle, Associate Administrator for Space Science and Applications, NASA, prepared statement in U.S. Congress, House of Representatives, Committee on Science and Astronautics, Subcommittee on Space Science and Applications, *Hearings: 1972 NASA Authorization, Pt. 3, March 1971* (Washington, DC: GPO, 1971), pp. 156–57; Leonard Jaffe, memorandum to Julian W. Scheer, Assistant Administrator for Public Affairs, NASA, January 10, 1967; Department of the Interior, Office of the Secretary, release, "Earth's Resources to Be Studied from Space," September 21, 1966; NASA, Program Review and Resources Management Office, "Chronology of NASA Earth Resources Program," December 13, 1971; George J. Vecchietti, Director of Procurement, NASA, memorandum to Philip N. Whittaker, Assistant Administrator for Industry Affairs, NASA, October 14, 1968.

ESA. *European Space Agency.*

escape tower. A *rocket*-powered framework designed to separate *spacecraft* modules from their *booster rockets* in case of accident. Escape towers are mounted atop the spacecraft and jettisoned after *launch*. (SP-6001, p. 32.)

escape velocity. The minimum speed needed to escape the gravitational pull of the Earth or another planet.

ESMC. *Eastern Space and Missile Center.*

ESRO. European Space Research Organization. Precursor to the *European Space Agency* (ESA). A 10-member Western European group formed to conduct scientific space research, ESRO came into formal existence in March 1964 (the ESRO Convention had been signed June 14, 1962).

Under a memorandum of understanding signed July 8, 1964, ESRO and NASA agreed to participate in a joint *satellite* project, also called ESRO. NASA's role in that cooperative venture was to provide *Scout launch vehicles*, conduct *launch* operations, provide supplemental *tracking* and data acquisition services, and train ESRO personnel. No funds were exchanged in the project. The agreement originally called for two cooperative satellites: ESRO 1 to investigate the polar ionosphere and ESRO 2 to study solar astronomy and cosmic rays. With development of the scientific *payloads*, it became apparent that ESRO 1 had a rather narrow *launch* opportunity and that it was important to launch it in the fall. Therefore ESRO 2 was moved up for first launch, although the number designations were not changed. After launch by NASA on October 3, 1968, ESRO 1 was assigned the name Aurorae by ESRO; it was designed to study the aurora borealis and related phenomena of the polar ionosphere. Its numerical designation later became ESRO IA when a duplicate *backup* satellite, ESRO IB, was launched October 1, 1969. ESRO IB was designated *Boreas* by ESRO. ESRO 2A, scheduled to be the first ESRO satellite, failed to reach *orbit* May 29, 1967. Its backup, ESRO 2B, was given the name *IRIS* (International Radiation Investigation Satellite) by ESRO after its successful launch on May 16, 1968.

Under a December 30, 1966, memorandum of understanding, ESRO became the first international space group to agree to pay NASA for launchings; it would reimburse NASA for launch vehicle and direct costs of equipment and services. The first satellite launched under this agreement, *HEOS* 1 (Highly Eccentric Orbit Satellite), was launched December 5, 1968. Later scientific and applications satellites planned by ESRO and to be launched by NASA were given functional names: *GEOS* (Geostationary Scientific Satellite, a different satellite from NASA's Geodetic Explorer or Geodynamic Experimental Ocean Satellite), to study cosmic *radiation* over a long period; EXOSAT, a high-energy astronomy satellite, for x-ray astronomy; METEOSAT, a *geostationary* meteorological satellite; OTS, a geostationary Orbital Test Satellite, a forerunner of the European Communications Satellite; AEROSAT, a joint Aeronautical Satellite developed with the U.S. Federal Aviation Agency and a U.S. contractor, to be launched in 1977 or 1978 for air traffic control, navigation, and communications; and MAROTS (Maritime Orbital Test Satellite), an adaptation of OTS funded principally by the United Kingdom for civil maritime communications and navigation.

SOURCES. SP-4402, p. 47, "Memorandum of Understanding between the European Space Research Organization and the United States National Aeronautics and Space Administration," attachment to NASA News Release 64-178; NASA News Release, "Press Briefing: ESRO II and NASA International Cooperative Programs," May 19, 1967; ESRO, *Europe in Space* (Paris: ESRO, March 1974), pp. 21–40; A. V. Cleaver, in *Spaceflight* 16, no. 6 (June 1974): 220–37.

ESSA. Environmental Science Services Administration. U.S. government environmental agency, precursor to the National Oceanic and Atmospheric Administration.

ESSA satellites. Meteorological satellites in the Tiros Operational Satellite (TOS) system that were financed and operated by the Environmental Science Services Administration *(ESSA)*. The name was selected by ESSA early in 1966 and was an acronym derived from Environmental Survey Satellite. It was also the abbreviation for the operating agency. Between 1966 and 1969 NASA procured, launched, and checked out in *orbit* the nine ESSA satellites, beginning with ESSA 1, launched February 3, 1966. On October 3, 1970, ESSA was incorporated into the new National Oceanic and Atmospheric Administration (NOAA). After *launch* by NASA, the subsequent series of satellites in the Improved TOS (ITOS) system were turned over to NOAA for operational use. The first ITOS spacecraft funded by NOAA, launched December 11, 1970, was designated NOAA 1 in orbit, following the pattern set by the ESSA series. See *Tiros, TOS, ITOS*. (SP-4402, p. 48–49; NASA-ESSA, ESSA 1 Press Kit, ES 66-7, January 30, 1966; NASA program office.)

ETI. Extraterrestrial intelligence. See also *SETI*.

ETR. *Eastern Test Range.*

European Space Agency (ESA). A 17-nation agency whose stated mission is "to shape the development of Europe's space capability and ensure that investment in space continues to deliver benefits to the citizens of Europe." ESA came into existence in 1975 out of *ESRO* and the European Launcher Development Organization (ELDO) after an attempt to develop a launching *rocket* for Europe had ended in failure.

Headquartered in Paris, ESA has developed its own launchers and *spacecraft* and is a partner in the *International Space Station*. In mid-2007 the member states were Austria, Belgium, Denmark, Finland, France, Germany, Greece, Ireland, Italy, Luxembourg, the Netherlands, Norway, Portugal, Spain, Sweden, Switzerland, and the United Kingdom. Canada, Hungary, Poland, and the Czech Republic also participate in some projects under cooperation agreements.

Europe's Spaceport. Official nickname for the *Centre Spatial Guyanais* located in *Kourou,* French Guiana.

EVA. *Extravehicular activity.*

the Excess Eleven. Self-imposed nickname of the *astronaut*s selected in mid-1967. They called themselves this because it was unlikely that they would go into space (most eventually did). They were the sixth group of astronauts and followed groups nicknamed the *Original Seven,* the *Next Nine,* and the *Fourteen.* (Compton, *Where No Man Has Gone Before,* SP-4214, p. 136.)

exobiology. The branch of biology that deals with life beyond the regions of the Earth. As a term and a discipline, it stemmed from early concerns by Nobel laureate and geneticist Joshua Lederberg, who believed that other planets could harbor life that could be contaminated by microorganisms carried from Earth via *spacecraft,* and that, on the other hand, Earth could fall victim to an unknown pathogen brought back from another planet, a pathogen to which Earth's inhabitants might have no immunity. Also known as xenobiology and, since the mid-1990s, as *astrobiology.* (Dick and Strick, *Living Universe.*)

ETYMOLOGY. The term was coined by Lederberg and gained immediate attention as it was linked to his urgent warnings about interstellar contamination and his call for the scientific study of life beyond Earth's *atmosphere.* He is quoted on the website for the National Library of Medicine (http://profiles.nlm.nih.gov/BB/) where his papers are archived: "I was the only biologist at the time who seemed to take the idea of extraterrestrial exploration seriously. People were saying it would be a hundred years before we even got to the moon." Lederberg, however, "was convinced that once the first satellite was up the timetable would be very short, and [his] fear was that the space program would be pushed ahead for military and political reasons without regard for the scientific implications." He collaborated with the well-known astronomer Carl Sagan in establishing exobiology as a scientific discipline. By publicly promoting exobiology, Lederberg almost singlehandedly gained a place for biologists in the *space program,* as well as a share of its ample research funds.

FIRST USE. The term was introduced by Lederberg at the First International Space Science Symposium in Nice, France, in January 1960 and was first mentioned in "Device Designed to Land on Mars and Send Reports of Life There," *New York Times,* January 14, 1960, p. 6.

exotic fuel. Any unusual fuel, especially with high energy content for added *thrust* (SP-6001, p. 33).

FIRST USE. "US Seeks to Bolster Rockets by 'Exotic' Fuel," *Los Angeles Times,* February 18, 1964, p. 10.

Explorer. First U.S. *satellite* to be successfully placed in Earth *orbit* (January 31, 1958). It was a bullet-shaped object 80 inches (2 m) long and weighing about 31 pounds (14 kg). Sixty-one satellites with the Explorer name were launched between 1958 and 1975.

When NASA was being formed in 1958 to conduct the U.S. civilian *space program,* responsibility for IGY *(International Geophysical Year)* scientific satellite programs was assigned to NASA. The decision was made by the National Advisory Committee for Aeronautics *(NACA)* to continue the name Explorer as a generic term for future NASA scientific satellites. Explorers were used by NASA to study (1) the *atmosphere* and ionosphere, (2) the *magnetosphere* and interplanetary space, (3) astronomical and astrophysical phenomena, and (4) the Earth's shape, *magnetic field,* and surface.

Many of the Explorer satellites had project names that were used before they were launched and then supplanted by Explorer designations once they were placed in *orbit.* Other Explorer satellites, particularly the early ones, were known before orbit simply by numerical designations. A listing of some of the Explorers' descriptive designations illustrates the variety of scientific missions performed by these satellites: Aeronomy Explorer, Air Density Satellite, Direct Measurement Explorer, Interplanetary Monitoring Platform (IMP), Ionosphere Explorer, Meteoroid Technology Satellite (MTS), Radio Astronomy Explorer (RAE), Solar Explorer, and Small Astronomy Satellite (SAS).

SAS-A, an X-ray Astronomy Explorer, became Explorer 42 when launched December 12, 1970, by an Italian crew from the *San Marco* platform off the coast of Kenya, Africa. It was also christened Uhuru (Swahili for freedom) because it was launched on Kenya's Independence Day. The small satellite, mapping the *universe* in x-ray wavelengths for four years, discovered x-ray *pulsar*s and evidence of *black holes.*

Geodetic Satellites (GEOS) were also called Geodetic Explorer Satellites and sometimes Geodetic Earth Orbiting Satellites. GEOS 1 (Explorer 29, launched November 6, 1965) and GEOS 2 (Explorer 36, launched January 11, 1968) refined knowledge of the Earth's shape and gravity field. GEOS-C, also known as GEOS-3 and to be launched in

1975 as a successor to GEOS 1 and 2, was renamed Geodynamic Experimental Ocean Satellite to emphasize its specific mission in NASA's Earth and ocean physics program while retaining the GEOS acronym. GEOS-C was to measure ocean currents, tides, and wave heights to improve the geodetic model of the Earth and knowledge of Earth-sea interactions. (The European Space Research Organization's *Geostationary* Scientific Satellite, also called GEOS, planned for 1976 *launch,* was not a part of the Geodetic Explorer series. See *ESRO.*)

The 52nd Explorer satellite, launched by NASA on June 3, 1974, was Hawkeye 1, a University of Iowa–built *spacecraft.* The university's *Injun* series had begun with Injun 1 on June 29, 1961, to study charged particles trapped in the Earth's magnetosphere. The first three Injuns were launched by the Air Force (Injun 2 failed to reach orbit; Injun 3 was launched December 13, 1962). NASA launched the next three, adding the Explorer name. Hawkeye 1 originally carried the prelaunch designation Injun F, but this was discarded; the Hawkeye name was approved by the NASA Project Designation Committee in June 1972. (Injun 4, November 21, 1964, was also named Explorer 25. Injun 5, August 8, 1968, was Explorer 40.)

Two International Sun-Earth Explorers—ISEE-A (sometimes called Mother) and ISEE-B (sometimes called Daughter)—were dual-launched in 1977, followed by ISEE-C (called Heliocentric) in 1978. The joint NASA-ESRO program earlier called the International Magnetosphere Explorer (IME) program was launched to investigate Sun-Earth relationships and solar phenomena.

An International Ultraviolet Explorer (IUE; originally designated SAS-D in the Small Astronomy Satellite series) was launched on January 26, 1978, as a cooperative NASA-U.K.-ESRO mission to gather high-resolution ultraviolet data on astronomical objects. It operated until 1996, when it was shut down for budgetary reasons despite still functioning at near-original efficiency.

ETYMOLOGY. The name Explorer, designating NASA's scientific satellite series, originated before NASA was formed. Explorer was used in the 1930s for the U.S. Army Air Service–National Geographic stratosphere balloons. On January 31, 1958, when the first U.S. satellite was launched by the U.S. Army as a contribution to the *International Geophysical Year* (IGY), Secretary of the Army Wilbur M. Bucker announced the satellite's name, Explorer 1. The name indicated the *mission* of this first satellite and its NASA successor to explore the unknown.

The Army Ballistic Missile Agency *(ABMA)* had previously rejected a list of the names of famous explorers for the satellite. The *Jet Propulsion Laboratory,* responsible for the fourth stage of the *Jupiter* C *rocket* (configured as the *Juno I launch vehicle*) and for the satellite, had called the effort Project Deal (see *Deal Project*): a loser in a poker game always called for a new deal, and this satellite was the answer to the Russian *Sputnik*. On the day of the launch, ABMA proposed the name *Top Kick,* which was not considered appropriate. The list of names was brought out again. All the names on the list had been crossed out, and only the heading Explorers remained. The late Richard Hirsch, a member of the National Security Council's Ad Hoc Committee for Outer Space, suggested that the first American satellite be called simply Explorer. The name was accepted and announced.

SOURCES. SP-4402, pp. 49–52; A. Ruth Jarrell, Historical Office, MSFC, letter to Historical Staff, NASA, December 16, 1963; R. Cargill Hall, *Project Ranger: A Chronology,* JPL/HR-2 (Washington, DC: NASA, 1971), p. 46; Eugene M. Emme, Historian, NASA, memorandum for the record (after conversation with Richard Hirsch, National Aeronautics and Space Council Staff), February 26, 1970; NASA Headquarters Preliminary Results Press Briefing (transcript), December 28, 1970; American Institute of Physics News Release, April 28, 1971; *NASA Facts* 3, no. 4 (Washington, DC: NASA, 1966); NASA News Releases 65-333, 65-354, 68-16, 70-108, 70-203; U.S. Congress, House of Representatives, Committee on Science and Astro-nautics, Subcommittee on Space Science and Applications, *Hearings: 1975 NASA Authorization, Pt. 3, February and March 1974* (Washington, DC: GPO, 1974), pp. 104–7, 189–190, 318; NASA, Historical Division, *Astronautics and Aeronautics 1968: Chronology on Science, Technology, and Policy,* SP-4010 (Washington, DC: NASA, 1969), p. 182.

extraterrestrial life. Life forms evolved and existing beyond the Earth. Despite robust programs in *exobiology, astrobiology,* and *SETI,* no such forms have yet been detected.

extravehicular activity (EVA). Activity conducted by an *astronaut* in space outside the *space vehicle,* requiring the astronaut to wear a life support suit and exit the vehicle. Also known as a *spacewalk.*

Extravehicular Mobility Unit (EMU). An independent anthropomorphic system that provides environmental protection, mobility, life support, and communications for a *Space Shuttle* or *International Space Station* crew member to perform *extravehicular activity* in Earth *orbit.*

eyeball instrumentation. Visual inspection.

eyeballs down. Positive gravitational pull. The acceleration stress that a human experiences as acting from above. See also *eyeballs up.* (SP-6001, p. 33.)

eyeballs in. Supine gravitational pull. The acceleration stress that a human experiences in the chest-to-back direction. (SP-6001, p. 33.)

ETYMOLOGY. John H. Glenn Jr. described a typical run of the centrifuge in

Mercury testing: "Then the gondola was turned so that you faced inward to let the *G* forces push toward the front of your body from front to back, and shoving you into the couch just as you would be pushed back by acceleration during launch. We called this riding 'eyeballs in' which meant that if our eyeballs were actually free to move that much—which they were not, of course—we would feel them being pushed clear back into their sockets." (Glenn in Carpenter et al., *We Seven,* p. 139.) See also *eyeballs out.*

eyeballs out. Prone gravitational pull. The acceleration stress that a human experiences in the chest-to-back direction. (SP-6001, p. 33.)
   ETYMOLOGY.   John H. Glenn Jr. explained: "If you suddenly had to abort and come to a rapid stop under these conditions, the *G* forces would reverse, and you would feel them going the other way. Since your eyes would tend to pop *out* of their sockets on a ride like that, we called this riding 'eyeballs out.'" (Glenn in Carpenter et al., *We Seven,* p. 139.) See also *eyeballs in.*

eyeballs up. Negative gravitational pull. The acceleration stress that a human experiences as acting from below. See also *eyeballs down.* (SP-6001, p. 33.)

F

Faith 7. *Mercury-Atlas.* Capsule carrying L. Gordon Cooper Jr. on May 15, 1963, a mission that examined the effects of extended spaceflight (Cooper circled the Earth 22 times in 34 hours). It was the capstone of the Mercury program.
   ETYMOLOGY.   Cooper recalled: "An awful lot of symbolism had gone into all those earlier names. I felt a certain responsibility. I selected the name Faith 7 to show my faith in my fellow workers, my faith in all the hardware so carefully tested, my faith in myself and my faith in God. The more you study, the more you know the scientific stuff it correlates. It confirms religious faith." (Quoted in Lattimer, *All We Did Was Fly to the Moon,* p. 17.)

Falcon. (1) *Call sign* for the *Apollo* 15 *Lunar Module,* which deployed the *Lunar Roving Vehicle,* an electric-powered four-wheel-drive car that traversed a total 17 miles (27 km). With the *rover* the astronauts were able to gather 169 pounds (77 kg) of lunar material. The upper portion of the Lunar Module became a small *sub-satellite* left in lunar *orbit* for the first time. (2) Former name of the Air Force base east of Colorado Springs which, among other things, is the center for command and control of all surveillance satellites. On June 5, 1998, Falcon AFB was renamed Schriever AFB in honor of retired Gen. Bernard A. Schriever, who pioneered the development of the nation's ballistic missile programs and is recognized as the father of the U.S. Air Force space and missile program. (3) Name of a launch vehicle developed by *SpaceX*. (4) Acronym for Force Application and Launch from CONUS, a small *launch vehicle* concept of the Defense Advanced Research Projects Agency. The agency describes Falcon as "a reusable Hypersonic Cruise Vehicle (HCV)."

ETYMOLOGY (definition 1). Because Apollo 15 was an all-USAF crew, its Lunar Module was named for that service's mascot. (Lattimer, *All We Did Was Fly to the Moon*, p. 85.)

fall-away section. Portion of *launch* apparatus jettisoned during flight (SP-6001, p. 35).

fallout / fall out (n., adj.). Early term, later replaced by *spinoff*, used to describe the secondary applications of space research: "Technological developments in the nation's aerospace program have led to any number of so-called 'fall out' items that have begun to appear in commercial and consumer markets" ("A Space Blanket for Earthlings," *New York Times*, August 21, 1967, p. 45).

family. NASA jargon used to characterize an anomalous event, as either *in family*, for one that was not unknown or unexpected, or *out of family*, for one that was unknown or unexpected.

USAGE. This way of describing safety issues was harshly criticized in the report of the *Columbia* Accident Investigation Board in 2003. "That's a real friendly way of talking about something that's not supposed to be happening in the first place," said a witness called by the commission. Gen. Kenneth Hess (USAF), a member of the investigation board, said, "You've heard NASA use the terminology 'in family' and 'out of family.' Well the family of foam loss just kept getting bigger and bigger. So you never got to a point where you could make a hard decision that something was unusual." (*Orlando Sentinel*, June 6, 2003, p. 1.)

Gen. Duane W. Deal (USAF), another member of the Columbia Accident Investigation Board, had this to say in his summary of the final report (under the heading "Avoid an Atrophy to Apathy"): "An organization should not invent clever ways of working around processes. For example, NASA created an ad hoc view of the anomalies as either 'in family' or 'out of family' depending on whether the anomaly had been previously observed. This led to 'a family that grew and grew'—until it was out of control. This ad hoc view led to an apathy and acceptance of items such as Challenger's solid rocket booster O-ring leakage and the foam strikes that had plagued the shuttle since its first mission but, until Columbia's demise, had never brought down an orbiter." (Deal, "Beyond the Widget: Columbia Accident Lessons Affirmed," *Air and Space Power Journal,* Summer 2004.)

Farside. See *Project Farside.*

faster, better, cheaper. An approach, pioneered by NASA Administrator Dan Goldin, that enabled NASA to cut costs while still delivering a wide variety of aerospace programs. The approach ultimately proved controversial, with the loss of vital missions to Mars, the loss of the Lewis *spacecraft,* and the cancellation of the Clark *satellite* due to project management failures.

ETYMOLOGY / FIRST USE. The term was used widely before its application to the management of NASA as an alternative to the status quo. For instance, mediation as a way of getting a divorce was seen as a "faster, better, cheaper" alternative to a court divorce. The first use in terms of NASA—albeit with the words in different order—dates back to the appointment of Daniel Goldin, at which time an unnamed White House official typified Goldin to William J. Broad of the *New York Times* as "a faster, cheaper, better kind of guy. He's obviously outside the NASA culture" ("Bush Names Aerospace Executive to Lead NASA in New Direction," *New York Times,* March 12, 1992, p. A1). At about this same time, according to Andrew J. Butrica, Vice President Dan Quayle in his role with the National Space Council talked of NASA adopting a policy of "faster, safer, cheaper and better" (Butrica, *Single Stage to Orbit: Politics, Space Technology, and the Quest for Reusable Rocketry* [Baltimore: Johns Hopkins University Press, 2003], p. 151).

Fat Albert *(sounding rocket).* See *Aries.*

ferret. Satellite using electromagnetic surveillance techniques.

ETYMOLOGY. From the notion of hunting with ferrets, which was the basis for the verb meaning "to force out of hiding."

firing window. Interval of time during which conditions are favorable for launching a *spacecraft* on a specific *mission.* Also called *launch window.*

first man. Name for the *astronaut* "first out" on the lunar surface—an honor achieved by Neil A. Armstrong. The phrase was also the title of James Hansen's 2005 biography of Armstrong.

fission. The nuclear reaction resulting from the splitting of atoms.

fixed satellite. A *satellite* that *orbits* the Earth from west to east at such a speed as to remain fixed over a given place on the Earth's equator at an altitude of approximately 22,300 miles (36,000 km). See *geostationary orbit* and *Syncom.*

flame bucket. A deep cavelike construction built beneath a launcher, open at the top to receive the hot gases of the *rocket* positioned above it, and open on one or three sides below, with a thick metal fourth side bent toward the open sides so as to deflect the exhausting gases. Also known as a flame trench. ("Glossary of Aerospace Age Terms"; SP-7, p. 15; SP-6001, p. 36.)

FIRST USE. "They have forgone, for example, the normal sand-blast cleaning of the steel flame bucket that diverts the flame from the Titan engines" (*Washington Post,* December 5, 1965, p. A1, reporting on the Gemini 7 launch and shortcuts being taken to prepare for the next one).

Flight Research Center (FRC). Renamed *Dryden Flight Research Center.*

flyby. (1, v.) To pass by in flight. (2, n.) A *mission* that calls for a *spacecraft* to observe a planet or other body from close proximity, without being placed in *orbit,* as opposed to a landing.

FIRST USE. "First Mariner Mission Will Be a Fly-by of Venus in 1962," *Missiles and Rockets,* October 31, 1953, p. 15. The term first came into widespread play in 1962 as the Mariner 2 *probe* came close the planet Venus. For example: "If nothing goes wrong—and NASA officials say they don't anticipate any trouble—historic data about the shrouded planet will be taken during a key half-hour period during the flyby" ("Mariner Sends Space Data 22.5 Million Miles," *Los Angeles Times,* November 26, 1962, p. 5).

USAGE. Although the term is almost universally written as a single word in the written media, it is hyphenated in some scientific references.

flying bedstead. Common nickname for the Lunar Landing Research Vehicle (LLRV), manufactured for NASA by Bell Aerosystems as a trainer for the Moon landing, simulating in free flight the actual landing on the Moon. Also known as the pipe rack, it was a complex combination of *rocket* motors and a vertical jet *engine* designed to accustom the astronauts to flying in the lower gravity of the Moon.

footprint. (1) The area within which a *spacecraft* is intended to land. (2) The area of the Earth's surface where a particular *satellite*'s signal can be received. A footprint can cover one-third of the globe but is usually less.

ETYMOLOGY / FIRST USE. The term appears to have been first applied to the *space program* in reference to the *Apollo* program: "Gemini is designed to be capable of landing within an area 500 miles by 200 miles. Apollo has what is called a 5,000- by 400-mile 'landing footprint.'" ("Dyna-Soar or Dinosaur," *New York Times,* March 16, 1963, p. 4.)

the Fourteen. Self-imposed nickname of third group of *astronaut* selectees, after groups that had come to be called the *Original Seven* and the *Next Nine* (Compton, *Where No Man Has Gone Before,* SP-4214, pp. 69, 136).

FR-1. The designation of the French *satellite* launched into *orbit* by NASA on December 6, 1965, in a cooperative U.S.-French program to investigate very-low-frequency electromagnetic waves. The name developed in 1964, when NASA and the French *space agency CNES*

(Centre National d'Études Spatiales) agreed, after preliminary *sounding rocket* experiments, to proceed with the satellite project. CNES provided the satellite and designated it FR-1 (for France or French satellite number one). The first flight unit was designated FR-1A and the *backup* unit, FR-1B. The second U.S.-French cooperative satellite, FR-2, was later renamed *Eole*. (SP-4402, pp. 52–53; NASA News Release 63-49; Homer E. Newell, Associate Administrator for Space Science and Applications, NASA, memorandum to NASA Headquarters and Field Centers, May 24, 1964.)

FRC. *Flight Research Center.* Renamed *Dryden Flight Research Center* (DFRC).

Freedom. See *International Space Station* (ISS).

Freedom 7. *Mercury-Redstone.* The *capsule* that took Alan B. Shepard into suborbital space on May 5, 1961. Shepard was the first American in space.

ETYMOLOGY. Named by Shepard, who explained: "Pilots have always named their planes. It's a tradition. It never occurred to me not to name the capsule. I checked with Dr. [Robert R.] Gilruth and I talked it over with my wife and with John Glenn, who was my backup pilot. We all liked it." (Quoted in Lattimer, *All We Did Was Fly to the Moon,* p. 7.) Shepard also added the 7 because the capsule to which he was assigned was model 7. Because the number also acknowledged the *Original Seven* astronauts, it was used for all Mercury missions.

free return. Also known as free return *trajectory.* Name for a key concept of the *Apollo* lunar flights for a flight path that allowed for the failure of the Service Module's main *engine.* If the Service Module's main engine failed to put the *spacecraft* into lunar *orbit,* it would loop around the Moon under the influence of lunar gravity and head back to Earth. (Compton, *Where No Man Has Gone Before,* SP-4214, p. 73.)

Friendship 7. *Mercury-Atlas.* The *capsule* that carried the first American to *orbit* the Earth, John H. Glenn Jr.

ETYMOLOGY. Glenn explained his choice of the name: "I put my wife Annie and the kids to work studying the dictionaries and a Thesaurus to come up with a list of suitable names on which the family could vote. We played around with Liberty, Independence, a lot of them. The more I thought about it, the more I leaned toward the name Friendship. Flying around the world, over all those countries, that was the message I wanted to convey. In the end that was the name the kids like best, too. I was real proud of them." (Quoted in Lattimer, *All We Did Was Fly to the Moon,* p. 11.)

FROST. Food Reserves on Space Trips. In *Apollo Terminology* this term is cross-referenced to *ice frost,* defined as "ice on the outside of a rocket." (SP-6001, p. 37.)

frustrum. The part of a solid *rocket booster* that houses the *recovery* parachute.

fusion. The combining of lighter elements into heavier ones. For lighter elements (e.g., hydrogen, helium) this process releases energy. Fusion is how stars produce energy, and it is being researched as a way to produce power on Earth.

fusion rocket. Conceptual nuclear-powered *booster* using the *fusion* process for the energy source as distinguished from the *fission* process, which was used in *NERVA* (NASA, *Glossary/Congressional Budget Submission*, p. 16).

G

G/G-force. The gravitational force of the Earth (SP 6001, p. 39).

galactic. Pertaining to any *galaxy*, including the *Milky Way* (SP-7).

Galactic Ghoul. See *Great Galactic Ghoul.*

galaxy. One of billions of systems, each including stars, nebulae, clusters of stars, gas, and dust, that make up the *universe.*

GALCIT. Guggenheim Aeronautical Laboratory, California Institute of Technology. Forerunner of the *Jet Propulsion Laboratory.* During World War II the GALCIT Rocket Research Project developed solid and liquid-*propellant* units to assist the *takeoff* of heavily loaded aircraft and began work on high-altitude rockets. They were known as JATO (Jet-Assisted Takeoff). Reorganized in November 1944 under the name Jet Propulsion Laboratory, the facility continued postwar research and development on tactical guided missiles, aerodynamics, and broad supporting technology for U.S. Army Ordnance.

Galileo. (1) *Probe* designed to perform in-depth studies of Jupiter's *atmosphere*, satellites, and surrounding *magnetosphere*. Launched on October 18, 1989, from *STS-34 Atlantis*, it remained in Jovian *orbit* for more than seven years, making it one of the most successful planetary exploration *spacecraft* ever. It also had an atmospheric *probe*. (NASA News Release 78-11; Memo for the Record, signed B. R. McCullar, December 8, 1977; NASA Names Files, record no. 17522.) (2) The as-yet undeployed European Union version of *GPS.*

ETYMOLOGY (definition 1). The *mission* to Jupiter was named in honor of Galileo Galilei, the Italian Renaissance scientist who was the first to observe the planet by telescope and discovered Jupiter's major moons in 1610. The name was decided at a meeting held on November 28, 1977, in which the

name bested the two top runners-up: *Juno* and Endeavor. A NASA news
release noted, "Galileo is the first planetary spacecraft to be named for a
person," although NASA's Orbiting Astronomical Observatory 3 *(OAO-3)*
was christened Copernicus after it was launched in 1972.

gantry. Tall cranelike structure with platforms on different levels used to
erect, assemble, and service large *rocket*s or *missile*s (SP-6001, p. 39).

gap filler. Ceramic coated-fabric strips used to prevent hot gas from
seeping into gaps between the *Space Shuttle*'s protective tiles.

garbage. (1) *Rocket* parts that go into *orbit* with a *satellite* ("Space Age
Slang," *Time* magazine, August 10, 1962, p. 12). (2) Miscellaneous
objects in orbit (SP-6001, p. 39).

GAS. Acronym for *Getaway Special,* which made for some nice punning
headlines. For example: "Here's a GAS of a Cargo Flight" (*Washington
Star,* May 14, 1978, p. A14).

GATV. Gemini Agena Target Vehicle. See *Gemini.*

Gemini. A series of two-man *spacecraft* capable of *docking* with another
vehicle in space.

In 1961, planning was begun on an Earth-orbital *rendezvous* program to follow the *Mercury* project and prepare for *Apollo* missions.
The improved or Advanced Mercury concept was designated *Mercury
Mark II* by Glenn F. Bailey, NASA Space Task Group Contracting Officer,
and John Y. Brown of McDonnell Aircraft Corporation. The two-man
spacecraft was based on the one-man Mercury *capsule,* enlarged and
made capable of longer flights. Its major purposes were to develop the
technique of rendezvous in space with another spacecraft and to
extend orbital flight time.

Project Gemini ended on November 15, 1966, after 12 missions
(2 uncrewed and 10 with crews). Its achievements had included
long-duration spaceflight, rendezvous and docking of two spacecraft
in Earth *orbit, extravehicular activity,* and precision-controlled *reentry*
and landing of spacecraft.

ETYMOLOGY. NASA Headquarters personnel were asked to suggest an
appropriate name for the project, and in a December 1961 speech at the
Industrial College of the Armed Forces, Dr. Robert C. Seamans Jr., then
NASA Associate Administrator, described Mercury Mark II, adding an offer
of a token reward to the person suggesting the name finally accepted. A
member of the audience sent him the name Gemini. Meanwhile, Alex P.
Nagy in NASA's Office of Manned Space Flight had also proposed Gemini
in a memo saying that the name "seems to carry out the thought nicely
of a two-man crew, a rendezvous mission, and its relation to Mercury."
Dr. Seamans recognized both as authors of the name. Gemini (Latin for
twins) was the name of the third constellation of the zodiac, made up of

the twin stars Castor and Pollux. These mythological twins were considered to be the patron gods of voyagers. The name was selected from several suggestions made in NASA Headquarters, including Diana, Valiant, and Orpheus. On January 3, 1962, NASA announced that the Mercury Mark II project had been named Gemini.

The crew of the first human Gemini mission, astronauts Virgil I. Grissom and John W. Young, nicknamed their spacecraft *Molly Brown.* The name came from the musical comedy "The Unsinkable Molly Brown" and was a facetious reference to the sinking of Grissom's *Mercury-Redstone* spacecraft after *splashdown* in the Atlantic Ocean on July 21, 1961. Molly Brown was the last Gemini spacecraft to have a nickname; after the Gemini 3 mission, NASA announced that "all Gemini flights should use as official spacecraft nomenclature a single easily remembered and pronounced name."

FIRST USE. The first major announcement of the name Gemini appears in Neal Stamford, "'Glenn Shot' Delay Put in Perspective," *Christian Science Monitor,* January 5, 1962, p. 5.

SOURCES. SP-4402, pp. 104–6; NASA Names Files, record no. 17523; Glenn F. Bailey, MSC, interview, December 13, 1966, reported by James M. Grimwood, Historian, MSC, May 23, 1968; Alex P. Nagy, Office of Manned Space Flight, NASA, memorandum to George M. Low, Office of Manned Space Flight, NASA, December 11, 1961; D. Brainerd Holmes, Director of Manned Space Flight Programs, NASA, memorandum to Associate Administrator, NASA, December 16, 1961; Holmes, memorandum to Associate Administrator, NASA, January 2, 1962; Robert C. Seamans Jr., Secretary, USAF, letter to Eugene M. Emme, Historian, NASA, June 3, 1969; desk calendar of Seamans, Associate Administrator, NASA, December 15, 1961; Bulfinch, *Mythology,* pp. 130–31; Holmes, memorandum to Associate Administrator, NASA, December 16, 1961; NASA, "NASA Photo Release 62-Gemini-2," January 3, 1962; Grimwood, *Project Mercury: A Chronology,* SP-4001 (Washington, DC: NASA, 1963), p. 133; Swenson, Grimwood, and Alexander, *This New Ocean,* SP-4201, pp. 491–92; Seamans, Associate Administrator, NASA, memorandum to George E. Mueller, Associate Administrator for Manned Space Flight, NASA, and Julian W. Scheer, Assistant Administrator for Public Affairs, NASA, May 4, 1965.

Gemini-Titan (GT). The program that would use *Titan booster*s to put *Gemini capsule*s in *orbit.* Comparable to the *Mercury-Atlas* designation.

Genesis rock. Name given to a sample of the lunar *crust* brought back by the crew of *Apollo* 15 as part of a haul of 173 pounds (78 kg) of Moon rocks it brought back on August 7, 1971. It is one of the oldest samples of lunar crust brought back to Earth.

GEO. Geostationary (or *geosynchronous*) Earth *orbit.*

Geodetic Earth Orbiting Satellite (GEOS). See *Explorer.*

Geodetic Explorer Satellite (GEOS). See *Explorer.*

Geodynamic Experimental Ocean Satellite (GEOS). See *Explorer.*

GEOS. (1) *ESRO* Geostationary Scientific Satellite. (2) NASA Geodetic Earth Orbiting Satellite. See *Explorer*. (3) NASA Geodetic Explorer Satellite. See *Explorer*. (4) NASA Geodynamic Experimental Ocean Satellite. See *Explorer.*

geostationary. Describing a *spacecraft* constrained to a constant latitude.

geostationary orbit. A circular *orbit* approximately 22,300 miles (36,000 km) above the Earth's surface in the plane of the equator. An object in such an orbit rotates at the same rate as the planet and therefore appears to be stationary with regard to any point on the surface. It is a specific type of *geosynchronous orbit.* (Paine Report, p. 198.)

geosynchronous. Referring to an Earth *orbit* with a period equal to almost a day: 23 hours 56 minutes 4 seconds. A *satellite* in geosynchronous orbit above the Earth's equator will stay over the same point on Earth at all times. Communications satellites are often put into geosynchronous orbits so that satellite dishes on Earth can remain oriented toward the same point in the sky at all times.

Getaway Special (GAS). Popular name for the Self-Contained Payload Program, an *STS* program that allowed organizations and individuals—both private and public—to place scientific and development experiments into space at a relatively modest cost.

    ETYMOLOGY. The term was coined by NASA Associate Administrator for Space Flight John F. Yardley (NASA News Release 77-4, "NASA Signs up Early 'Getaway' Shuttle Payloads," January 7, 1977).

Giant Leap. Term for the first human steps on the lunar surface, from the words uttered by Neil Armstrong: "That's one small step for [a] man—one giant leap for mankind."

Giant Leap Tour. From the name for the 25-nation tour staged by NASA to honor the *Apollo* 11 astronauts. According to Leon Wagener, the tour was staged as if the three were a rock band, and "included a dozen ticker-tape parades and dinners with presidents, kings and movie stars" (Wagener, *One Giant Leap,* p. 238).

Giotto. *European Space Agency mission* to the *comets* Halley and Grigg-Skjellerup, 1985. Encountered Halley March 12–15, 1986, and Grigg-Skjellerup July 10, 1992. Named after Italian artist Giotto di Bondone (1266–1337).

GISS. *Goddard Institute for Space Studies.*

Glenn Research Center (GRC). Officially the John H. Glenn Research Center at Lewis Field. GRC is composed of a main campus adjacent to the Cleveland (Ohio) Hopkins International Airport and the

6,400-acre (2,590-hectare) *Plum Brook Station* in Sandusky, Ohio. The research and technology development work conducted at the center focuses on aeronautical propulsion, space propulsion, space power, *satellite* communications, and *microgravity* sciences in combustion and fluid physics.

Congress authorized a flight-propulsion laboratory for the *NACA* on June 26, 1940, and in 1942 the new laboratory began operations adjacent to the Cleveland Municipal Airport. It was known as the Aircraft Engine Research Laboratory. On September 28, 1948, the NACA renamed it Lewis Flight Propulsion Laboratory in honor of Dr. George W. Lewis (1882-1948). Dr. Lewis not only was a leading aeronautical engineer, whose work in flight research has been termed "epochal contributions to aeronautics," but also made his mark as an administrator, serving as the NACA's Director of Aeronautical Research from 1919 to 1947. He was responsible for the planning and building of the new flight-propulsion laboratory that was later to bear his name.

Upon the formation of the National Aeronautics and Space Administration on October 1, 1958, the facility became Lewis Research Center. The center's research and development responsibilities concentrated chiefly on advanced propulsion and space power systems. It had management responsibilities for the *Agena* and *Centaur launch vehicle* stages.

GRC today consists of over 300 acres (121 hectares) and has 150 buildings (including 31 major research facilities) and 3,500 civilian and contract employees at its Cleveland location. The latest additions to the center have been the Research and Analysis Center and the Power Systems Facility. The center was renamed once more on March 4, 1999. At the dawn of the new millennium it became the NASA John H. Glenn Research Center at Lewis Field in tribute to Ohio *astronaut* and U.S. senator, Glenn.

The Plum Brook Station, which is a component of GRC and contains the Welcome Center, is located on Lake Erie near Sandusky. The site, formerly a U.S. Army Ordnance plant, was acquired from the Army through a gradual process beginning in 1956 and completed in 1963. The name Plum Brook Station derived from the Army's name of the former ordnance facility, Plum Brook Ordnance Works, after a small stream running through the site. It had a nuclear research reactor and a wide range of propulsion test facilities.

Nuclear propulsion program cutbacks to adjust to NASA budget reductions in 1973 prompted a decision to phase down most of the Plum Brook facilities. The facility was shut down and placed in a safe, dry storage mode. It is presently being decommissioned.

The Space Power Facility—one of the world's largest space environment simulation chambers, equipped with a solar simulation system,
instrumentation, and data-acquisition facilities—was kept in operation for use by other government agencies. The Air Force, Navy, National Oceanic and Atmospheric Administration, and Energy Research
and Development Administration indicated possible interest in using
the facility. It is now being used to test the *Orion Crew Exploration
Vehicle.*

SOURCES. SP-4402, p. 152-53; Hunsaker, *Forty Years of Aeronautical
Research,* p. 242; Rosholt, *Administrative History of NASA,* p. 21; Vice Adm.
Emory S. Land, "George William Lewis: An Address," in George William
Lewis Commemoration Ceremony [program], Cleveland, NACA Lewis
Flight Propulsion Laboratory, September 28, 1948; Emme, *Aeronautics
and Astronautics,* p. 102; NASA News Release 99-12. For more information
about Glenn Research Center, see Virginia P. Dawson, *Engines and Inno-
vation: Lewis Laboratory and American Propulsion Technology,* SP-4306
(Washington, DC: GPO, 1991).

global dimming. Phenomenon by which diminished amounts of
sunlight are reaching the Earth's surface because of pollution and
thicker clouds.

Global Positioning System. See *GPS.*

global warming. A gradual warming of the Earth's *atmosphere*
reportedly caused by the burning of fossil fuels and industrial
pollutants.

Goddard Institute for Space Studies (GISS). A center for theoretical
research established in 1961 as the New York office of the Theoretical
Division of *Goddard Space Flight Center* (GSFC). In July 1962 it was
separated organizationally from the Theoretical Division and renamed
Goddard Institute for Space Studies. Located at Columbia University in
New York City, it is today a laboratory of the Earth-Sun Exploration
Division of NASA's GSFC and a unit of the Columbia University Earth
Institute. Research at GISS emphasizes a broad study of global climate
change.

Goddard Space Flight Center (GSFC). Research center named for
Dr. Robert H. Goddard, a pioneer in *rocket* research, established in
1959 in Greenbelt, Maryland. Since that time, GSFC has played a major
role in space and Earth science.

In August 1958, before NASA officially opened for business,
Congress authorized construction of a NASA "space projects center" in
the "vicinity of Washington, D.C." The site selected was in Maryland on
land then part of the Department of Agriculture's Beltsville Agricultural
Research Center. On January 15, 1959, NASA designated four divisions

of NASA Headquarters as the Beltsville Space Center. Project *Vanguard* personnel, transferred by Executive Order of the President from the Naval Research Laboratory to NASA in December 1958, formed the nucleus of three of the four divisions and hence of the center. On May 1, 1959, NASA renamed the facility Goddard Space Flight Center in honor of the father of modern rocketry, Dr. Robert H. Goddard (1882–1945). A rocket theorist as well as a practical inventor, Dr. Goddard achieved a list of "firsts" in rocketry that included the first *launch* of a liquid-*propellant* rocket in March 1926.

In its early years Goddard Space Flight Center was responsible for uncrewed *spacecraft* and *sounding rocket* experiments in basic and applied research; it operated the worldwide Space Tracking and Data Acquisition Network (STADAN), which later became the *Spaceflight Tracking and Data Network* (STDN); and it managed development and launch of the *Thor-Delta launch vehicle.*

The Goddard facility at Greenbelt covers 1,270 acres (514 hectares) and maintains the adjacent Magnetic Test Facility and Propulsion Research site. The outlying sites include the Antenna Performance Measuring Range and the Optical Tracking and Ground Plane Facilities. The unique Magnetic Test Facility is a magnetic-quiet area and is operated from a single control building. Other unique facilities at Goddard include the Flight Dynamics Facility and a high-capacity centrifuge capable of rotating 5,000-pound (2,270-kg) payloads at up to 30 rpm. In 1990 the Space Systems Development and Integration Facility opened, including an enormous *clean room* for the testing of spacecraft hardware. The *Wallops Flight Facility* (WFF; Wallops Island, Va.) was consolidated with NASA Goddard in 1982 and quickly became NASA's primary facility for suborbital programs. Wallops continues to support projects using sounding rockets, balloons, and scientific aircraft to facilitate research about the Earth, all regions of the atmosphere, and the near-Earth and space environment.

Today GSFC remains a key player in Earth and *space science.* For more information about Goddard Space Flight Center, see Rosenthal, *Venture into Space,* SP-4301; Wallace, *Dreams, Hopes, Realities,* SP-4312.

GOES. Geostationary Operational Environmental Satellite (meteorology). See *SMS.*

go fever. Term of the *Space Age* through the *Apollo* period signifying a willingness to go aloft with a notion of calculated risk; resolution to proceed as the moment of *liftoff* approaches. With two *STS* catastrophes behind it, NASA has seen "go fever" giving way to "safety fever." In 2001, CNN space correspondent Miles O'Brien reported, "Fifteen years ago, the space agency that put men on the moon was beset with an epidemic of 'go fever.' Today, the man in charge of the $3 billion shuttle program says he guards against 'go fever' as if it were a plague. His name is Ron Dittemore, and during a conference call with a cadre of reporters on the space beat the other day, he offered some insights into NASA's 'P-C' thinking (that's Post-Challenger)." (O'Brien, "Fifteen Years after Challenger, NASA Inoculates against 'Go Fever,'" January 18, 2001; http://archives.cnn.com/2001/TECH/space/01/18/downlinks.40/.)

golden slippers. The gold-plated foot restraints designed to hold an *Apollo* astronaut in position as he worked just outside the entryway of the *Lunar Module* ("Glossary of Terms Used in Space Flight," *New York Times,* March 4, 1969, p. 15). "Golden Slippers" was a spiritual popularized in the years following the American Civil War by the Fisk Jubilee Singers.

go/no-go. Term used to characterize a part, component, or system that can have only two parameters: go (functioning properly) or no-go (not functioning properly). For instance, the *Apollo* 11 *mission* contained nine separate go/no-go points, starting with a decision to *launch.* (SP-6001, p. 40.)

Goodwill rock. Name given to an *Apollo* 17 lunar rock sample that, after it had been examined for its scientific value, was broken into many pieces, allowing President Richard M. Nixon to give away 135 samples, mostly to heads of state.

GoreSat/GoreScan. See *Triana.*

GOX. Gaseous oxygen.

GPS. Global Positioning System. A worldwide radio-navigation system developed by the U.S. Department of Defense.

Grand Tour / Grand Tour of the Planets. (1) Proposed *probe*s to the outermost planets of the solar system that would have exploited the short-term alignment of Jupiter, Saturn, Uranus, Neptune, and Pluto in the late 1970s, an alignment that would not recur for 176 years. One of the proposed *mission* designs had two probes. One would fly by Jupiter, Uranus, and Neptune. The other would fly by Jupiter, Saturn, and Pluto. The Grand Tour missions, as well as later proposals for a "mini grand tour," were canceled owing to lack of funds. However, most elements of the Grand Tour were added to *Voyager* missions. (2) A term that has come to be used specifically of Voyager 2's mission.

ETYMOLOGY. Originally, the name for a European travel itinerary that reached its height of popularity in the 18th century, particularly among the British aristocracy. The tour served an educational purpose for wealthy university graduates, its primary value lying in exposure to the cultured artifacts of antiquity and the Renaissance. According to the *OED,* the first use of the term (and perhaps its introduction into the English language) was made by Richard Lassels in *An Italian Voyage* (1670).

FIRST USE. The term was applied to the planetary *flyby*s in 1965 at the *Jet Propulsion Laboratory* during a study of the mechanics of the mission (NASA Names Files, record no. 17525, letter of February 23, 1970, from H. J. Stewart to George H. Henry on the history of the project).

gray mice. NASA insider terminology, ca. 1976 and later, for new projects with low visibility and minimal impact on the agency's budget. In contrast to *purple pigeons,* which were new projects with high visibility and substantial budgetary impact. See also *wild turkey.* (NASA Names Files, record no. 17540, memo from John E. Naugle of November 12, 1976, suggests that these terms were Naugle's creation.)

Great Galactic Ghoul. (1) Name for a mythical monster responsible for early failures in space, a latter-day incarnation of the gremlins that bedeviled fighter pilots during World War II. (2) An area of space about 35 million miles (56 million km) from Earth close to the *orbit* of Mars containing a high concentration of meteoroids and cosmic dust. The term has been applied most commonly to the period between 1960 and 1997, when America and Russia sent 29 *spacecraft* to Mars, only 8 of which returned useful data. *New York Times* writer John Noble Wilford discussed the creature with scientists at the *Jet Propulsion Laboratory* (JPL): "It was only a joke, one recalled with a despairing sigh last week, but a joke born of hard experience and rooted in culture. Mars seems to lie at some intersection of rational inquiry and wondering romance, which can by turns inspire hope or dread." (Wilford, "O Brave New World, That Has Such a Silence around It," *New York Times,* August 29, 1993.)

ETYMOLOGY / FIRST USE. Definition 1: "The Great Galactic Ghoul is the mythical monster who gets the blame for every unexplained failure in the space business" (Abigail Brett, "Galactic Ghoul Clicks off Lunar TV," *Washington Post,* August 6, 1971, p. A8). Definition 2: *Washington Post* writer Thomas O'Toole suggested that the notion of the Ghoul, in the sense of a haunted region, was born at JPL in 1969 when *Mariner* 7's battery exploded. "It's uncanny," John Casani of JPL told O'Toole. "The Ghoul always seems to know when we're coming its way." (O'Toole, "'Galactic Ghoul' Strikes," *Washington Post,* January 25, 1976, p. 3.)

greenhouse effect. Warming of the surface and lower *atmosphere* of a planet. The greenhouse effect tends to intensify with any increase in atmospheric carbon dioxide (CO_2).

FIRST USE. The term appears to have first been used in a public forum in a lecture given in September 1955 by Dr. John G. Hutton of the General Electric Company to the Cleveland section of the American Institute of

Electrical Engineers. Hutton claimed that the CO_2 belt around the Earth may be "having a greenhouse effect on our climate" ("Why Earth Warms," *New York Times,* September 25, 1955, p. E11).

ground truth data. Geophysical parameter data, measured or collected by other means than the instrument itself, used as correlative or calibration data for that instrument data. Such data may include data taken on the ground or in the *atmosphere.* Ground truth data are another measurement of the phenomenon of interest; they are not necessarily more "true" or more accurate than the instrument data.

GSFC. *Goddard Space Flight Center.*

g-suit/G-suit. A suit that exerts pressure on the abdomen and lower parts of the body to prevent or retard the collection of blood below the chest under positive acceleration (SP-7; SP-6001, p. 39).

GT- . *Gemini-Titan* (followed by *mission* number).

g-tolerance. A tolerance in a person or other animal, or in a piece of equipment, to an acceleration of a particular value.

Guggenheim Aeronautical Laboratory, California Institute of Technology (GALCIT). Forerunner of the *Jet Propulsion Laboratory.*

guillotine. A cutting device, sometimes powered by explosives, employed to cables, water lines, and wires during the separation of spacecraft modules from one another or from launch towers (called "umbilical" towers). In the Apollo days, the umbilical was severed before the astronauts came back to Earth by a guillotine that cut through all the tubes and wires between the command and service modules.

Gumdrop. *Call sign* for the *Apollo* 9 Command Module, March 3-13, 1969. The *mission* was crewed by James A. McDivitt, David R. Scott, and Russell L. Schweickart. It was the first *crewed* flight of all lunar hardware in Earth *orbit* and the first crewed flight of the *Lunar Module* (*LM*), which had the call sign *Spider.*

ETYMOLOGY. From the appearance of the *spacecraft* when transported on Earth when it was wrapped in a blue shroud, giving it the appearance of a wrapped gumdrop. "This was the first mission in which the use of the names for the spacecraft was again authorized," McDivitt recalled. "There was no way that you could fly two spacecraft and call them by the same call sign. So we went to the names again and picked 'Spider' and 'Gumdrop' . . . not very glamorous but they fit the picture." (Quoted in Lattimer, *All We Did Was Fly to the Moon,* p. 11.)

H

H-II Transfer Vehicle. See *HTV*.

Habitat. See *Project Habitat*.

Ham. Name of the chimpanzee carried in *Mercury* 2, a test *mission* of the *Mercury-Redstone capsule–launch vehicle* combination of January 31,1961. Ham and his capsule were successfully recovered after a 16½ minute flight in suborbital space. Named for Holloman Aeromedical Laboratory, where Ham's training took place.

hard landing. Collision landing as opposed to a *soft landing*.

Hawk *(sounding rocket)*. A low-cost sounding rocket developed by NASA in 1974–75 using surplus motors from the Army's Hawk antiaircraft missiles. The research *rocket* inherited the Army's name, an acronym for Homing All the Way Killer, although the new uses would be far removed from the purposes of the weapon system.

To be flown as a single-stage Hawk or in two-stage combination as the *Nike*-Hawk, for a variety of research projects, the 14-inch (35.6-cm) diameter rocket provided a large volume for payloads. Both stages of the Nike-Hawk would use surplus Army equipment. Development testing was proceeding under *Wallops Flight Center* management. By December 1974, two flight tests of the single-stage Hawk sounding rocket had been launched, the first one lifting off successfully on May 29, 1974.

The single-stage Hawk could carry a 100-pound (45-kg) *payload* to a 49-mile (80-km) altitude, or 200 pounds (90 kg) to 35 miles (57 km). Engineers were working toward a performance capability for the Nike-Hawk of 100 pounds (45 kg) to 130 miles (210 km), or 200 pounds (90 kg) to 100 miles (160 km). The Nike-Hawk had its last launch September 9, 1975. (SP-4402, p. 131.)

Hayabusa (Japanese for peregrine falcon). Japanese sample-return *mission* launched on May 9, 2003, and *soft-land*ed on asteroid Itokawa on November 20, 2005, with a return to Earth planned for June 2010. Originally called MUSES-C.

HCMM. Heat Capacity Mapping Mission. *Explorer* applications *satellite*.

HEAO. High Energy Astronomy Observatory. In September 1967 NASA established the Astronomy Missions Board (AMB) to consult the scientific community and submit for consideration a long-range program for the 1970s. The board's X-ray and Gamma-ray Panel completed its report in September 1968, recommending an *Explorer*-class *spacecraft* with a larger *payload* capability, designated

High Energy A by the panel and Heavy Explorer in other sections of the AMB position paper. The spacecraft was alternately referred to as the Super Explorer, but all three names were later dropped because of objections to the resulting acronyms (HEX and SEX). HEAO, an acronym for High Energy Astronomy Observatory, first appeared in June 1969 and was officially adopted as the spacecraft developed into an observatory-class *satellite*.

HEAO was originally planned to be the largest uncrewed spacecraft put into orbit by the United States, weighing almost 22,000 pounds (10 metric tons) and capable of carrying the larger instruments required to investigate high-energy electromagnetic *radiation* from space including x-rays, gamma rays, and high-energy cosmic rays.

In January 1973 the project was suspended because of budget cuts. A scaled-down project was substituted in fiscal year 1975, calling for three spacecraft instead of four, to be launched by an *Atlas-Centaur* vehicle instead of the *Titan* IIIE, in 1977, 1978, and 1979. With the smaller *launch vehicle*, HEAO was revised to carry fewer instruments and to weigh about 7,000 pounds (3,200 kg). The first *mission* was to make an x-ray survey, the second to make detailed x-ray studies, and the third to make a gamma and cosmic ray survey of the sky. Launches of spacecraft from NASA's *Space Shuttle* after 1980 would carry heavier gamma and cosmic ray experiments to complete the scientific objectives.

The first satellite in the series, HEAO-A or HEAO 1, was launched on August 12, 1977, and remained in operation until January 9, 1979. The second of NASA's three High Energy Astrophysical Observatories, HEAO-2, renamed *Einstein* after *launch* on November 12, 1978, was the first fully imaging x-ray telescope put into space. HEAO-3 was launched September 20, 1979.

SOURCES. SP-4402, p. 54; NASA, Office of Technology Utilization, *A Long-Range Program in Space Astronomy*, SP-213 (Washington, DC: NASA, 1969), pp. 16–26; NASA News Releases 73-40, 74-240.

Heat Capacity Mapping Mission (HCMM). *Explorer* applications *satellite*.

heat shield. Device that protects people, animals, or equipment from heat, such as a shield in front of a *reentry capsule*.

heliocentric. Sun-centered.

Heliocentric *(satellite)*. International Sun-Earth Explorer ISEE-C. See *Explorer*.

Helios. (1) Name used by NASA for the Advanced Orbiting Solar Observatory (AOSO) project, canceled in 1965. It was to have performed experiments similar to the Helios probes described in definition 2. (2) Two solar *probes* developed by West Germany in cooperation with NASA. In June 1969 NASA and the German Ministry for Scientific Research (BMwF) agreed to a joint project for launching two probes to study the interplanetary medium and explore the near-solar region. The probes would carry instruments closer to the Sun than any previous *spacecraft,* approaching to within 28 million miles (45 million km). The probes were to be launched on *Titan III–Centaur* vehicles. NASA launched the West German–built Helios 1 into solar *orbit* on December 10, 1974, and a second was launched on January 15, 1976. (SP-4402, p. 84; NASA News Release 69-86; NASA, Historical Staff, *Astronautics and Aeronautics, 1963: Chronology on Science, Technology, and Policy,* SP-4004 [Washington, DC: NASA, 1964], p. 73, and *Astronautics and Aeronautics, 1965,* SP-4006 [1966], p. 554.)

ETYMOLOGY (definition 2). The joint U.S.-German project was designated Helios, the name of the ancient Greek god of the Sun, by German Minister Karl Kaesmeier. The name had been suggested in a telephone conversation in August 1968 between him and *Goddard Space Flight Center*'s Project Manager, Gilbert W. Ousley. NASA had previously used the name for the Advanced Orbiting Solar Observatory (AOSO) (see definition 1).

heliosphere. The large region of space influenced by the Sun's *solar wind* and the interplanetary *magnetic field.* This vast sea of electrical plasma activity may extend as far as 10 billion miles (16.5 billion km) from the Sun, affecting the *magnetosphere,* ionosphere, and *upper atmosphere* of Earth and other solar system bodies. (Paine Report, p. 197.)

HEOS. Highly Eccentric Orbit Satellite. *Satellite* built and named by the European Space Research Organization *(ESRO).* HEOS 1 was launched December 5, 1968, to investigate interplanetary *magnetic field*s and to study solar and cosmic ray particles outside the *magnetosphere.* Nine scientific groups in five countries provided experiments on board the satellite. Under a December 30, 1966, memorandum of understanding and a March 8, 1967, contract with ESRO, the *mission* was the first cost-reimbursed NASA launch of a foreign scientific satellite. HEOS 2, the second satellite in the series, was launched by NASA on January 31, 1972, to continue the study of the interplanetary medium. See *ESRO.* (SP-4402, p. 54; ESRO Communiqué no. 41, March 1967; Ellen T. Rye, Office of International Affairs, NASA, telephone interview, April 20, 1967; NASA News Release 68-20; Oscar E. Anderson, Director, International Organizations Division, Office of International Affairs, NASA, memorandum to Eugene M. Emme, Historian, NASA, December 20, 1968.)

high Earth orbit. Flight path above *geosynchronous orbit,* between 22,300 miles (36,000 km) and 50,000–60,000 miles (80,000–97,000 km) above the surface of the Earth.

High Energy Astronomy Observatory. See *HEAO.*

Highly Eccentric Orbit Satellite. See *HEOS.*

Hiten. Japanese *flyby* and orbiter *mission* to the Moon. The *spacecraft,* built by the Japanese Space Agency (ISAS), was launched on January 24, 1990. The spacecraft entered a circumlunar *orbit* and released a small orbiter, Hagomoro, into lunar orbit. This was the first lunar flyby for Japan. Hiten intentionally crashed into the Moon on April 10, 1993. The transmitter on Hagomoro failed, rendering it scientifically useless.

ETYMOLOGY. The spacecraft was named for a Buddhist angel who plays music in heaven, and Hagomoro was named for the veil worn by Hiten. The spacecraft was originally called MUSES-A.

hold. (1, n.) Unscheduled delay or pause in the launching sequence or *countdown* for a *missile* or *space vehicle.* (2, v.) During a countdown, to halt the sequence of events until an impediment has been removed so that the countdown can be resumed.

FIRST USE. Introduced to the public in a press conference following the failure of the first *Vanguard launch:* "We started the count during the night, as you know, and the count-down ran smoothly. There were only a few 'holds' (delays)." (Quoted from news conference with J. Paul Walsh, Vanguard Deputy Director, *New York Times,* December 7, 1957, p. 8.)

EXTENDED USE. Like other space terms, the "hold" of *Cape Canaveral* became the "hold" of everyday life: "When anything is stymied or stalled a missile man is 'holding.' When things are going well he is 'counting.' If plans are changed, he 'recycles.' If plans are canceled, he 'scrubs.' If he gives instructions to someone, he 'programs' that person." (Shelton, *Countdown.*)

Horizon. See *Project Horizon.*

housekeeping. Routine tasks required to maintain a *spacecraft* in habitable and operational condition during flight. By extension, the tasks required to keep any system functioning.

FIRST USE. "The space agency said it plans to select another contractor to perform housekeeping chores" ("Chrysler to Get Saturn Contract of $200 Million," *Wall Street Journal,* November 20, 1961, p. 30).

Houston. *Call sign* for *mission control* at the Manned Spacecraft Center / *Johnson Space Center* in Houston. Traditional form of address used by *astronauts* in space.

HST. *Hubble Space Telescope.*

HTV. H-II Transfer Vehicle. A Japanese-built autonomous logistical resupply vehicle designed to berth to the *International Space Station.* See also *ATV.*

Hubble Space Telescope (HST). A telescope in Earth *orbit* taking sharp optical images of very faint objects. An international cooperative project between NASA and the *European Space Agency,* the Hubble was launched in 1990 from the *Space Shuttle Discovery,* with an expected 15-year *mission.* Its optics were flawed and were corrected by a servicing mission by *Space Shuttle* astronauts in December 1993. Over the next decade three more servicing missions followed, adding new instruments and replacing batteries and gyroscopes. A fifth servicing mission, scheduled for 2009, will allow HST to remain functional until its successor, the *James Webb Space Telescope,* is placed in orbit around 2013. As of April 25, 2005, HST had taken more than 750,000 photos of the cosmos, images that have awed, astounded, and confounded astronomers and the public. The Space Telescope Science Institute in Baltimore conducts Hubble science operations. The institute is operated for NASA by the Association of Universities for Research in Astronomy.

ETYMOLOGY. Originally the Large Space Telescope (LST), in the spring of 1983 it was renamed the Edwin P. Hubble Space Telescope, after the American astronomer (1889–1953).

human spaceflight. Preferred alternative to *manned spaceflight.*

USAGE. The *Style Guide* for the NASA History Series updated to January 26, 2006, contains this directive: "All references referring to the space program should be non-gender specific (e.g. human, piloted, un-piloted, robotic). The exception to the rule is when referring to the Manned Space Craft Center, the predecessor to the Johnson Space Center in Houston, or any other official program name or title that included 'manned' (e.g. Associate Administrator for Manned Space Flight)."

human spaceflight program. NASA's first four human spaceflight projects were *Mercury, Gemini, Apollo,* and *Skylab.*

As the first U.S. human spaceflight project, Project Mercury, which included two suborbital flights and four orbital flights, "fostered Project Apollo and fathered Project Gemini."

The third human spaceflight project at NASA was Apollo, the lunar exploration program. The national goal of a human lunar landing in the 1960s was set forth by President John F. Kennedy on May 25, 1961: "I believe that this nation should commit itself to achieving the goals, before this decade is out, of landing a man on the moon and returning him safely to earth. No single space project in this period will be more impressive to mankind, or more important for the long-range exploration of space; and none will be so difficult or expensive to accomplish . . . But in a very real sense, it will not be one man going to the moon if we make this judgment affirmatively, it will be an entire nation."

The interim Project Gemini, completed in 1966, was conducted to provide spaceflight experience, techniques, and training in preparation for the complexities of Apollo lunar-landing missions.

Project Skylab was originality conceived as a program to use hardware developed for Project Apollo in related missions. It evolved into the Orbital Workshop program with three record-breaking missions in 1973–74 to man the laboratory in Earth *orbit,* producing new data on the Sun, Earth resources, materials technology, and the effects of space on human beings.

The *Apollo-Soyuz Test Project* was a one-time icebreaking effort toward international cooperation during the early human spaceflight program. The United States and the Soviet Union flew a joint *mission* in 1975 to test new systems that permitted their *spacecraft* to dock with each other in orbit, for space rescue or joint research.

As technology and experience broadened our ability to explore and use space, post-Apollo planning called for ways to make access to space more practical, more economical, more nearly routine. Early advanced studies grew into the *Space Shuttle* program. Development of the reusable space transportation system, meant to be used for most of the nation's human and robotic space missions in the 1980s, became the major focus of NASA's program for the 1970s. European nations cooperated by undertaking development of *Spacelab,* a pressurized reusable laboratory to be flown in the Shuttle. Beginning in the late 1990s the Shuttle was used to deliver materials and crew for the *International Space Station* (ISS), the most complex space project ever undertaken. Plans are for the Space Shuttle to be retired by 2010, following completion of the ISS. *Project Constellation,* announced by President George W. Bush in January 2004, was created to follow the Shuttle era and intended to take humans back to the Moon and beyond to Mars.

SOURCES. Swenson, Grimwood, and Alexander, *This New Ocean,* SP-4201, p. 105; John F. Kennedy, Special Messages to the Congress, May 25, 1961; General Services Administration, National Archives and Records Service, Office of the Federal Register, *Public Papers of the Presidents of the United States: John F. Kennedy, 1961* (Washington, DC: 1962), p. 404.

Huntsville Facility. Unit that became the *Marshall Space Flight Center* (MSFC).

Huygens. *Probe* portion of Cassini-Huygens, a NASA–*European Space Agency mission* to Saturn, 1997. Named after the Dutch astronomer Christiaan Huygens. The Cassini-Huygens spacecraft was launched on October 15, 1997, and entered Saturn's *orbit* on July 1, 2004.

hypersonic. Pertaining to speeds of Mach 5 or greater.

I

ICE. International Cometary Explorer. *Explorer* 59 (ISEE-3). NASA *mission* to *Comet* Giacobini-Zinner (launched 1978; encounter 1985). Renowned for the fact that it confirmed the belief that comets and asteroids were *dirty snowballs*.

ice frost. Ice on the outside of a *rocket* vehicle over surfaces super-cooled by liquid oxygen. (Compare and contrast to *FROST,* SP-6001, p. 46.)

ICESAT. Ice Cloud and Land Elevation Satellite. Part of NASA's Earth Observing System *mission.* The *spacecraft* was launched by a *Delta* II *rocket* from Vandenberg AFB on January 13, 2003. It carried a single instrument, GLAS (Geoscience Laser Altimeter System), which enables accurate surface-level measurements of ice sheets. Ice surface variations in Greenland and Antarctica are important predictors of *global warming.*

igniter. Device that starts the fire in the combustion chamber of a *rocket engine*; comparable to a sparkplug.

ignition. When fuel and oxygen mix and a sustained chemical reaction begins; usually applied to *rocket engine*s.

IGY. *International Geophysical Year.*

IMAGE. Imager for Magnetopause-to-Aurora Global Exploration. *Mission* to study the global response of the *magnetosphere* to the changes in *solar wind.*

IMP. Interplanetary Monitoring Platform. See *Explorer.*

impactor object. Object designed to crash into another. The impactor deployed from a *flyby spacecraft* that slammed into *Comet* Tempel 1 as part of the *Deep Impact mission* of 2005 had a copper core and weighed 820 pounds (372 kg).

Improved Tiros Operational Satellite (ITOS). See *TIROS, TOS, ITOS.*

in family. NASA jargon used to characterize an anomalous event that was known or expected. See *family.*

in-house. Referring to something within NASA, albeit sometimes using outside support (NASA, *Glossary/Congressional Budget Submission,* p. 21).

Injun. *Satellite* launched on June 29, 1961, to gather data on Earth's *radiation belts.* See *Explorer.*

    USAGE. A derisive term for American Indian used in Western movies of the 1930s and 1940s, the word is now widely considered a racial slur.

inner solar system. The part of the solar system between the Sun and the main asteroid belt. It includes the planets Mercury, Venus, Earth, and Mars, as distinct from the outer planets Jupiter, Saturn, Uranus, and Neptune and the dwarf planet Pluto. (Paine Report, p. 198.)

insertion. The point at which a *spacecraft* acquires centrifugal force equal to the gravitational force and achieves *orbit*. The term came into popular use during the *Mercury* orbital missions.

INTASAT. Spanish communications *satellite*. In May 1972 NASA and the Spanish Space Commission CONIE (Comision Nacional de Investigacion del Espacio) signed a memorandum of understanding on a joint research program in which NASA would *launch* Spain's first satellite. CONIE named the satellite INTASAT (INTA + satellite, INTA being an acronym for Instituto Nacional de Tecnica Aeroespacial, the government laboratory responsible for development of the satellite). Designed and developed in Spain to measure the total electron count in the ionosphere and ionospheric irregularities, INTASAT was launched *piggyback* in a three-satellite launch (with NOAA 4 and *OSCAR* 7) on November 15, 1974. The 33-pound (15-kg) satellite beamed data for two years to scientists around the world from its Sun-synchronous *polar orbit*. (SP-4402, pp. 54–55; GSFC, *Goddard News* 20, no. 6 [September 1972]: 1; NASA News Releases 72-275, 75-19; NASA program office.)

INTEGRAL. International Gamma Ray Astrophysical Laboratory (satellite).

Intelsat. Series of *satellites* owned and operated by *INTELSAT,* the International Telecommunications Satellite Organization. They were launched and tracked, on a reimbursable basis, by NASA for the Communications Satellite Corporation, the U.S. representative in and manager of INTELSAT. The INTELSAT consortium's method of designating its satellites went through numerous changes as new satellites were launched, producing alternative names for the same satellite and varying the numbering system.

The first of the INTELSAT satellites, Intelsat I, was named Early Bird because it was the satellite in the early capability program, the program to obtain information applicable to selection and design of a global commercial system and to provide experience in conducting communications satellite operations. Early Bird, the world's first commercial communications satellite, was launched by NASA on April 6, 1965, and placed in synchronous orbit over the Atlantic Ocean. Intelsat II-A, also known as Lani Bird, was the first communications satellite of the consortium's Intelsat II series. Lani Bird was launched in October 1966 to transmit transpacific communications but failed to achieve synchronous orbit. (It was named by the Hawaiian press, Lani meaning bird of heaven). Intelsat II-B, or Pacific 1, the second in the Intelsat II series, was launched in January 1967 and placed in orbit to provide transpacific service. Intelsat II-C (later redesignated Intelsat-II F-3, for flight 3 in series II), or Atlantic 2, was the second INTELSAT satellite to provide transatlantic service. The United Press International wire service nicknamed it Canary Bird because of the association with the Canary Islands Earth station. The name Canary Bird appeared widely in the press but was not adopted by INTELSAT. It was placed in synchronous orbit over the Atlantic in March 1967.

Subsequent satellites followed the same sequences: Intelsat II-D, or Pacific 2, was launched in September 1967 and later renumbered Intelsat-II F-4; Intelsat III-A (later Intelsat-III F-1) failed to achieve orbit in September 1968; Intelsat-III F-2, or Atlantic 3, was launched in December 1968. Satellites in the Intelsat IV series were numbered according to a different system, beginning with Intelsat-IV F-2, launched January 25, 1971. Intelsat-IV F-2 was the first in the IV series to be launched, with the designation F-2 meaning the second fabrication—the second satellite built rather than the second flight in the series. Other satellites in the series followed this pattern, with Intelsat-IV F-8 being launched into orbit November 21, 1974.

Each successive series of satellites increased in size and communications capacity. Satellites in the Intelsat II series were improved versions of Early Bird; Intelsat III satellites had five times the communications capacity of the II series; and Intelsat IV satellites not only had an increased capacity more than five times that of the III series but also were nearly ten times as heavy. Intelsats V, VII, and IX have continued this very successful series of communications satellites with increasing sophistication and global coverage.

SOURCES. SP-4402, p. 56–57; COMSAT Corp., *Prospectus* (Washington, DC: COMSAT, 1964), p. 14; Larry Hastings, COMSAT Corp., telephone interview, April 21, 1967; COMSAT News Releases 67-45, 67-48, 69-53; NASA News Releases 68-195, 69-6; COMSAT Corp., Public Relations Office, telephone interview, December 22, 1971; NASA program office.

INTELSAT. International Telecommunications Satellite Organization, the world's largest commercial satellite communications services provider.

interface. (1, n.) Surface that forms a shared boundary, especially in electronic data systems, and where some exchange of energy (e.g., information) takes place. (2, v.) To connect two systems—computer and machine, man and machine, even two federal agencies: "Interface is another word that was recast at the space center. Defined in the dictionary as a surface that lies between two parts of matter or space and forms their common boundary, it grew to encompass any kind of interaction at KSC. Perhaps this was subliminal recognition that Kennedy Space Center was the Great Interface where the many parts and plans that went into the moon launch had to be fitted together." (Benson and Faherty, *Moonport,* SP-4204.)

ETYMOLOGY. Latin *inter* (between) + *facies* (shape, form, or face).

FIRST USE. The technical use of the term dates to the 1880s. As for its use in a social context, a *Boston Globe* article devoted to the history of the term claims, "A first appearance was in Marshall McLuhan's *Gutenberg Galaxy* [1962], the book in which he began to expound his theories of communication. The interface of the Renaissance was the meeting of medieval pluralism and modern homogeneity and mechanism." (*Boston Globe,* "Glossary: Interface," August 18, 1982). The McLuhan book mentioned was one of his characteristically mind-stretching exercises in drawing connections between concepts and disciplines. The *Globe* noted, "In that same year, 1962, a story in the *Washington Star* speculated that the word seems to mean the liaison between two different agencies." The fashion industry had picked up the word somewhat earlier, using it interchangeably with "interlining," while a 1969 *New Yorker* piece called the word "a space-age verb meaning, roughly, to coordinate."

USAGE. The term has not always been uttered with a straight face and can be used to express sarcasm. Senator John Glenn once speculated that the Pentagon did not support a mobile MX *missile* system because of "the unacceptable social interface."

International Geophysical Year (IGY). By international agreement, a period during 1957–58 when greatly increased observation of worldwide geophysical phenomena would be undertaken through the cooperative efforts of participating nations. A precedent for the project had been set by the International Polar Years of 1882 and 1932.

Officially deemed the "greatest scientific research program ever undertaken," the IGY involved more than 5,000 scientists participating in an effort to find out as much as possible about the Earth, the Sun, and outer space during the "year" (which actually ran for 18 months, from July 1, 1957, to the end of 1958). The period covered a time of maximum activity in solar flares. Millions of facts would be collected, and major questions, such as whether or not the Earth's climate was changing, would be posed and presumably answered. Two participating nations, the United States and the Soviet Union, had each earlier told the world on separate occasions that they would put a small Earth-circling *satellite* into *orbit* as part of its contribution to the IGY.

International Radiation Investigation Satellite. See *IRIS.*
International Satellites for Ionospheric Studies. See *ISIS.*

International Space Station (ISS). The largest and most complex international scientific project in history. When assembly is complete the ISS will be composed of about 925,000 pounds (420,000 kg) of hardware brought into orbit in about 40 separate launches over the course of more than a decade. To date there have been over 50 flights to the ISS, including flights for assembly, crew rotation, and logistical support. The first flight was in November 1998 using the Proton rocket, and the second followed in December using the *Space Shuttle*. The first crew arrived on November 2, 2000 (Expedition 1). As of the end of 2007 the ISS had been continuously crewed for more than six years and was more than 50 percent complete. It is planned to be completed with the retirement of the Space Shuttle in 2010.

The ISS draws upon the resources of 16 nations led by the United States: Canada, Japan, Russia, Brazil, and 11 nations of the *European Space Agency*. The American, European, and Japanese laboratories *Destiny, Columbus*, and *Kibo* are at the hub of the station's science research, with other key elements including Canada's *Remote Manipulator System*, and Russia's *Zarya* (Sunrise) and *Zvezda* (Star) Service Modules. These elements are connected by U.S.-built *nodes*.

Transportation logistics for the ISS are undertaken by the U.S. Space Shuttle, Russia's *Soyuz* and Proton rockets, the European Space Agency's *Ariane* rocket, and Japan's H-II. Russia's Progress is a resupply vehicle used for cargo and *propellant* delivery to the ISS, and for boosting it to higher altitudes. The Space Shuttle ferries in its cargo bay laboratory experiments and supplies in three Italian-built modules known as Multi-Purpose Logistics Modules (MPLMs). They are named Donatello, Leonardo, and Raffaelo. (*Reference Guide to the International Space Station*, SP-2006-557.)

ETYMOLOGY. Initially named Space Station Freedom by President Ronald Reagan from a list of over 700 nominations. The decision to name it Freedom was announced on July 18, 1988, by Marlin Fitzwater, Assistant to the President, who stated, "The name Freedom is tied to the President's earliest statements on the program. When the President announced his decision to build a space station in his January 1984 State of the Union Address, he noted that he was inviting our friends and allies to join us so 'we can strengthen peace, build prosperity, and expand freedom for all who share our goals.'" Because of political and economic pressures, in 1993 Freedom (briefly renamed Alpha), Mir-2, and the European and Japanese modules were combined into the International Space Station.

International Space Year (ISY). Year of international space exploration, 1992, proposed as an effort to study the Earth from space.

International Sun-Earth Explorer (ISEE). See *Explorer*.

International Ultraviolet Explorer (IUE). Originally *SAS*-D. See *Explorer*.

International Years of the Quiet Sun. A full-scale follow-up during the period January 1, 1964–December 31, 1965, to the *International Geophysical Year*. It was an intensive effort to collect geophysical observations during a period of minimum solar activity.

Interplanetary Monitoring Platform (IMP). See *Explorer*.

Intrepid. *Call sign* for the *Apollo* 12 *Lunar Module*. This *mission* was crewed by an all-Navy crew of Charles P. Conrad Jr., Richard F. Gordon Jr., and Alan L. Bean. The Command Module for the mission was the *Yankee Clipper*.

ETYMOLOGY. This name and that of the Command Module, Yankee Clipper, were picked by the Apollo 12 crew from more than 2,000 names submitted by the companies that built the two *spacecraft*: North American Rockwell's Space Division and the Grumman Aerospace Corporation. At a press conference in Houston on October 12, 1969, the *astronauts* said that the name was picked for all that it implied in the way of "courage and determination" and because it was in keeping with Navy traditions. Since the Revolutionary War, seven USN ships have carried the name Intrepid. ("Apollo Crew Names Spacecraft 'Yankee Clipper' and 'Intrepid,'" *Washington Post,* October 12, 1969, p. 3; "Apollo Spaceships Have Names with Salty Ring," *New York Times,* November 15, 1969, p. 24.)

Ionosphere Explorer. See *Explorer*.

Iris *(sounding rocket)*. Name of a sounding rocket used briefly (1960–62) by NASA. See *Arcas*.

IRIS. International Radiation Investigation Satellite (ESRO 2B). *Satellite* designed, developed, and built by the European Space Research Organization *(ESRO)*. ESRO assigned the name to ESRO 2B, a *backup* satellite to ESRO 2A, which had been launched May 29, 1967, but had failed to achieve *orbit*. Under an agreement with ESRO, NASA launched IRIS 1 on May 16, 1968, to study solar astronomy and cosmic ray particles. See *ESRO*.

ISIS. International Satellites for Ionospheric Studies. A cooperative *satellite* project of NASA and the Canadian Defence Research Board, designed to continue and expand the ionospheric experiments of the *Alouette* 1 topside sounder satellite.

The first ISIS *launch,* known as ISIS-X, was achieved November 28, 1965, when NASA launched Alouette 2 and *Explorer* 31 from the Western Test Range with a single *Thor-Agena* B *booster*. The Canadian topside sounder and the U.S. Direct Measurement Explorer were designed to complement each other's scientific data on the ionosphere. Both ISIS 1 (launched January 29, 1969) and ISIS 2 (launched March 31, 1971) carried experiments to continue the cooperative investigation of the ionosphere.

In 1969 the Canadian government proposed the substitution of an experimental communications satellite for the last of the projected ISIS *spacecraft* (ISIS-C). The satellite was redesignated CAS-C, an acronym used by NASA to denote an international Cooperative Applications Satellite. In April 1971 a memorandum of understanding was signed by NASA and the Canadian Department of Communication providing for the launch of CAS-C, which later was again redesignated CTS-A, for Communications Technology Satellite. (SP-4402, p. 58; NASA News Release 71-72.)

ETYMOLOGY. The name was devised in January 1963 by John Chapman, project manager of the Canadian team; Dr. O. E. Anderson, NASA Office of International Affairs; and other members of the topside sounder Joint Working Group. They selected Isis because the acronym was also the name of an ancient Egyptian goddess.

ISS. *International Space Station.*

ITOS. Improved Tiros Operational Satellite (meteorology). See *TIROS, TOS, ITOS.*

IUE. International Ultraviolet Explorer. See *Explorer.*

ivory tower. Vertical test stand (SP-6001, p. 50).

J

James E. Webb Memorial Rocket. See *Webb's Giant.*

James Webb Space Telescope (JWST). A large, infrared-optimized space telescope scheduled for *launch* in 2013. It was originally referred to as the *Next Generation Space Telescope* (NGST). In some ways it is a successor to the *Hubble Space Telescope,* though it is optimized for a different wavelength regime.

ETYMOLOGY. Named for James E. Webb (1906–92), the man who ran the fledgling *space agency* NASA from February 1961 to October 1968 as its second Administrator. While Webb is best known for leading *Apollo* and a series of lunar exploration programs that landed the first humans on the Moon, he also initiated an ambitious *space science* program, responsible for more than 75 launches during his tenure, including America's first interplanetary explorers. "It is fitting that Hubble's successor be named in honor of James Webb. Thanks to his efforts, we got our first glimpses at the dramatic landscapes of outer space," said NASA Administrator Sean O'Keefe when the name was announced. (www.ngst.nasa.gov/faq.html.)

Japan Aerospace Exploration Agency (JAXA). Agency formed on October 1, 2003, through the merger of three previously independent space organizations: the Institute of Space and Astronautical Science, the National Aerospace Laboratory of Japan, and the National Space Development Agency of Japan.

Jason *(sounding rocket).* See *Argo.*

Javelin *(sounding rocket).* See *Argo.*

JAXA. Common short form of *Japan Aerospace Exploration Agency,* giving it a look and sound parallel to NASA.

Jet Propulsion Laboratory (JPL). NASA research center located in Pasadena, California. Staffed and managed for the government by a leading private university, Caltech, as a federally funded research and development center. Along with *Goddard Space Flight Center,* JPL is NASA's center for Earth and *space science spacecraft* development and operations.

In 1936, students at the *Guggenheim Aeronautical Laboratory* of the California Institute of Technology *(GALCIT),* directed by Dr. Theodore von Kármán, began design and experimental work with liquid-*propellant rocket engine*s. During World War II the *GALCIT* Rocket Research Project developed solid and liquid-propellant units to assist the *takeoff* of heavily loaded aircraft and began work on high-altitude rockets. Reorganized in November 1944 under the name Jet Propulsion Laboratory, the facility continued postwar research and development on tactical guided missiles, aerodynamics, and broad supporting technology for U.S. Army Ordnance.

JPL participated with the Army Ballistic Missile Agency *(ABMA)* in the development and operation of the first U.S. *satellite, Explorer* 1, the succeeding Explorer missions, and the *Pioneer* 3 and 4 lunar *probe*s. On December 3, 1958, shortly after NASA came into existence, the functions and facilities of JPL were transferred from the U.S. Army to NASA. Operating in government-owned facilities, JPL remained a laboratory of Caltech under contract to NASA and continued planetary exploration programs such as *Ranger, Surveyor,* and the *Mariner* series, conducted supporting research, and founded and operated the worldwide *Deep Space Network* (DSN) for communication with lunar and planetary spacecraft. Since that time it has been responsible for numerous robotic space missions.

The 1960s saw the construction of a series of "space-related" buildings: the Spacecraft Assembly Facility, Control Systems Laboratory, and Celestial Simulator, all in 1961; the Space Simulator Facility in 1962; and the Space Flight Operations Command Facility in 1963. Later in the decade the emphasis moved toward laboratory space, with the construction of the Earth Space Science, Physical Sciences, and Spectroscopy Laboratory; the Gyro Laboratory; the Magnetic

Laboratory; and the Environmental Laboratory (1965–67). The primary administration building was completed in 1964. A number of new facilities have since been built: a Robotics Laboratory in 1971; an Isotope Thermoelectric Systems Lab in 1972; an Earth and Space Science laboratory in 1985; a Microdevices Laboratory in 1986; and the Observational Instruments Lab in 1989. The newest buildings are the In-Situ Instruments Lab (2001) and the Optical Interferometery Development Lab (2002).

The laboratory now covers some 177 acres (72 hectares) adjacent to the site of von Kármán's early rocket experiments. Jet propulsion is no longer the focus of JPL's work, but the well-known name remains the same.

SOURCES. SP-4402, pp. 143-45; Emme, *Aeronautics and Astronautics,* pp. 34, 48; President Dwight D. Eisenhower, Executive Order no. 10793, December 3, 1958, in Rosholt, *Administrative History of NASA,* p. 47. For more information about the Jet Propulsion Laboratory, see Clayton R. Koppes, *JPL and the American Space Program: A History of the Jet Propulsion Laboratory* (New Haven: Yale University Press, Haven, 1982; Peter Westwick, *Into the Black: JPL and the American Space Program, 1976–2004* (New Haven: Yale University Press, 2007).

jettisonable. Describing a *space vehicle* component that is intended to be ejected as soon as it is no longer useful. For example, the *solid rocket boosters* of the *Space Shuttle* are jettisonable.

John H. Glenn Research Center at Lewis Field. Full and formal name for the *Glenn Research Center* (GSC).

Johnson Space Center (JSC). NASA installation located in Houston with primary responsibility for human spaceflight. From the *Gemini, Apollo,* and *Skylab* projects to today's *Space Shuttle* and *International Space Station* (ISS) programs, the center continues to lead NASA's efforts in human space exploration. JSC's famed Mission Control Center, or MCC, has been the operational hub of every American human space mission since Gemini IV. The MCC manages all activity onboard the ISS and directs all Space Shuttle missions, including station assembly flights and *Hubble Space Telescope* servicing. Today JSC serves as the lead NASA center for the ISS, a U.S.-led collaborative effort of 16 nations, and the largest, most powerful, complex human facility ever to operate in space. JSC is also home to NASA's Astronaut Corps and is responsible for training space explorers from the United States and our Space Station partner nations. As such, it is the principal training site for both Space Shuttle crews and ISS expedition crews. JSC leads NASA's flight-related scientific and medical research efforts. Additionally, the center manages the development, testing, production, and delivery of hardware supporting spacecraft functions including life support systems, power systems, crew equipment, electrical power generation and distribution, navigation and control, cooling systems, structures, flight software, robotics, and spacesuits and spacewalking equipment.

On January 3, 1961, NASA's *Space Task Group*—an autonomous subdivision of *Goddard Space Flight Center* that managed Project *Mercury* and was housed at *Langley Research Center*—was made an independent NASA field installation. Following congressional endorsement of President Kennedy's decision to accelerate the U.S. *human spaceflight program,* Congress in August 1961 appropriated funds for a new center for such a program. On September 9, 1961, NASA announced that the Manned Spacecraft Center (MSC) would be built at Clear Lake near Houston, Texas, and on November 1, 1961, Space Task Group personnel were told that "the Space Task Group is officially redesignated the Manned Spacecraft Center."

Known as the Manned Spacecraft Center for 11½ years, the facility was responsible for design, development, and testing of crewed spacecraft; selection and training of astronauts; and operation of human spaceflights—including the Mercury, Gemini, Apollo, and *Skylab* programs and the U.S.-Soviet *Apollo-Soyuz Test Project.* It was lead center for management of the Space Shuttle program.

ETYMOLOGY. Following the January 22, 1973, death of former President Lyndon B. Johnson, a supporter of the U.S. *space program* from its earliest beginnings, Senator Lloyd M. Bentsen (D-Tex.) proposed that MSC be renamed the Lyndon B. Johnson Space Center. Senator Robert C. Byrd (D-W.Va.) introduced Senate Joint Resolution 37 on Senator Bentsen's behalf on January 26, and House joint resolutions were introduced within the next few days. Support from NASA Headquarters and MSC was immediate. The Senate and House acted February 6 and 7, and President Richard Nixon signed the resolution on February 17, 1973.

As Senator, Johnson had drafted and helped enact the legislation that created NASA. As Vice President he had chaired the National Aeronautics and Space Council during the early years of the space program, when the decision was made to place a man on the Moon. As President he continued to offer strong support.

Signing the resolution renaming MSC, President Nixon said, "Lyndon Johnson drew America up closer to the stars, and before he died he saw us reach the moon—the first great platform along the way."

SOURCES. SP-4402, pp. 149–50; Rosholt, *Administrative History of NASA,* pp. 124, 214; Swenson, Grimwood, and Alexander, *This New Ocean,* SP-4201, p. 392; *Congressional Record,* January 26, 1973, p. S1344; January 29, 1973, p. H553; February 7, 1973, p. H77; February 2, 1973, p. D71;

February 6, 1973, pp. S2229–30; February 7, 1973, pp. H838–39; February 20, 1973, p. D117; NASA Notice 1132, February 17, 1973; U.S. Congress, Senate, Committee on Aeronautics and Space Sciences, Tenth Anniversary, 1958–1968, S. Doc. 116, July 19, 1968; Statements by Presidents of the United States on International Cooperation in Space, S. Doc. 92-40, September 24, 1971, pp. 55–90; JSC, transcript, "Dedication of Lyndon B. Johnson Space Center," August 27, 1973; White House Release, Key Biscayne, Fla., February 19, 1973. For more information about Johnson Space Center, see Dethloff, *Suddenly, Tomorrow Came,* SP-4307.

Jovian. Of or pertaining to the planet Jupiter. Associated with Jupiter or similar to Jupiter, as in Jovian mission.

JPL. *Jet Propulsion Laboratory* (Pasadena, Calif.).

JSC. *Johnson Space Center* (Houston, Tex.).

Juno *(launch vehicle).* Juno I and II were early launch vehicles adapted from existing U.S. Army missiles by the Army Ballistic Missile Agency *(ABMA)* and the *Jet Propulsion Laboratory* (JPL). Juno I, a four-stage configuration of the *Jupiter C,* launched the first U.S. *satellite, Explorer* 1, on January 31, 1958. The "UE" painted on the *Redstone* first stage of that Juno I indicated that the Redstone was number 29 in a series of launches. The ABMA code for numbering Redstone *boosters* was based on the word Huntsville, with each letter (except for the second "L") representing a number:

H U N T S V I L E
1 2 3 4 5 6 7 8 9

Later that year, at the request of the Department of Defense's *Advanced Research Projects Agency,* ABMA and JPL designed the Juno II, which was based on the Jupiter intercontinental ballistic *missile* and had the *upper stage*s of the Juno I. Responsibility for Juno II was transferred to NASA after its establishment on October 1, 1958. Juno II vehicles launched three *Explorer* satellites and two *Pioneer space probe*s. Juno V was the early designation of the launch vehicle that became the *Saturn I.*

ETYMOLOGY. The ancient Roman goddess Juno, queen of the gods, was the sister and wife of Jupiter, king of the gods. Since the new launch vehicle was the satellite-launching version of the Jupiter C (Jupiter Composite Reentry Test Vehicle), the name Juno was suggested by Dr. William H. Pickering, JPL Director, in November 1957. Army officials approved the proposal and the name was adopted.

SOURCES. SP-4402, p. 14; William H. Pickering, Director, JPL, teletype message to MIG John B. Medaris, Commanding General, ABMA, November 18, 1957; Medaris, teletype message to Gen. James M. Gavin, Chief of Research and Development, Hq. USA, November 20, 1957; Wernher von Braun,

"The Redstone, Jupiter, and Juno," in *The History of Rocket Technology,* ed. Eugene M. Emme (Detroit: Wayne State University Press, 1964), pp. 107–21; R. Cargill Hall, JPL, letter to Eugene M. Emme, Historian, NASA, October 9, 1965.

Jupiter. (1) Army intermediate-range ballistic missile (IRBM) that followed the *Redstone,* becoming the *Jupiter C* and *Juno* for space missions. It was the forerunner of the *Saturn* rockets that went to the Moon. (2) Fifth planet from the Sun, a gas giant or Jovian planet.

Jupiter C. Jupiter Composite Reentry Test Vehicle.

JWST. *James Webb Space Telescope.*

K

Kennedy Space Center (KSC). Dubbed America's Gateway to the Universe, one of the world's prime *launch* sites for *crewed* and uncrewed missions into space. Formally named John F. Kennedy Space Center, NASA, the installation at *Cape Canaveral (known as Cape Kennedy* 1963–73) evolved through a series of organizational changes and redesignations.

In 1951 the Experimental Missiles Firing Branch of the Army Ordnance Guided Missile Center in Huntsville, Alabama, was established to supervise test flights of the U.S. Army's *Redstone* intermediate-range ballistic *missile* at the Long Range Proving Ground at Cape Canaveral, Florida. In January 1953, when its responsibilities were expanded, the Army facility was renamed the Missile Firing Laboratory.

On July 1, 1960, the Missile Firing Laboratory became part of NASA's *Marshall Space Flight Center* (MSFC)—the nucleus of which was the laboratory's parent organization at Huntsville—and was absorbed organizationally into MSFC's Launch Operations Directorate. The other basic element of the Launch Operations Directorate was a NASA unit known as AMROO (Atlantic Missile Range Operations Office). AMROO had functioned as NASA's liaison organization with the military-operated Atlantic Missile Range (formerly Long Range Proving Ground) at Cape Canaveral. Together, the Missile Firing Laboratory and AMROO formed MSFC's Launch Operations Directorate.

The Launch Operations Directorate was discontinued as a component of MSFC on March 7, 1962, and the Launch Operations Center was established as a separate NASA field installation, officially activated July 1, 1962.

Adjacent to Cape Canaveral was the 125-square-mile (324-sq-km) Merritt Island. In the autumn of 1961 NASA had selected it for launches in the *Apollo* lunar program. On January 17, 1963, the Launch Operations Center became the executive agent for management and operation of the Merritt Island Launch Area (usually called MILA). Headquarters of Kennedy Space Center moved to new facilities on Merritt Island July 26, 1965, and NASA discontinued the MILA designation, calling the entire NASA complex the Kennedy Space Center. The center was responsible for overall NASA launch operations at the Eastern Test Range (formerly Atlantic Missile Range), Western Test Range, and KSC itself, including launches of satellites, *probes*, human space missions, and the *Space Shuttle*.

SOURCES. SP-4402, pp. 149–50; Frank E. Jarrett Jr. and Robert A. Lindemann, *Historical Origins of NASA's Launch Operations Center to July 1, 1962,* KHN-1 (Cocoa Beach: KSC, 1964), pp. 21–22, 32, 54; Rosholt, *Administrative History of NASA,* pp. 123, 214–15; President Lyndon B. Johnson, Executive Order no. 11129, November 29, 1963, in Angela C. Gresser, *Historical Aspects Concerning the Redesignation of Facilities at Cape Canaveral,* KHN-1 (Cocoa Beach: KSC, 1964), p. 15. For more information about Kennedy Space Center, see Benson and Faherty, *Moonport,* SP-4204; Kenneth Lipartito, Orville R. Butler, and Gregg A. Buckingham, *A History of the Kennedy Space Center* (Gainesville: University Press of Florida, 2007).

Kibo. Also known as the Japanese Experiment Module (JEM), a key element of the *International Space Station.* Kibo's pressurized module will be used mainly for *microgravity* experiments, and its Exposed Facility can accommodate experiments that require direct exposure to the environment of outer space. It was successfully launched in 2008. (*Reference Guide to the International Space Station,* SP-2006-557).

kick in the apogee. Raising a *satellite* into Earth *orbit* by firing a *rocket engine* at its point of maximum altitude ("Space Age Slang," *Time* magazine, August 10, 1962, p. 12).

Kitty Hawk. *Apollo* 14 Command Module piloted by Stuart Roosa while Alan Shepard and Edgar Mitchell went to the lunar surface in the Antares *Lunar Module.* The Apollo 14 *mission* began on January 31 and ended on February 9, 1971.

ETYMOLOGY. Named by Roosa: "I did it to honor the place where it all began with the Wright Brothers and felt that this name was most appropriate" (quoted in Lattimer, *All We Did Was Fly to the Moon,* p. 82).

Kiwi reactor. Program to develop and test nuclear reactors on the ground to be used in flight engines for nuclear propulsion systems (NASA, *Glossary/Congressional Budget Submission,* p. 24).

Kourou. Location of *Europe's Spaceport* in northeastern South America in

French Guiana, an overseas department of France. In 1964 the French government chose Kourou, from among 14 sites, as a base from which to launch its satellites. When the *European Space Agency* (ESA) came into being in 1975, the French government offered to share its *Centre Spatial Guyanais* (CSG). For its part, ESA approved funding to upgrade the launch facilities at the CSG to prepare the Spaceport for the *Ariane* launchers under development.

KSC. *Kennedy Space Center.*

L

LAGEOS. Laser Geodynamic Satellite. In 1971 NASA was considering the possibility of launching a passive *satellite,* Cannonball, on a *Saturn launch vehicle* left over from the *Apollo* program. Definition and documentation were completed in late 1971. Subsequently the Office of Applications began defining a similar but less costly satellite as a new project to begin in fiscal year 1974. The redefined satellite was given the functional name Laser Geodynamic Satellite. LAGEOS was to be the first in a series of varied satellites within NASA's Earth and Ocean Physics Applications Program (EOPAP), including some launched on uncrewed vehicles in 1976 and 1977 and later ones on the *Space Shuttle.*

Approved as a "new start" for fiscal year 1974, with a 1976 *launch* date, the *terrestrial* reference satellite was to be a very heavy ball weighing 906 pounds (411 kg), although less than 3 feet (1 m) in diameter, and covered with laser reflectors to permit highly accurate measurements of the Earth's rotational movements and movements of the Earth's *crust.* The *orbit* and the weight of the simple, passive satellite were planned to provide a stable reference point for decades. The high, 3,666-mile (5,900-km) orbit would permit simultaneous measurements by laser ranging from *Earth stations* a continent apart. LAGEOS was launched in 1976 and LAGEOS II in 1992. Data was used in earthquake prediction and other applications.

SOURCES. SP-4402, pp. 58–59; Eberhard Rees, Director, MSFC, memorandum to Dale D. Myers, Associate Administrator for Manned Space Flight, NASA, January 25, 1973; U.S. Congress, House of Representatives, Committee on Science and Astronautics, Subcommittee on Space Science and Applications, *Hearings: 1975 NASA Authorization, Pt. 3, February and March 1974* (Washington, DC: GPO, 1974), pp. 104–6; NASA Fiscal Year 1974

Budget News Conference (transcript), January 23, 1973; MSFC News Releases 73-184, 75-49.

Lagrangian point. See *libration point.*

Laika (Russian for barker). The name of the mixed-breed female terrier sent into space on *Sputnik* 2.

lander. *Spacecraft* or part of one that is designed to land on the surface of a planet or the Moon.

Landsat. New name for Earth Resources Technology Satellite. See *ERTS.* A series of unpiloted Earth-orbiting NASA satellites that acquire multi-spectral images in various visible and infrared (IR) bands. The first Landsat was launched in 1972, and the latest, Landsat 7, was launched April 15, 1999.

Langley Research Center (LaRC). NASA research center that concentrates on aeronautics and atmospheric sciences.

Construction of *NACA*'s first field station began at Langley Field near Hampton, Virginia, in 1917. In April 1920 President Wilson concurred with NACA's suggestion that the facility be named Langley Memorial Aeronautical Laboratory in honor of Dr. Samuel P. Langley (1834–1906). Dr. Langley was the third Secretary of the Smithsonian Institution, an "inventor, brilliantly lucid writer and lecturer on science, original investigator in astronomy and especially of the physics of the Sun, pioneer in aerodynamics." He was "all this and more." His persistent investigation of mechanical flight led to successful flights by his steam-powered, heavier-than-air "aerodromes" in 1896. On May 6 his model made two flights, each close to a mere 0.6 mile, and on November 28 his aerodrome achieved a flight of more than 0.7 mile.

The facility was dedicated July 11, 1920, marking "the real beginning of NACA's own program of aeronautical research, conducted by its own staff in its own facilities." It was the only NACA laboratory until 1940. On October 1, 1958, it became a component of the National Aeronautics and Space Administration and was renamed Langley Research Center.

The center conducted basic research in a variety of fields for aeronautical and space flight and had management responsibility for the *Lunar Orbiter* and *Viking* projects and the *Scout launch vehicle*s. The supercritical wing, an improved airfoil, was developed at Langley. The Lunar Landing Research Facility came into operation at the center in July 1965. The huge structure—250 feet (76.2 m) high and 400 feet (121.9 m) long—would be used to explore techniques and to forecast various problems of landing on the Moon. The facility enabled a test vehicle to be operated under one-sixth g conditions. Originally, the 250-foot-high frame was designed to test the Apollo Lunar Lander and to train Apollo astronauts. By 1972, however, the structure was converted to an Impact Dynamics Facility, where crashes are conducted under controlled circumstances. The impact runway can be modified to simulate other crash environments, such as packed dirt, to meet a specific test requirement. Additional historic additions to the Langley Research Center include the Eight-Foot High-Speed Tunnel (built 1935) and the Rendezvous Docking Simulator (1963).

On the Langley grounds today, an anechoic chamber serves for noise-reduction tests, and a transonic wind tunnel is used to study problems that occur in transonic speed ranges. In 1985 the U.S. Department of the Interior designated the Variable Density Tunnel, Full-Scale Tunnel, Eight-Foot High-Speed Tunnel, Lunar Landing Research Facility, and Rendezvous Docking Simulator as National Historic Landmarks. Today the focus of Langley's research is aviation safety, quiet aircraft technology, small aircraft transportation, and aerospace vehicles system technology. It supports NASA space programs with atmospheric research and technology testing and development.

SOURCES. SP-4402, pp. 150–52; Hunsaker, *Forty Years of Aeronautical Research*, pp. 250–251; Bessie Zaban Jones, *Lighthouse of the Skies: The Smithsonian Astrophysical Observatory, Background and History, 1846–1955*, Smithsonian Publication 4612 (Washington, DC: Smithsonian Institution, 1965), pp. 105, 155–58; Emme, *Aeronautics and Astronautics*, p. 102. For more information about Langley Research Center, see James R. Hansen, *Engineer in Charge: A History of the Langley Aeronautical Laboratory, 1917–1958*, SP-4305 (Washington, DC: GPO, 1987); Hansen, *Spaceflight Revolution: NASA Langley Research Center from Sputnik to Apollo*, SP-4308 (Washington, DC: GPO, 1995); James Schultz, *Crafting Flight: Aircraft Pioneers and the Contributions of the Men and Women of NASA Langley Research Center*, SP-2003-4316 (Washington, DC: GPO, 2003).

Lani Bird (Hawaiian name for bird of heaven). Intelsat II-A communications *satellite*. See *Intelsat*.

LaRC. *Langley Research Center*.

Large Space Telescope (LST). Project that became the *Hubble Space Telescope*.

Laser Geodynamic Satellite. See *LAGEOS*.

launch. (1, v.) To send forth a *rocket* or *missile* under its own power. (2, n.) The result of this action, i.e., the transition from static repose to dynamic flight by the *rocket*. (3, n.) The event itself.

launch complex. The site, facilities, and equipment used to *launch* a *rocket* vehicle.

Launch Complex 39. Area at *Kennedy Space Center* created to *launch* American astronauts to the Moon, and still used in the *human space-*

flight program today. Placed on the National Register of Historic Places in 1973 and classified as a historic site. (*Historic Properties,* KSC Archives, John F. Kennedy Space Center, Florida, August 2005, NP-2005-08-KSC.)

launch crew. A group of technicians that prepares and *launch*es a *rocket* (SP-7).

launch pad / launchpad. The load-bearing steel and concrete base or platform from which a *rocket* vehicle is *launch*ed. Also known as the launching pad, launch area, or pad.

> FIRST USE. The term first enters the popular vocabulary with preparation for the launch of the *Vanguard* satellite in the fall of 1957: "J. Paul Walsh, deputy director of Project Vanguard, told a news conference that the test of the important second section was carried out on a launch pad at Patrick Air Force Base, Fla." ("US Rocket Test is Success," *Washington Post,* November 21, 1957, p. A4).

launch point. The geographical position from which a *rocket* vehicle is *launched* (SP-7).

launch profile. The shape of a vehicle's *trajectory* with reference to the surface of the Earth represented graphically.

launch vehicle. A *rocket* or other vehicle used to place a *satellite* in *orbit* or to *launch* a *space probe.* See, for example, *Atlas, Delta, Saturn,* and *Titan.* For more information, see Launius and Jenkins, eds., *To Reach the High Frontier.*

> ETYMOLOGY. "In the early days of the U.S. civilian space program the term 'launch vehicle' was used by NASA in preference to the term 'booster' because 'booster' had been associated with the development of the military missiles. 'Booster' now has crept back into the vernacular of the Space Age and is used interchangeably with 'launch vehicle.'" (SP-4402.)

launch window. Opening in time or space through which a *spacecraft* must be *launch*ed in order to achieve a desired orbital position. Also called *firing window.* (SP-7, p. 158; NASA, *Glossary/Congressional Budget Submission,* p. 24.)

> ETYMOLOGY / FIRST USE. The term first comes into popular use as the United States prepared to launch the *Ranger* 3 in 1962 to take the first close-up television images of the lunar surface. The window established for the launch was to take advantage of maximum lighting for the *mission.* The *Los Angeles Times* used the term in a report on the launch: "This permitted extension of the launch 'window' or time restriction to about one hour in midafternoon either next Friday or Saturday" ("Crews Rush to Ready Ranger for Moon Shot," *Los Angeles Times,* June 22, 1962. p. 9).

LC. *Launch complex.*

the leans. Illusion of a *spacecraft* being tilted, with a corresponding leaning of the crew in the opposite direction, caused by a false reaction uncorrected by visual cues (Pitts, *The Human Factor,* SP-4213, p. 264).

learning failures. Approach to spaceflight testing that accepted a certain number of failed missions as part of the development process. Because NASA was an open program, this approach was dropped in favor of one that sought success on the first *mission,* a policy that became known as *all up.* This is discussed by Homer Newell in *Beyond the Atmosphere*: "By any reasonably objective measures—certainly by previously accepted standards—Vanguard was a successful development. Yet the early launch failures made the entire program a symbol for failure in the public mind. The lesson of Vanguard was plain: NASA could not afford to regard failure under any guise. Success had to be sought on the first try, and every reasonable effort bent toward achieving that outcome." Newell goes on to say that this policy was epitomized in George Mueller's all-up approach. (Newell, *Beyond the Atmosphere,* SP-4211, pp. 159–60.)

LEM. *Lunar Excursion Module.* An abbreviation changed officially to *LM,* for *Lunar Module,* in the early days of the *Apollo* program. The pronunciation "lem" was retained even after the abbreviation was changed to LM.

LEO. *Low Earth orbit.*

Leonardo. One of the Italian-built Multi-Purpose Logistics Modules *(MPLMs)* that are carried in the cargo bay of the *Space Shuttle* and ferried to the *International Space Station.* They contain experiments and supplies. (*Reference Guide to the International Space Station,* SP-2006-557.)

LeRC. *Lewis Research Center.*

Lewis and Clark. Two small Earth-scanning satellites announced in June 1994 and named for American explorers Meriwether Lewis and William Clark ("For NASA '*Smallsats,*' a Commercial Role," *Washington Post,* June 9, 1994, p. A7). The Lewis *mission* failed, and the Clark mission was canceled.

Lewis Research Center (LeRC). Now *Glenn Research Center.*

Liberty Bell 7. *Mercury capsule* flown for 16 minutes on July 21, 1961, by Virgil I. "Gus" Grissom.

ETYMOLOGY. Grissom chose the name because the capsule implied American independence in space and was shaped like a bell. "John Glenn felt that the symbolic number '7' should appear on our capsules in honor of the team, so this was added," said Grissom. "Then one of the engineers got the bright idea that we ought to dress up Liberty Bell by painting a crack on it just like the real one. No one seemed quite sure what the crack looked like, so we copied it from the 'tails' side of a fifty-cent piece. Ever since my flight, which ended with my capsule sinking to the bottom of the Atlantic, there has been a joke around the Cape that that was the last capsule that we would ever launch with a crack in it." (Grissom in Carpenter et al., *We Seven,* pp. 217–18.)

libration point. Point in space at which a small body, under the gravitational influence of two larger ones, will remain approximately at rest relative to them. Libration points are also called Lagrange (or Lagrangian) points, after Joseph-Louis Lagrange, the French mathematician who calculated their existence in 1772. (Paine Report, p. 198.)

LIDAR. Light detection and ranging. Laser counterpart of radar.

lifting body. Configuration wherein the body itself produces lift. It is related to, but the opposite of, a flying wing, an aircraft whose fuselage is contained by the wing.

liftoff. The beginning of the flight into space. Informally, *blastoff.* See also *takeoff.*

light the candle. To be *launch*ed into space. From a *New York Times* article on John Glenn's return to space in 1998 as a member of a Shuttle crew: "But nothing he has achieved in politics has come close to rivaling the worldwide acclaim he received as an astronaut, and there was speculation today that Mr. Glenn wanted to cap his career the way he began it—by lighting the candle, in the jargon of astronauts" ("Glenn to Slip the Bonds of Earth," *New York Times,* January 17, 1998, p. A8). Neal Thompson gave the title *Light This Candle* to his biography of astronaut Alan Shepard (New York: Crown, 2004).

light-year. The distance light travels in space (technically, in a pure vacuum) in one year, approximately 6 trillion miles (9.5 trillion km).

Lincoln Near-Earth Asteroid Research (LINEAR). Telescope in Socorro, New Mexico, that searches the skies for new asteroids.

LINEAR. *Lincoln Near-Earth Asteroid Research.*

Little Joe *(launch vehicle).* A relatively simple and inexpensive launch vehicle designed specifically to test the *Mercury spacecraft abort* system in a series of suborbital flights. Based on a cluster of four solid-*propellant rocket* motors, as conceived by Langley Research Center's Maxime A. Faget and Paul E. Purser, the *booster* acquired its name in 1958 as Faget's nickname for the project was gradually adopted. The configuration used in the tests added four Recruit rockets, but the original concept was for four Pollux rocket motors, fired two at a time. There were eight Little Joe flights. Little Joe II was similar in design and was used to test the *Apollo* spacecraft abort system.

ETYMOLOGY. Since their first cross-section drawings showed four holes up, the engineers called the project Little Joe, from the craps game for a hard four roll of the dice: a pair of twos. The appearance on engineering drawings of the four large stabilizing fins protruding from the airframe also helped to perpetuate the name.

SOURCES. SP-4402, p. 10; Swenson, Grimwood, and Alexander, *This New Ocean,* SP-4201, pp. 123-24; Paul E. Purser, MSC, handwritten note to James M. Grimwood, Historian, MSC, October 1963; Robert W. Mulac, LaRC, letter to Historical Staff, NASA, December 10, 1963.

LM. Common term for the *Lunar Module,* the *Apollo* module designed to land on the Moon. It was changed from the earlier acronym *LEM,* for *Lunar Excursion Module,* but was still pronounced "lem." A *New York Times* "Glossary of Terms," published in connection with the Apollo 9 *mission,* defines LM as "the Lunar Module," or "lem" (*New York Times,* March 4, 1969, p. 15).

LOI. Lunar orbit insertion; going into Moon *orbit.*

Long Range Proving Ground. Former name of USAF facility that become the Atlantic Missile Range and then the Eastern Test Range. Now the *Eastern Space and Missile Center.* See *Kennedy Space Center.*

los/LOS. *Loss of signal.*

loss of signal (los/LOS). The end of "real time" in ground *tracking* and communication—especially relevant to early human *spaceflight.* "The initial contact is called acquisition . . . and real time ends at loss of signal" (Kranz, *Failure Is Not an Option,* p. 87).

low Earth orbit (LEO). Circling the globe at an altitude not high enough to sustain a spaceflight of lengthy duration (owing to atmospheric drag). A decaying *orbit.* (NASA, *Glossary/Congressional Budget Submission,* p. 26.)

lox/LOX. Liquid oxygen. A translucent bluish magnetic liquid obtained by compressing gaseous oxygen and then cooling it to below its boiling point. Used as an oxidizer in *rocket propellants.*

ETYMOLOGY / FIRST USE. Defined in 1945 in the lexicon attached to G. Edward Pendray's *The Coming Age of Rocket Power.* Early definers had fun with playing this off against the lox of bagels and lox (Nova Scotia salmon).

loxing. Also loxing up. Process of loading liquid oxygen. See *lox.*

loxogen. Liquid oxygen. See *lox.*

LRV. *Lunar Roving Vehicle.*

LST. Large Space Telescope. Project that became the *Hubble Space Telescope.*

Luna. Series of Soviet uncrewed lunar *probes.* The Luna series were the first human-made objects to attain *escape velocity* from Earth's gravity, to impact on the Moon, to photograph the far side of the Moon, to *soft-land* on the Moon, to retrieve and return lunar samples to the Earth, and to deploy a lunar *rover* on the Moon's surface. The program ran from 1958 to 1976. The last of the Luna series of *spacecraft,* Luna 24, was the third *mission* to retrieve lunar ground samples. See *Lunik.*

Lunar Excursion Module (*LEM*). Early name for what would become the

Lunar Module or *LM*. A *spacecraft* that carries *astronaut*s from the Command Module (CM) to the surface of the Moon and back. (NASA, *Glossary/Congressional Budget Submission,* p. 25; SP-6001, p. 54.)

Lunar Module (*LM*). Formerly *Lunar Excursion Module.* The section of the *Apollo spacecraft* designed to land on the Moon.

lunarnaut. Also lunanaut. Short-lived term for one who travels or has traveled to the Moon. Though little used today, it was applied enough during the *Apollo* era to have a full entry in the *OED* and other dictionaries.

Lunar Orbiter *(space probe).* A literal description of the *mission* assigned to each *probe* in that project: to attain lunar *orbit,* whence it would acquire photographic and scientific data about the Moon. Lunar Orbiter supplemented the *Ranger* and *Surveyor* probe projects, providing lunar data in preparation for the *Apollo* landings and the Surveyor *spacecraft soft landing*s. The name evolved informally through general use. NASA had under consideration plans for a Surveyor spacecraft to be placed in orbit around the Moon. This Surveyor was called Surveyor Orbiter to distinguish it from those in the lunar-landing series. When the decision was made to build a separate spacecraft rather than use Surveyor, the new probe was referred to simply as Orbiter or Lunar Orbiter. Five Lunar Orbiter flights launched in 1966 and 1967 made more than 6,000 orbits of the Moon and photographed more than 99 percent of the lunar surface, providing scientific data and information for selecting the Apollo piloted landing sites. *Tracking* data increased knowledge of the Moon's gravitational field and revealed the presence of the lunar *mascon*s. (SP-4402, pp. 83–85; U.S. Congress, House of Representatives, Committee on Appropriations, Subcommittee on Independent Offices, *Hearings: Fiscal Year 1966 Independent Offices Appropriations, Pt. 2* [Washington, DC: GPO, 1965], p. 858; NASA News Release 68-2.)

Lunar Prospector. The third NASA *Discovery mission.* Lunar Prospector was designed to determine the origin of the Moon, the Moon's evolution, and the current state of lunar resources by performing low-altitude mapping of the lunar surface's composition and of its *magnetic field*s, gravity fields, and gas release events. The mission (1998) provided a new map of the Moon. This map shows, in unprecedented detail, the Moon's chemical composition its well as its magnetic and gravitational fields. (http://aerospacescholars.jsc.nasa.gov/HAS/cirr/glossary.cfm.)

Lunar Roving Vehicle (LRV). Four-wheel transporter for two astronauts on surface of the Moon for *Apollo* missions 15, 16, and 17. The LRV greatly expanded the area which could be explored by astronauts. Also known as a *rover.*

lunar-tic/lunartic. Nickname applied to those who proclaim their belief that the Moon landing was a hoax.

ETYMOLOGY. Play on word lunatic, given great impetus by Jack Hitt's *New York Times* piece of the same title (February 9, 2003, pp. 52–56).

Lunik. Nickname adopted by Western media for the first Soviet *Luna probe* launched in January 1959.

ETYMOLOGY. "Jubilant Moscow citizens called the new space traveler Lunik—a combination of luna for moon and Sputnik" ("Red Rocket Passes Moon," *Chicago Tribune,* January 4, 1959, p. 1).

FIRST USE. Headline in *New York Times,* January 4, 1959, p. E1: "Now 'Lunik.'"

Lunokhod (Russian for Moon walker). A pair of robotic lunar *rover*s landed on the Moon by the Soviet Union in 1970 and 1973. They were in operation in parallel with the *Zond* series of *flyby mission*s. The Lunokhod missions were primarily designed to explore the surface and return pictures. They complemented the *Luna* series of missions, which were intended to be sample-return missions and orbiters.

MA-. *Mercury-Atlas* (followed by *mission* number).

MAF. *Michoud Assembly Facility.*

Magellan. NASA Venus radar mapping *mission,* 1989–94. The first Shuttle-launched interplanetary *spacecraft.* Following its *launch* from *Atlantis* using an Inertial Upper Stage (IUS) on May 4, 1989, during the *STS*-30 mission, Magellan began a 15-month voyage to map Earth's planetary twin. Magellan mapped almost the entire surface of Venus using synthetic aperture radar and conducted measurements of the planet's gravitational field. Its radar was able to penetrate the cloud cover and map Venus's surface, paved with lava flows and pocked with craters and volcanic mountain escarpments. The mission was America's first uncrewed interplanetary expedition after an 11-year hiatus.

ETYMOLOGY. Named for the Portuguese navigator and explorer Magellan (ca. 1480–1521), who sailed under the flags of Portugal (1505–12) and Spain (1519–21). From Spain he sailed around South America, discovering the strait that is named for him, and across the Pacific. Although Magellan was killed by natives in the Philippines, his ships continued on to Spain and completed the first circumnavigation of the world.

The *NASA Daily Activities Report for the Week Ending December 27, 1985,* discusses the naming of Magellan as well as the plan for naming planetary missions: "Magellan has been approved by the Director, Public Affairs Division, as the official name for the Venus Radar Mapper (VRM) Project. The name Magellan was the unanimous choice of the Special Project Name Committee established by Dr. Edelson in September 1982 to solicit and recommend names for the VRM mission. Approximately 1,470 name suggestions were proposed to the committee" (p. 1). The announcement goes on to say that the name was consistent with the general plan to name major planetary missions after noted historic persons, such as scientists, astronomers, mathematicians, and explorers. Magellan was deemed appropriate because the name connoted exploration, discovery, and circumnavigation—all of which were elements of the mission. (NASA Names Files, record no. 17530.)

magnetic field. (1) Region in space with a detectable magnetic force at every point within the region. (2) Specifically, the region of magnetic force that surrounds the Earth.

magnetic storm. A large particle disturbance of Earth's *magnetic field* due to solar flares.

magnetosphere. A region surrounding a planet, extending out thousands of miles and dominated by the planet's *magnetic field* so that charged particles are trapped in it (Paine Report, p. 198).

Malemute *(sounding rocket upper stage)*. A *rocket* second stage developed in 1974 in an interagency program with NASA, Sandia Laboratories, and the Air Force Cambridge Research Laboratories as sponsoring agencies. Designed to be flown with either the *Nike* or the *Terrier* first stage, the Malemute began flight tests in 1974. The Nike-Malemute sounding rocket could lift a 200-pound (90-kg) *payload* to 310 miles (500 km). The Terrier-Malemute could lift the same payload to 435 miles (700 km). The new vehicles were intended to replace the *Javelin* and *Black Brant* IV rockets in the NASA inventory.

ETYMOLOGY. According to *Origins of NASA Names,* SP-4402, "It was named for the Alaskan Eskimo people by the contractor, Thiokol Corporation, in Thiokol's tradition of using Indian-related names." This may be misleading as the term means a breed of sled dog developed in Alaska—according to the *American Heritage Dictionary*, "Any of a breed of powerful dog developed in Alaska as a sled dog and having a thick gray, black, or white coat."

SOURCES. SP-4402, pp. 132–33; William J. Bolster, Flight Performance Branch, Sounding Rocket Division, GSFC, information sent to Historical Office, NASA, January 30, 1975.

maneuver. The ability of a *spacecraft* to change directions.

Man in Space Soonest (MISS). U.S. Air Force program to put a human in space and the parallel program to the Army's *Project Adam*. Both MISS and Project Adam were superseded by NASA's *Mercury* program. The program was created after the *launch* of *Sputnik* as an outgrowth of Air Force Project 7969, a piloted ballistic *rocket* system.

manned. Adjective used officially and unofficially to describe *human spaceflight* for most of the *Space Age.* It is gradually being superseded by the preferred terms *crewed, human,* or *piloted.*

Manned Orbiting Laboratory (MOL). Part of the U.S. Air Force's *human spaceflight program,* a successor to the canceled X-20 *Dyna-Soar* project. It was announced to the public on the same day that Dyna-Soar was canceled, December 10, 1963. The MOL program was canceled in June 1969.

Manned Spacecraft Center (MSC). Former name for *Johnson Space Center.*

Manned Spacecraft Operations Building (MSOB). Located at the *Kennedy Space Center.* "The training area that the Apollo astronauts used to practice the mechanics of ALSEP [Apollo Lunar Surface Experiment Package] deployments and other EVA activities was located behind the MSOB. Other training activities, such as those involving a Lunar Module mock-up, were conducted in the building" (*ALSJ* Glossary).

manned spaceflight. See *human spaceflight, human spaceflight program.*

Manned Space Flight Network (MSFN). System that provided reliable, (usually) continuous, and instantaneous communication with the astronauts, *launch* vehicle, and *spacecraft* from *liftoff* to *splashdown* (*ALSJ* Glossary).

mare (plural, maria). Any of several mostly flat dark areas on the lunar surface. Also formerly applied to certain dark features on Mars, and in 2007 officially approved to apply to certain features on Saturn's moon Titan (where they are large expanses of dark material thought to be liquid hydrocarbon). Maria are the largest topographical features on the Moon and can be seen from Earth with the unaided eye. In fact, they form the face of the "man in the Moon." (http://aerospacescholars .jsc.nasa.gov/HAS/cirr/glossary.cfm.)

ETYMOLOGY. Latin for sea. The term was mistakenly applied to the Moon by 17th-century lunar observers who thought the areas might be real seas. The 19th-century Italian astronomer Schiaparelli applied the term to features on Mars. Though the International Astronomical Union adopted his Martian nomenclature in 1958, much of it was later superseded (1973) with better resolution made possible by Space Age technology, which showed that Mars has no seas.

Mariner (interplanetary *probe*). Series of space probes designed to investigate the Earth's planetary neighbors Venus and Mars, and eventually Mercury, Jupiter, and Saturn. The name was adopted in May 1960 as part of the Cortright system of naming planetary missions after nautical terms.

Mariner *spacecraft* performed a number of record-setting missions from the early years of the project. On December 14, 1962, Mariner 2 came within 22,000 miles (35,000 km) of Venus, climaxing a four-month space flight that provided new scientific data on interplanetary space and Venus. On July 14, 1965, after seven months of interplanetary flight, Mariner 4 took the first close look at Mars from outside the Earth's *atmosphere,* returning high-quality photographs and scientific data. It provided the first indication of craters on Mars. On October 19, 1967, Mariner 5 flew within 2,500 miles (4,000 km) of Venus, obtaining additional information on the nature and origin of the planet and on the interplanetary environment during a period of increased solar activity. During 1969, Mariner 6 and 7 continued the investigation of the Martian atmosphere, flying within 2,200 miles (3,500 km) of the planet.

Following the unsuccessful Mariner 8 *launch* attempt, Mariner 9 was launched May 30, 1971, and put into *orbit* around Mars on November 13, 1971, the first man-made object to orbit another planet. Mariner 9 photographed the moons of Mars, mapped 100 percent of the planet, and returned data showing that it had been geologically and meteoro-logically active in the past, as evidenced by possible ancient river beds. Mariner 10, launched November 3, 1973, flew past Venus in February 1974 to a March 1974 encounter with Mercury, for the first exploration of that planet. The *spacecraft's trajectory* around the Sun swung it back for a second encounter with Mercury in September 1974 and returned for a third in March 1975. Venus data gave clues to the planet's weather system, suggested the planet's origin differed from the Earth's, and confirmed the presence of hydrogen in its *atmosphere.* Mercury data revealed a strong *magnetic field,* a tenuous *atmosphere* rich in helium, a cratered *crust,* and possibly an iron-rich core; it brought new insight into the formation of the *terrestrial* planets.

USAGE. Mariner H was designated Mariner 8 by NASA Associate Administrator John E. Naugle because of pressure from the press for easier identification. This designation was a departure from past precedent of assigning a number to spacecraft only after a successful launch.

SOURCES. SP-4402, p. 86–87; Edgar M. Cortright, Assistant Director of Lunar and Planetary Programs, NASA, memorandum to NASA Ad Hoc Committee to Name Space Projects and Objects, 17 May 1960; NASA Ad Hoc Committee to Name Space Projects and Objects, minutes of meeting May 19, 1960; NASA News Release 75-19.

Mariner Jupiter-Saturn. *Probe* later renamed *Voyager.*

Mars 96. Russian *spacecraft* launched into Earth *orbit* on November 17, 1996. At 14,800 pounds (6.7 metric tons), Mars 96 was the heaviest and most ambitious planetary *mission* launched by any country. The spacecraft reentered the Earth's *atmosphere* and crashed into the Pacific Ocean because the planned second burn of the Block D-2 fourth stage failed to take place.

Mars Climate Orbiter. NASA orbiter mission to Mars, December 11, 1998–September 23,1999. This *mission*—designed to study the Martian weather, climate, water, and carbon dioxide budget, in order to understand the reservoirs, behavior, and atmospheric role of volatiles and to search for evidence of long-term and episodic climate changes—was *launch*ed on a *Delta* II from Pad A of *Launch Complex* 17 at *Cape Canaveral* Air Station, Florida. The *spacecraft* reached Mars and executed a 16-minute 23-second *orbit* insertion main *engine burn* on September 23, 1999. The *spacecraft* passed behind Mars and was to reemerge and establish radio contact with Earth. However, contact was never reestablished, and no signal was ever received from the spacecraft. Findings of the Failure Review Board indicate that a navigation error resulted from some spacecraft commands having been sent in English units instead of metric.

Mars Environmental Survey (MESUR) Pathfinder. Name for project that became simply *Mars Pathfinder.*

Mars Exploration Rovers. See MERs.

Mars Express. *European Space Agency* Mars *orbiter* and *lander,* 2003. *Spacecraft launch*ed June 2, 2003, from *Baikonur Cosmodrome* in Kazakhstan, using a *Soyuz*-Fregat *rocket,* which began its interplanetary voyage. See also *Beagle 2.*

ETYMOLOGY. The use of the word express in this *mission*'s name refers originally to the relatively short duration of its voyage, which was initiated when Earth and Mars were in closer proximity than ever in recorded history.

Mars Global Surveyor *(space probe).* A probe *launch*ed in 1996 and placed in *orbit* around Mars in September 1997, detecting a *magnetic field* there. The existence of a planetary magnetic field has important implications for the geological history of Mars and for the possible development and continued existence of life on Mars. The probe continued mapping the Martian surface in detail , taking hundreds of photographs, until contact was lost on November 2, 2006.

Marshall Space Flight Center (MSFC). One of NASA's largest and most diversified installations, with a history of contributions dating from the 1961 flight of the first U.S. *astronaut* into space to the *Apollo* missions to the Moon; the development and operation of America's *Space*

Shuttle fleet; and the construction of and pursuit of scientific discovery aboard the *International Space Station.*

In April 1950 the U.S. Army established a team of *rocket* specialists headed by Dr. Wernher von Braun as the Ordnance Guided Missile Center at *Redstone* Arsenal, Huntsville, Alabama. This center was the origin of what eventually became the George C. Marshall Space Flight Center. On February 1, 1956, the Army Ballistic Missile Agency *(ABMA)* was formed at Redstone Arsenal. ABMA was a merger and expansion of existing agencies there; its team of scientists formed the nucleus of the Development Operations Division.

Early in 1960 President Eisenhower submitted a request to Congress for the transfer of ABMA's space missions to NASA, including certain facilities and personnel, chiefly the Development Operations Division. The transfer became effective March 14, 1960, and NASA set up its Huntsville Facility in preparation for formal establishment of the field center later that year. The next day, March 15, President Eisenhower proclaimed that the NASA facility would be called the George C. Marshall Space Flight Center.

MSFC officially began operation with the formal mass transfer of personnel and facilities from ABMA on July 1, 1960. The center's primary *mission* responsibility was development of the *Saturn* family of *launch vehicles,* used in the Apollo lunar-landing program, in the *Skylab* experimental *space station* program, and in the U.S.-Soviet *Apollo-Soyuz Test Project.* MSFC also held responsibility for development of the Skylab Orbital Workshop and Apollo Telescope Mount, as well as integration of the Skylab cluster of components. It was responsible for three major elements of the Space Shuttle: the solid-fueled *rocket booster,* the Space Shuttle main *engine,* and the external tank. Today, Marshall manages Space Shuttle propulsion elements and science aboard the International Space Station, as well as other research. Marshall focuses on the development of transportation and propulsion systems, large complex systems, space infrastructure, applied materials and manufacturing processes, and scientific spacecraft research and instruments.

ETYMOLOGY. The name chosen by President Eisenhower honored George C. Marshall, General of the Army, who was Chief of Staff during World War II, Secretary of State 1948–49, and author of the Marshall Plan. General Marshall was the only professional soldier to receive the Nobel Peace Prize, awarded to him in 1954. He was close to Eisenhower both personally and professionally.

SOURCES. SP-4402, p. 154–56; David S. Akens, *Historical Origins of the George C. Marshall Space Flight Center,* MHM-1 (Huntsville, AL: MSFC, 1960),

p. 77; Frank E. Jarrett Jr. and Robert A. Lindemann, *Historical Origins of NASA's Launch Operations Center to July 1, 1962,* KHM-1 (Cocoa Beach: KSC, 1964), p. x; Rosholt, *Administrative History of NASA,* p. 120; President Dwight D. Eisenhower, Executive Order no. 10870, March 17, 1960. For more information about Marshall Space Flight Center, see Andrew J. Dunar and Stephen P. Waring, *Power to Explore: A History of Marshall Space Flight Center, 1960–1990,* SP-4313 (Washington, DC: GPO, 1999).

Mars Observer. NASA *mission* to Mars, 1992. The first and last of NASA's Planetary Observer series. It was launched in 1992 and was lost shortly before entering *orbit* around Mars in 1993.

Mars Odyssey. NASA orbiter mission to Mars, 2001. The *spacecraft* successfully achieved *orbit* in October 2001 around Mars following a six-month, 286-million-mile (460-million-km) journey. Following *aerobraking* operations, Odyssey entered its science-mapping orbit in February 2002 and began characterizing the composition of the Martian surface at unprecedented levels of detail, finding evidence of ice-rich soil very near the surface in the arctic regions. See also *Odyssey.*

Marsokhod (Russian for Mars walker). The Mars *rover* Marsokhod was designed in Russia (formerly the Soviet Union) in 1992 and used in feasibility tests by the Russian Space Agency and NASA. The name recalls the *Lunokhod,* the Russian lunar rover used in the early 1970s. Marsokhod has thus far been used only for testing. For example, NASA Ames Research Center undertook field tests of Marsokhod in the late 1990s. While these were robotic in nature, *crewed* versions have also been designed.

Mars Pathfinder. NASA environmental survey mission to Mars, 1996. Formerly known as the *Mars Environmental Survey* (MESUR) Pathfinder, Mars Pathfinder is one of the first two *Discovery* missions and was *launch*ed in December 1996. The *mission* consists of a *lander* and a surface *rover, Sojourner.* It landed on Mars on July 4, 1997. There had not been a landing on Mars in 21 years, since the *Viking spacecraft.* The main objective of this mission was to test new ideas in spacecraft engineering and to study the rocks. The Mars Pathfinder was the second of NASA's low-cost planetary Discovery missions to be launched.

Mars Polar Lander. A failed *spacecraft* that was to land on Mars December 3, 1999, and deploy two Deep Space 2 microprobes named *Amundsen* and *Scott.* All contact with the spacecraft was lost at the point of *separation* of the *lander* and microprobes. Subsequent investigations pointed to shortcomings in project management and preflight testing. The microprobes, which were to have deployed near the planet's south pole, were named after prominent explorers of the Earth's south pole. The names were chosen by a public contest won by Paul Withers, a graduate student at the University of Arizona in Tucson. (NASA News Release 99–135, "Mars Penetrator Probes Named for Pioneering Polar Explorers," November 15, 1999.)

Mars Reconnaissance Orbiter. NASA orbiter *mission* to Mars, 2005.

Mars Sample Return Missions. Originally proposed as 2003 and 2005 *mission*s to return Mars rock samples (drill cores) and soils to Earth in 2008. These missions have been postponed several times and are now scheduled for the next decade.

Mars Surveyor. A continuing line in NASA's budget dedicated to the *launch* of two small- to intermediate-sized Mars missions at every launch opportunity (every 26 months) between 1996 and 2005. The first Surveyor mission, *Mars Global Surveyor,* was launched in November 1996. The second was the *Mars Pathfinder,* launched a month later.

mascon. Any of the high-density regions below the surface of the Moon's seas (maria; see *mare*) that exert gravitational pull on the motion of *spacecraft* in lunar *orbit*. At first these concentrations were something of a mystery, but it was concluded after studying the lunar samples brought back to Earth by *Apollo* 11 astronauts that it was the maria themselves that were the mascon rather than some mysterious and unseen mass beneath the lunar surface. (Victor Cohn, "Moon May Be Like a Glassy, Shattered Ball," *Los Angeles Times,* September 1, 1969, p. A1.)

ETYMOLOGY / FIRST USE. Mass + concentration. According to the *OED,* the term appears to have been coined in an article in *Science* magazine by P. M. Muller and W. L. Sjogren (August 16, 1968, p. 680): "The Urey-Gilbert theory of lunar history has predicted such large-scale high-density mass concentrations below these maria, which, for convenience, we shall call mascons." The term reached a wider audience on June 7, 1969, when the Apollo 10 astronauts revealed that their *Lunar Module* had been drawn off course by subtle variations in lunar gravity and by, in the words of Thomas O'Toole of the *Washington Post,* "a single massed concentration (mascon) of matter below the moon's surface." O'Toole added, "So dense was this lunar mascon, [Thomas P.] Stafford said, that it caused the [Lunar Module] to dip suddenly toward the moon at a speed 14 miles-an-hour faster than it had been going." (O'Toole, *Washington Post,* June 8, 1969, p. 1.)

mass-driver. An electromagnetic accelerating device for propelling solid or liquid material, for example, from the Moon into space, or for providing propulsion by ejecting raw lunar soil or asteroidal material as reaction mass.

max q. Slang for the most dangerous part of *liftoff,* which comes at about the first minute when the *craft* meets the most resistance when the combination of increasing speed and decreasing density is maximum. During the *launch* of *space vehicle*s, the launch *crew* can be heard talking about "passing max q." Vehicles are designed to withstand only a certain maximum q before they will suffer structural damage. According to Canadian *Shuttle astronaut* Chris Hadfield, "During launch that's the moment when the vehicle is getting the maximum shaking, rattling and rock-and-rolling." ("He'll Have His Nose to the Glass: Canada's Top Astronaut Is Ready for an Amazing Experience as Endeavor Blasts off to Install Canadarm2," *Canada Post,* April 16, 2001, p. A13.) Max-Q is also the name of an all-astronaut rock band, of which Hadfield was a member. The Houston-based band with a changing lineup was formed in early 1987 by a group of astronauts that included Robert L. Gibson, who joked that, like the Shuttle, the band "makes lots of noise but no music."

ETYMOLOGY / FIRST USE. Max q is an aeronautical term, short for "maximum quotient": the point of maximum stress on an aircraft during flight that it can sustain without failing. It was part of the jargon of test pilots that transferred to the astronaut corps. The term has been in use in human spaceflight since the *Mercury* flights. Regarding John Glenn's *Friendship 7* flight, Marvin Miles wrote in the *Los Angeles Times,* "At about the time the Atlas wrote its contrail, Glenn rode through the greatest vibration of the flight, the brief period of max 'Q' or maximum dynamic pressure, a function of speed and air density" (*Los Angeles Times,* February 21, 1962, p. 1).

MCC. (1) *Mercury* Control Center at *Cape Canaveral,* 1960–65. (2) *Mission Control Center* at Houston, 1965–72.

MCC-H. *Mission Control Center,* Houston.

MCC-M. Mission Control Center, Moscow.

MCO. *Mars Climate Orbiter.*

the Meatball. Nickname for the NASA insignia or logo as it was first designed in 1959. It was proposed by James Modarelli, the head of the Lewis Research Center Reports Division. After an official seal was designed, this emblem was used as the more informal of the two. The logo includes a sphere representing a planet, stars symbolizing space, and a red chevron (in 1959, the latest in hypersonic wing design) signifying aeronautics; a *spacecraft* is shown *orbit*ing the wing. In 1992, Administrator Dan Goldin brought NASA's Meatball out of retirement to evoke memories of glory days of the *Apollo* Project and to show that "the magic is back at NASA." (Lewis Press Release, December 1, 1997, "The NASA Meatball," http://history.nasa.gov/meatball.htm.) See also *the Worm.*

Mechta (Russian for dream). Soviet name for the *probe* that was called *Lunik* 1 in the West. Launched on January 2, 1959, it orbited the Sun.

Mercury. Project Mercury, NASA's human *space program.*

On April 9, 1959, NASA announced selection of the seven men chosen to be the first U.S. space travelers, or *astronauts.* Robert R. Gilruth, head of the *Space Task Group,* proposed the name Project Astronaut to NASA Headquarters, but the suggestion lost out in favor of Project Mercury, largely because it was thought that the former name might lead to an overemphasis on the personality of the man inside the *spacecraft.*

In Project Mercury the United States acquired its first experience in conducting human space missions and its first scientific and engineering knowledge of man in space. After two suborbital and three orbital missions, Project Mercury ended with a fourth orbital space flight, a full-day mission by L. Gordon Cooper Jr., May 15–16, 1963.

In each of Project Mercury's human spaceflights, the assigned astronaut chose a *call sign* for his spacecraft just before his *mission.* The choice of *Freedom 7* by Alan B. Shepard Jr. established the tradition of the numeral 7, which came to be associated with the team of seven Mercury astronauts. When Shepard chose Freedom 7—"Freedom because it was patriotic," according to Shepard (Shepard and Slayton, *Moon Shot,* p. 95)—the numeral seemed significant to him because it appeared that "capsule No. 7 on booster No. 7 should be the first combination of a series of at least seven flights to put Americans into space." The lead astronaut for the second flight, Virgil I. Grissom, named his spacecraft *Liberty Bell 7* because "the name was to Americans almost synonymous with 'freedom' and symbolical numerically of the continuous teamwork it represented." John Glenn, assigned to take the nation's first orbital flight, named his Mercury *spacecraft Friendship 7.* Scott Carpenter chose *Aurora 7* because, he said, "I think of Project Mercury and the open manner in which we are conducting it for the benefit of all as a light in the sky. Aurora also means dawn in this case the dawn of a new age. The 7, of course, stands for the Original Seven astronauts." Walter M. Schirra selected *Sigma 7* for what was primarily an engineering flight, a mission to evaluate spacecraft systems; sigma is an engineering symbol for summation. In selecting sigma, Schirra also honored "the immensity of the engineering effort behind him." L. Gordon Cooper's choice of *Faith 7* symbolized, in his words, "my trust in God, my country, and my teammates."

ETYMOLOGY. Traditionally depicted wearing a winged cap and winged shoes, Mercury was the messenger of the gods in ancient Roman mythology (to the ancient Greeks he was known as Hermes). The symbolic associations of the name appealed to Abe Silverstein, NASA's Director of

Space Flight Development, who suggested it for the human spaceflight project in the autumn of 1958. Dr. T. Keith Glennan, NASA Administrator, and Dr. Hugh L. Dryden, Deputy Administrator, agreed upon the name Mercury on November 26, 1958, and Dr. Glennan announced the name to the public on December 18. (Bulfinch, *Mythology,* p. 18; Swenson, Grimwood, and Alexander, *This New Ocean,* SP-4201, pp. 131–32, 160; NASA Names Files, record no. 17531.)

FIRST USE. The public announcement of the code name for NASA's man-in-space program was reported widely on December 18, 1958, including a two-page edition of the strikebound *New York Times* published as an incidental part of the main story ("Big Rocket Engine for Space Flights Is Ordered by US," *New York Times,* December 18, 1958).

Mercury-Atlas (MA). The program that would use *Atlas booster*s to put *Mercury capsules* in *orbit*. Comparable to the *Gemini-Titan* designation.

Mercury Mark II. One of the names for *Gemini* before that name was bestowed on the two-person *spacecraft.*

Mercury-Redstone (MR). The program employing *Redstone* boosters to put *Mercury capsules* in *orbit.*

Mercury Surface, Space Environment, Geochemistry and Ranging. See *MESSENGER.*

Merritt Island Launch Area (MILA). See *Kennedy Space Center.*

MERs. Mars Exploration Rovers. The two *rovers*, Spirit and Opportunity, were also known as MER-A and MER-B (pronounced mur-ay and mur-bee).

MESSENGER/Messenger. Mercury Surface, Space Environment, Geochemistry and Ranging. NASA orbiter mission to Mercury, 2004. The first *spacecraft* to *orbit* the planet Mercury is designed to study the characteristics and environment of the planet. Specifically, the scientific objectives of the *mission* are to characterize the chemical composition of Mercury's surface, the geological history, the nature of the *magnetic field,* the size and state of the core, the volatile inventory at the poles, and the nature of Mercury's exosphere and *magnetosphere* over a *nominal* orbital mission of one Earth year. Messenger was *launch*ed on August 3, 2004, on a *Delta* 7925H (a Delta II heavy *launch vehicle* with nine strap-on solid *rocket booster*s). The spacecraft was injected into solar orbit 57 minutes later.

USAGE. Even prior to launch, the popular press dropped the upper-case spelling of the name announced in the NASA press releases: "This week a probe called Messenger will take off from Cape Canaveral" (*U.S. News & World Report,* August 9, 2004, p. 54).

MESUR. Mars Environmental Survey Pathfinder. Early name for *Mars Pathfinder.*

MET. *Mission* elapsed time; the time since *liftoff* (Kranz, *Failure Is Not an Option*, p. 395).

meteor. A piece of solid matter from space, burning up in Earth's *atmosphere* due to friction with the air. (The luminous streaks meteors trace across the sky are commonly called shooting stars, although they have nothing to do with stars.) Any part of the object that survives the fiery passage through the air and hits the ground is called a *meteorite*. See also *meteoroid*. (ASP Glossary.)

meteorite. A *meteor* that survives entry into the Earth's *atmosphere* (with a typical speed of about 25,000 mph, or 40,000 kph) and reaches the ground.

meteoroid. A piece of solid matter in outer space too small to be called an *asteroid*. A meteoroid that enters the Earth's atmosphere is called a *meteor*.

Michoud Assembly Facility (MAF). An 832-acre (337-hectare) site owned by NASA, located in eastern New Orleans, Louisiana. The primary product from the facility is the external tank for the *Space Shuttle*. The Michoud Assembly Facility is a government-owned, contractor-operated component of the George C. *Marshall Space Flight Center* (MSFC).

On September 7, 1961, NASA selected the then-unused government-owned Michoud Ordnance Plant at Michoud, Louisiana, as the site for industrial production of *Saturn launch vehicle* stages under the overall direction of the MSFC. NASA called the site Michoud Operations. On July 1, 1965, Michoud Operations was redesignated Michoud Assembly Facility to "better reflect the mission" of the facility. Following construction of the first stages of the *Saturn IB* and *Saturn V* launch vehicles for *Apollo, Skylab,* and *Apollo-Soyuz Test Project* missions, Michoud was selected in 1972 as the site for the manufacture and final assembly of the *Space Shuttle*'s external *propellant* tank. During Hurricane Katrina in August 2005 a ride-out team of 37 employees at Michoud risked their lives to stay at the site to keep the generators running, to keep the pumps going, and to protect the facilities and the flight hardware entrusted to them.

SOURCES. SP-4402, p. 155–56; NASA News Release 61–201; Robert C. Seamans Jr., Associate Administrator, NASA, letter to MSFC, in MSFC, Historical Office, *History of the George C. Marshall Space Flight Center: July 1–December 31, 1961*, MHM-4, vol. 1 (Huntsville, AL: MSFC, 1962), p. 38; MSFC News Release 65–167.

microgravity. Extremely low level of gravity. As experienced by Shuttle crews, for example, one millionth the level of gravity on Earth's surface. USAGE. Preferred term to *weightlessness* or *zero-gravity*, which are both considered misnomers in the sense that there is always some gravity in force.

milk stool. Physical arrangement of the three storable *propellant rocket* engines located below the main pressure vessel of the *Lunar Excursion Module* (SP-6001, p. 61).

Milky Way. Band of hazy light that can be seen from clear, dark locations and that stretches all the way around the sky. When looked at using binoculars or a small telescope, it is seen to be composed of vast numbers of faint individual stars. It is actually the disk of our own *galaxy* seen from our perspective within the disk. The flat lens shape of the galaxy appears to surround us. Astronomers often use the term Milky Way to refer to our entire galaxy, rather than to just its appearance in our sky. (ASP Glossary.)

MilSatCom. Military Satellite Communications.

Minitrack. Network of U.S. stations placed at different points around the world in order to track the flight of artificial *satellites*. Initially created for the *Vanguard* program during the *International Geophysical Year* (IGY), Minitrack got its first workout *tracking Sputnik*. (Noted in NASA, *Glossary/Congressional Budget Submission,* p. 26, that it had become the basic network for tracking small scientific *Earth satellites.*)

Mir (pronounced meer). A highly successful Soviet (and later Russian) *space station.* It was humanity's first permanently inhabited long-term research station in space. Through a number of collaborations, it was made internationally accessible to *cosmonauts* and *astronauts* of many different countries. Mir was assembled in *orbit* by successively connecting seven separate modules—three instrument and life support modules (Mir, Kvant, and Kvant 2), three science modules (Kristall, Spektr, and Priroda), and a *docking* module that was added to allow the U.S. *Space Shuttle* to dock with the station—each *launched* separately from February 19, 1986, to 1996. In 2001 Russia decommissioned the aging space station and allowed it to burn up in the Earth's *atmosphere.*

ETYMOLOGY. Russian for peace. In Russian the name also denotes a peasant commune in which serfs lived apart from their owners. "The mir was an institution of both peasants and serfs in which they enjoyed a continuity of culture and less interference from owners" (*New York Review of Books,* November 17, 1987, p. 43).

missile. An object or a weapon that is fired, thrown, dropped, or otherwise projected at a target; a projectile.

missile gap. The belief that the United States was behind the Soviet Union in the deployment of both long- and medium-range missiles. A missile-gap panic occurred during 1957–61 following the Soviet *launch* of *Sputnik* in 1957. This perceived technological imbalance between the United States and the Soviet Union, coupled with the Gaither Report (1957, which discredited U.S. military preparedness and urged a 50 percent increase in military spending), led to a high infusion of money into U.S. weapons research and development. A "bomber-gap" panic had occurred between 1954 and 1957, based on an erroneous U.S. intelligence report that the Soviets had more long-range bombers than the United States.

mission. (1) The objectives of a *space program.* (2) A single large operation or task.

mission control. Those responsible for directing a *spacecraft* and, in the case of human missions, directing its crew.

Mission Control Center. The complex in Houston, Texas, created initially to direct the *Gemini* and *Apollo* flights.

> **FIRST USE.** The term first appears in the popular press in an advertisement placed in the *New York Times*: "Project Officer to have technical review and coordination responsibility in the Washington Program Office for the design, development, and operation of the new Mission Control Center in Houston, Texas" ("Electronic Engineers for Apollo & Gemini," *New York Times,* December 9, 1962, p. 107).

mission critical. Describing any operation that cannot tolerate intervention, compromise, or shutdown during the performance of its critical function.

> **FIRST USE.** The term was first invoked in public during the *Skylab* program in response to questions about malfunctioning equipment including a solar array: "In days of shrinking budgets, NASA decided that a number of tests and studies not mission critical could be dispensed with" ("Skylab: Jigsaw in the Sky," *Christian Science Monitor,* May 15, 1973, p. 10).

mission profile. Plot of the flight plan of a vehicle indicating altitude, distance, and speed (NASA, *Glossary/Congressional Budget Submission,* p. 26).

Mission to Planet Earth (MTPE). (1) Name for space-based monitoring of the Earth to chart environmental change using a sequence of ever more sophisticated *EOS* satellites. The first MTPE *mission* was *launch*ed on *STS*-48, flown by *Space Shuttle Discovery,* which deployed the Upper Atmosphere Research Satellite (UARS). In 1985 the name Mission to Planet Earth replaced *Project Habitat,* the program's name since its inception in 1982. Budget cuts severely restricted the program, and in November 1997 NASA's Office of Mission to Planet Earth was renamed the Earth Science Enterprise, whose centerpiece was still the Earth Observing System (EOS). (2) Term used in 1987 by Dr. Sally Ride in her report *Leadership and America's Future in Space* (commonly known as the Ride Report) as one of four major initiatives that could take NASA into the 21st century. See *Ride Report.*

Mobile Equipment Transporter. Sometimes referred to as Modular Equipment Transporter. The *Apollo* 14 transporter was a two-wheeled rickshaw vehicle used to carry tools, containers, spare film, etc.

mock-up. A full-size replica or dummy of a vehicle, e.g. a *spacecraft,* often made of some substitute material such as wood to assess design features.

module. A self-contained unit of a *launch vehicle* or *spacecraft* that serves as a building block for its overall structure and usually designated by its primary function: command, lunar landing, or service.

ETYMOLOGY. According to Benson and Faherty (*Moonport,* SP-4204), "Apollo scientists and engineers were establishing a terminology for new things; no one had defined them in the past because such things did not exist." Module is an example. As late as 1967 the *Random House Dictionary of the English Language* gave as the fifth definition of module under computer technology: "A readily interchangeable unit containing electronic components, especially one that may be readily plugged in or detached from a computer system." The space world was well ahead of the dictionary because, as every American television viewer knew, a module— command, service, or lunar—was a unit of the *spacecraft* that went to the Moon.

Moffett Field Laboratory. Forerunner of *Ames Research Center.*

MOL. *Manned Orbiting Laboratory.*

Molly Brown. Unofficial name for the *Gemini* 3 *spacecraft* launched on March 23, 1965, and flown by former *Mercury astronaut* Gus Grissom in command of the *mission,* with John W. Young, a Naval aviator chosen as an astronaut in 1962, accompanying him.

ETYMOLOGY. "During Project Mercury, each pilot had named his own spacecraft, although Cooper had some trouble selling NASA on Faith 7 for the last spacecraft in the program. Grissom and Young now had the same difficulty with 'Molly Brown.' Grissom had lost his first ship, Liberty Bell 7, which sank after a faulty circuit blew the hatch before help arrived. 'Molly Brown,' the 'unsinkable' heroine of a Broadway stage hit, seemed to Grissom the logical choice for his second space command. NASA's upper echelons thought the name lacking in dignity; but since Grissom's second choice was 'Titanic,' they grudgingly consented, and the name remained 'Molly Brown,' though only quasi-officially. Later spacecraft were officially referred to by a Roman numeral, although a few had nicknames as well." (Hacker and Grimwood, *On the Shoulders of Titans,* SP-4203, p. 233.)

Molniya orbit. Highly elliptical *orbit* inclined at an angle of about 65 degrees to the plane of the equator, such as employed by the Russian

Molniya communications *satellites*. The orbit allows the *spacecraft* to pass close to the Earth near the south pole, traveling quickly, and at a great distance from the Earth near the north pole, traveling slowly. This results in the satellite spending most of each 12-hour orbit over the northern hemisphere, making it more useful for northern hemisphere applications.

moon. A small natural body that orbits a larger one. A natural *satellite*.

USAGE. According to a glossary published by the Astronomical Society of the Pacific, "Some astronomers frown on this use of the word 'moon' which they feel should be reserved for Earth's natural satellite exclusively. Probes we launch into orbit around the Earth are called 'artificial satellites'" (ASP Glossary). However, the popular use of "moon" as a synonym for satellite has a long and distinguished history in science fiction as well as the popular press. The May 1949 cover of *Popular Science* magazine depicts a satellite labeled "US Orbiter" and poses the question, "Is the US Building a 'New Moon?'"

Moon buggy. Vehicle for astronauts to use to explore the surface of the Moon. Even after that vehicle had been given the formal name *Lunar Roving Vehicle (LRV),* the term was still used widely by the public and even the astronauts. Also called the lunar *rover*.

ETYMOLOGY. A buggy was a light one-horse (sometimes two-horse) vehicle, for one or two persons. The term was appropriated as affectionate American slang for one's automobile and later for the dune buggy, a go-anywhere off-road vehicle.

FIRST USE. "But in time astronauts will undoubtedly make many return trips, landing at different sites, staying longer, roaming the surface in special 'moon buggies'" ("At the Next Blast-off the Order May Be: 'Destination Moon,'" *New York Times,* October 27, 1968, p. E12).

Moondoggle (Moon + boondoggle). A derisive reference to the *Apollo* program. It was employed by those who opposed the program on strictly fiscal arguments or who saw it as taking away from the War on Poverty and other public assistance programs. Amitai Etzioni used it as the title for his book *The Moondoggle: Domestic and International Implications of the Space Race* (Garden City, NY: Doubleday, 1964).

Moonport / Moonport, USA. Nickname for *Cape Canaveral* throughout the *Apollo* era. It provided the title for Benson and Faherty's history, *Moonport: A History of Apollo Launch Facilities and Operations,* SP-4204 (1978).

FIRST USE. Washington (UPI): "The space agency said Thursday that Cape Canaveral will be enlarged about five times to serve as Moonport, USA" ("Cape Canaveral Picked as Moon Flight Base," *Los Angeles Times,* August 25, 1961, p. 2).

moonquake. A seismic disturbance on the Moon.

Moonshot / Moon shot. (1) The *launching* of a *spacecraft* to the Moon. 2) A Moon-bound *spacecraft.*

> **ETYMOLOGY.** Although used in science fiction for attempts to reach the Moon, the term had broad common use in pari-mutual racetrack betting for an extreme longshot. For example: "Today's program wasn't conducted at Fort Knox, but there was plenty of gold around—that is, for a select few who happened on some moonshot winners" ("Rex Romanus Wins Graw Race at $195," *Washington Post,* April 19, 1947, p. 9).

moon suit. Inflatable coveralls worn for high-altitude tests and space tests.

Moon walker. Human who walks on the lunar surface.

> **FIRST USE.** "Search for the Spacemen," *New York Times,* June 25, 1961, p. SM28, where the term appears in a caption to a photo of a man in a self-contained space suit.

MOOSE. Man out of Space Easiest. *Apollo* emergency space escape system. (SP-6001, p. 63.)

Mother *(satellite).* International Sun-Earth Explorer ISEE-A. See *Explorer.*

motorman's pal. A tube attached to the lining of a *space suit* used for collecting urine samples during flight. The *Time* magazine account of John Glenn's *Mercury mission* contained this line: "He found he could urinate easily into the 'motorman's pal' which was attached to the lining of his space suit" (*Time,* March 2, 1962, p. 4).

> **ETYMOLOGY.** Originally a device strapped to the leg of a male streetcar worker on a long run.

MOUSE. Minimum Orbital Unmanned Satellite of the Earth. A small, 100-pound (45-kg) *satellite* proposed in a study published in 1953 by Fred Singer of the University of Maryland.

MPL. Mars Polar Lander.

MPLMs. Multi-Purpose Logistics Modules. These Italian-built high-tech containers, named Donatello, Leonardo, and Raffaelo, are carried in the cargo bay of the *Space Shuttle* and ferried to the *International Space Station.* They contain experiments and supplies. (*Reference Guide to the International Space Station,* SP-2006-557.)

MR-. *Mercury-Redstone* (followed by *mission* number).

MSC. Manned Spacecraft Center. See *Johnson Space Center.*

MSFC. *Marshall Space Flight Center.*

MSFN. *Manned Space Flight Network.* Pronounced Miss Finn, according to *ALSJ* Glossary.

MSOB. *Manned Spacecraft Operations Building* at *Kennedy Space Center.*

MTF. Mississippi Test Facility. See *Stennis Space Center.*

MTS. Meteoroid Technology Satellite. See *Explorer.*

Multi-Purpose Logistics Modules. See MPLMs.

multistage rocket. A *rocket* having two or more stages that operate in succession, each being discarded as its job is done.

MUSES-A. See *Hiten.*

MUSES-C. See *Hayabusa.*

Muttnik. One of the many punning nicknames for the Soviet *Sputnik* 2, which carried the dog *Laika.*

Mylar. Name for a clear thin polyester film 0.0005 inch (0.5 mm) thick manufactured by DuPont with a long and strong association with the *space program* dating back to 1960 and the first *Echo satellite,* an aluminum metallized Mylar sphere that was placed into Earth *orbit* to act as a reflector for a series of passive communications experiments. (NASA News Release 60–237, "Project Echo Payload and Experiment.") Extensive use of the material for insulation, especially during the *Apollo* program, led to the commercial development of the *space blanket,* a 56- by 84-inch (142- by 213-cm) sheet that folded into a package that would fit into a person's hand. Because it reflects up to 80 percent of the user's body heat, it can sustain life for prolonged periods of time even in subzero temperatures. The same rectangle was also sold and promoted as a suntan blanket in ads that promised, "This Moon Mission Blanket Will 'Toast' You to a Delicious Tan."

ETYMOLOGY. Trademarked by E. I. DuPont de Nemours and Co., Wilmington, Delaware, in 1954. Like many DuPont names, it has no meaning other than a mellifluous sound.

N

NACA. The National Advisory Committee for Aeronautics. The 43-year-old-group that was absorbed along with its 8,000 employees by the new NASA on September 30, 1958.

nanosatellites. Very small, lightweight (under 44 lb/20 kg) satellites containing mems (microelectronic mechanical systems) equipment, components, and payloads.

NASA. *National Aeronautics and Space Administration.* Agency in charge of all civilian *space programs* for the United States.

USAGE. Few terms better illustrate the economy of effort in the *space program* as words were shortened, blended, and turned into acronyms. The same urge that turned liquid oxygen into *lox* ensured that hardly anyone ever said the full name of NASA—although in the early days it was often called "the NASA" just as people still say "the Pentagon" or "the FBI." By contrast, *NACA* was always called "the N-A-C-A" when not referred to by its full name.

NASA Earth Observatory. An online publishing organization of the *National Aeronautics and Space Administration*. It is the principal source of free *satellite* imagery and other scientific information about our planet for consumption by the general public. It is focused on the climate and the environment.

NASAese. Also known as NASA-speak. Slang term for the jargon, penchant for initialisms, and overstated language that often characterizes the language of NASA: "The only significant problem the astronauts encountered was a balky 'text and graphics system'— NASAese for a text machine" ("Magellan Probe Zips toward Venus," *Los Angeles Times,* May 9, 1989, p. A2). See also *space-speak*.

NASA Meatball logo. See *the Meatball*.

NASA-speak. See *NASAese*.

NASC. National Aeronautics and Space Council. Established in 1958 to serve as the President's advisory group on space. President Nixon abolished it in 1973.

National Advisory Committee for Aeronautics. See *NACA*.

National Aeronautics and Space Administration (NASA). The official name for the U.S. civilian *space agency*. Fifty years after its founding, NASA arguably leads the world in space exploration, standing on the shoulders of a long line of explorers. Its astronauts have circled the world, walked on the Moon, piloted the first winged spacecraft, and constructed the *International Space Station*. Its robotic spacecraft have studied the Earth, visited all the planets (and will soon visit the dwarf planet Pluto), imaged the universe at many wavelengths, and peered back to the beginnings of time. Its *scramjet* aircraft have reached the aeronautical frontier, traveling 7,000 miles per hour (11,300 kph), ten times the speed of sound, setting the world's record.

NASA's birth was directly related to the launch of *Sputnik* and the ensuing race to demonstrate technological superiority in space. Driven by the competition of the Cold War, on July 29, 1958, President Dwight D. Eisenhower signed the National Aeronautics and Space Act, providing for research into the problems of flight within the Earth's atmosphere and in space. After a protracted debate over military versus civilian control of space, the act inaugurated a new civilian agency designated the National Aeronautics and Space Administration. The agency began operations on October 1, 1958. At the moment of its creation it consisted of only about 8,000 employees and an annual budget of $100 million. In addition to a small headquarters staff in

Washington that directed operations, NASA had at the time three major research laboratories inherited from the National Advisory Committee for Aeronautics (the Langley Aeronautical Laboratory established in 1918, the Ames Aeronautical Laboratory activated near San Francisco in 1940, and the Lewis Flight Propulsion Laboratory built at Cleveland, Ohio, in 1941) and two small test facilities (one for high-speed flight research at Muroc Dry Lake in the high desert of California and one for sounding rockets at Wallops Island, Virginia). It soon added several other government research organizations.

In addition to its headquarters in Washington, DC, NASA's facilities include ten centers around the country: *Ames Research Center, Dryden Flight Research Center, Glenn Research Center, Goddard Space Flight Center, the Jet Propulsion Laboratory, Johnson Space Center, Kennedy Space Center, Langley Research Center, Marshall Space Flight Center,* and *Stennis Space Center.* NASA is staffed by nearly 19,000 civil servants and many more contractors. Its budget for fiscal year 2007 was nearly $17 billion.

For more information about NASA, see Glennan, *Birth of NASA;* Roger Launius, *A History of the U.S. Civil Space Program* (Malabar, FL: Krieger, 1994); *Legislative Origins of the National Aeronautics and Space Act of 1958,* Monographs in Aerospace History no. 8 (Washington, DC: NASA, 1998); Logsdon, ed., *Exploring the Unknown,* vol. 1; Howard McCurdy, *Inside NASA: High Technology and Organizational Change in the U.S. Space Program* (Baltimore: Johns Hopkins University Press, 1993); Walter McDougall, *The Heavens and the Earth.*

ETYMOLOGY. The name changed from Agency to Administration as the House of Representatives reacted to the Eisenhower proposal. "Administration" was first used in the context of the House of Representatives, where a specially convened Democratic-led House committee revised and strengthened the bill. The name did not become official until the signing into law (July 29, 1958) of the National Aeronautics and Space Act of 1958.

FIRST USE. The first newspaper to name the new agency appears to have been the *Washington Post:* "The House yesterday voted to set up a new National Aeronautics and Space Administration, under a single civilian administrator to direct United States efforts to 'leapfrog' Russian scientific achievements in outer space" ("House Votes to Set up Civilian Space Unit," *Washington Post,* June 3, 1958, p. A1). The term appeared in the *Los Angeles Times* the next day; it did not appear in the *New York Times* until June 17.

National Aeronautics and Space Agency. The name for the new entity—first proposed by President Dwight D. Eisenhower in a letter sent to Congress on April 2, 1958—that would incorporate and then replace the National Advisory Committee for Aeronautics, *NACA*

("Text of Message on Space Agency," *Los Angeles Times,* April 3, 1958, p. 22). The word Administration replaced Agency in the final congressional bill—the National Aeronautics and Space Act, passed July 29, 1958.

National Aeronautics and Space Council. See *NASC.*

National Commission on Space. See *Paine Report.*

National Defense Education Act (NDEA). Created "to meet critical national needs," this was the primary educational reaction to the *Sputnik* "surprise." NDEA appropriated $47.5 million in student loans for 1958, with expenditures on loans budgeted to exceed $100 million by 1962. Also, over four years nearly $300 million dollars went to fund the purchase of scientific equipment and the establishment of National Defense Fellowships for graduate students.

On September 2, 1958, President Dwight D. Eisenhower signed into law the National Defense Education Act, "an emergency undertaking to be terminated after four years to bring American education to levels consistent with the needs of our society." When President Eisenhower signed the Act into law, it was noted that this was the first time since 1917 that serious attention had been paid to school reform. NDEA poured billions of dollars into the educational system over the next decade to pay for language labs, the "new math," and the broad curriculum overhaul of the late 1950s and early 1960s. The funds were not limited to colleges and universities but were also poured into high schools. And the aid for science and math spurred increased support for liberal arts. English majors could get loans provided by NDEA, part of which could be "forgiven" if they went into teaching. The federal government had committed itself to shoring up the public schools and giving a major boost to colleges and universities.

Between 1958 and 1968, NDEA also provided loan money for more than 1.5 million individual college students—fellowships directly responsible for producing 15,000 Ph.D.'s a year. NDEA allocated approximately $1 billion over four years to supporting research and education in the sciences. Federal support for science-related research and education increased between 21 and 33 percent per year through 1964, representing a tripling of science research and education expenditures over five years. States were given money to strengthen schools on a 50–50 matching basis, thousands of teachers were sent to NDEA-sponsored summer schools, and the National Science Foundation sponsored no fewer than 53 curriculum development projects. By the time of the lunar landing in 1969, NDEA alone had pumped $3 billion into American education.

National Space Transportation System (NSTS). Name for the *Space Shuttle* program from its inception until March 1990, when it was officially renamed the Space Shuttle by Program Director Robert Crippen. One of the main reasons stated by Crippen was that the name Space Shuttle was better known to those outside NASA. ("'Space Shuttle' Now Officially Program's Title," *Space News Roundup*, March 9, 1990, p. 10.)

NEAR. *Near Earth Asteroid Rendezvous.*

Near Earth Asteroid Rendezvous (NEAR). The first of NASA's *Discovery missions* and the first mission ever to go into *orbit* around an asteroid (Eros) and to touch down on its surface. The *spacecraft, launch*ed on February 16, 1996, was equipped with an x-ray / gamma ray spectrometer, a near-infrared imaging spectrograph, a multispectral camera fitted with a CCD (charge-coupled device) imaging detector, a laser altimeter, and a magnetometer.

Near Earth Asteroid Rendezvous–Shoemaker (NEAR Shoemaker). New name for the *NEAR spacecraft*, renamed in honor of Dr. Eugene M. Shoemaker, the renowned geologist who influenced decades of research on the role of asteroids and *comets* in shaping the planets. The renaming took place in March 2000 when the *craft* was in its third year in space conducting the first close-up study of the asteroid Eros. The tribute was announced on March 14, 2000, at the Lunar and Planetary Science Conference in Houston by Dr. Carl B. Pilcher, NASA Science Director for Solar System Exploration. The spacecraft landed gently on the asteroid Eros on February 12, 2001, and continued to send signals for a week.

near-Earth asteroids (NEAs). See *near-Earth objects.*

near-Earth objects (NEOs). Asteroids and *comets* in solar *orbit* with a closest approach to the Sun of less than 1.3 times that of the Earth (i.e., less than 120 million miles, or 193 million km).While many of these objects pose no threat to the Earth, a subset known as Earth-crossing asteroids (ECAs) and potentially hazardous asteroids (PHAs) have orbits with the potential for a close encounter or collision with the Earth. The Earth is bombarded by small meteorites every day, but most of these objects are less than 160 feet (50 m) in size and burn up in the *atmosphere*. NEO surveys focus primarily on near-Earth asteroids (NEAs). The terms NEO and NEA are thus often used interchangeably. (U.S. Congress, House of Representatives, Committee on Science, Subcommittee on Space and Aeronautics, *The Threat of Near-Earth Asteroids: Hearing . . . October 3, 2002* [Washington, DC: GPO, 2003], p. 1.)

NEAs. Near-Earth asteroids. See *near-Earth objects.*

NEOs. *Near-Earth objects.*

NERV. Nuclear Emulsion Recovery Vehicle. A *nosecone* sent into space in September 1960 to determine the effects of *radiation* on astronauts going into space. The name of the second NERV was changed to BIOS I.

NERVA. Nuclear Engine for Rocket Vehicle Application. A U.S. rocket program to develop a nuclear propulsion system for journeys to the Moon and the planets. Begun in the early 1960s, the project's funding was cut by the Nixon administration in the 1970s, and it was terminated in 1973.

neutron star. A compact star consisting predominantly of neutrons. Neutron stars have masses in the range of about one to three solar masses and sizes around 12 miles (19 km). Their density is comparable to that of atomic nuclei (i.e., about 100 to 1,000 trillion times the density of water). (http://observe.arc.nasa.gov/nasa/space/ stellardeath/stellardeath_6.html.)

New Horizons. NASA *spacecraft* designed to fly by Pluto and its moons and transmit images and data back to Earth. It was launched January 19, 2006, aboard an *Atlas* V–*Centaur* vehicle. After encountering the Pluto system it will continue on into the Kuiper belt, where it may fly by a Kuiper belt object and return further data.

New Millennium Program (NMP). Program jointly established in 1995 by NASA's Office of Space Science (OSS) and Office of Earth Science. It was intended to speed up space exploration through the development and testing of leading-edge technologies. Managed by the *Jet Propulsion Laboratory* / California Institute of Technology, NMP provides a critical bridge from initial concept to exploration-*mission* use. Through NMP, selected technologies are demonstrated in the "laboratory" of space under conditions that can't be replicated on Earth. (http://nmp .nasa.gov/PROGRAM/program-index.html.)

NEXRAD. Next Generation Radar. A network of Doppler radars operated by the National Oceanic and Atmospheric Administration (NOAA) for weather surveillance purposes.

Next Generation Space Telescope (NGST). Now the *James Webb Space Telescope* (JWST).

the Next Nine. NASA's second *astronaut* group, selected in September 1962 for *Gemini* and *Apollo* flights. They are Neil Armstrong, Frank Borman, Charles Conrad, Jim McDivitt, Jim Lovell, Elliott See, Tom Stafford, Ed White, and John Young.

NGST. *Next Generation Space Telescope.*

Nike (*sounding rocket* first stage). The Nike, a solid-*propellant* first stage, was an adaptation of the Nike antiaircraft *missile* developed, beginning in 1945, by the Hercules Powder Company for U.S. Army Ordnance. In NASA's sounding rocket program, Nike was used with *Apache, Cajun, Tomahawk, Hawk,* or *Malemute upper stages,* as well as with the *Aerobee* 170, 200, and 350. (Emme, *Aeronautics and Astronautics,* p. 49; Peter T. Eaton, Office of Space Science and Applications, NASA, letter to

Historical Staff, NASA, May 2, 1967; Edward E. Mayo, Flight Performance Branch, Sounding Rocket Division, GSFC, information sent to Historical Office, NASA, January 30, 1975.)

ETYMOLOGY. In ancient Greek mythology, Nike was the winged goddess of victory.

Nimbus (meteorological *satellite*). Second-generation research satellite following the first meteorological satellite series, Tiros. Nimbus 1 was launched August 28, 1964, and provided photographs of much higher resolution than those provided by Tiros satellites until it ceased transmission September 23, 1964. Nimbus 2 (1966) and 3 (1969) operated for a few years, followed by Nimbus 4 (1970) and 5 (1972) to continue providing meteorological data and testing a variety of weather-sensing and measuring devices. See *ESSA* and *TIROS, TOS, ITOS.*

ETYMOLOGY. Latin for rainstorm or cloud; the meteorological term meaning precipitating clouds. The name was suggested in late 1959 by Edgar M. Cortright, Chief of NASA's Advanced Technology Programs, who directed the formation of NASA's meteorological satellite programs, including Nimbus and TIROS.

SOURCES. SP-4402, p. 60; Robert F. Garbarini, Director of Applications, Office of Space Science and Applications, NASA, letter to Historical Staff, NASA, December 30, 1963; William K. Widger Jr., *Meteorological Satellites* (New York: Holt, Rinehart and Winston, 1966), p. 153.

nodes. U.S. modules that connect the elements of the *International Space Station*. Node 1, called Unity, was the first U.S.-built element launched. Node 2 will connect the U.S., European, and Japanese laboratories. Node 3 will provide additional habitation functions, including hygiene and sleeping compartments.

no-go. The decision to cancel. One result of the *go/no-go* decision to continue or *abort* a *mission* (Kranz, *Failure Is Not an Option,* p. 395).

nominal. Functioning acceptably, normally. *NASAese* for "no problem." It is often used as an antonym to "anomalous."

FIRST USE. This is an interesting application of a term that has vastly different meanings in other contexts. The *OED* lists the aerospace meaning as the sixth distinct sense of the word, with a 1966 example as its earliest citation: "The mission is to launch the 800-lb. prime vehicle to effect a nominal re-entry at 400,000 ft. following injection at 26,000 fps" (*Aviation Week & Space Technology,* December 5, 1966, pp. 30–31).

nose cone / nosecone. Leading edge of a *rocket* vehicle, consisting of a chamber or chambers in which a *satellite,* instruments, animals, plants, or auxiliary equipment may be carried. It protects the instrumentation package, satellite, or warhead during *launch* and *reentry.* (SP-6001, p. 66.)

Nozomi (Japanese for hope). Japanese orbiter mission to Mars, launched 1998. Planned as a Mars orbiting aeronomy *mission* to study the Martian upper *atmosphere* and its interaction with the *solar wind,* and to develop technologies for use in future planetary missions. Efforts to put the Nozomi *spacecraft* into Martian *orbit* were abandoned when it was unable to achieve Martian orbit and went into orbit around the Sun. (http://nssdc.gsfc.nasa.gov/database/MasterCatalog?sc= 1998-041A.)

NRL. Naval Research Laboratory.

NSTS. *National Space Transportation System.* Early name for the *Space Shuttle.*

nudging. Deflecting asteroids on collision course with Earth.

OAO (Orbiting Astronomical Observatory). Name for a series of four astronomy *satellites.* The first satellite of the program, OAO 1, was *launch*ed into almost perfect *orbit* on April 8, 1966, but its power supply failed. OAO 2, launched December 7, 1968, took the first ultraviolet photographs of stars, returning data previously unobtainable. OAO 3, launched August 21, 1972, contained the largest telescope put into orbit by the Unites States to that date. After launch it was given the additional name Copernicus in honor of the Polish astronomer, as part of the international celebration of the 500th anniversary of his birth.

ETYMOLOGY. The term Orbiting Astronomical Observatory was first mentioned in writing by Dr. James E. Kupperian Jr. in a December 1958 draft project outline, and NASA project officials approved this name as a working designation. The question of a new name arose in March 1959 when NASA was preparing the first official project document. The long name had been shortened in common usage to OAO. The NASA officials Kupperian, Dr. G. F. Schilling, and Dr. Nancy Roman decided to keep the long title, with OAO as a short title. The intent at the time was to keep a meaningful name, one that was short, descriptive, and professional.

SOURCES. SP-4402, pp. 60–62; James E. Kupperian Jr., Office of Space Flight Development, NASA, draft project outline, December 1958, with approval indicated by Gerhardt F. Schilling, Chief, Astronomy and Astronomy Programs, Office of Space Flight Development, NASA; NASA, "Proposed National Aeronautics and Space Administration Project" (first official OAO project document), March 12, 1959; Kupperian, letter to Historical Staff, NASA, November 18, 1963; NASA News Releases 72-141, 72-156.

Odyssey. (1) *Call sign* for the *Apollo* 13 Command Module, crewed by James A. Lovell Jr., John L. Swigert Jr., and Fred W. Haise Jr. This was the third lunar landing attempt, aborted after rupture of the Service Module oxygen tank, which damaged several components of the CM including life support systems. The *Aquarius LM*—a self-contained *spacecraft* unaffected by the accident—was used as a "lifeboat" to provide austere life support for the return trip. (2) A NASA orbital mission to Mars. See *Mars Odyssey*.

ETYMOLOGY. From the title of the epic Greek poem attributed to Homer about the wanderings of Odysseus. The use of the name (and of Aquarius) for the Apollo mission (definition 1) was announced by James E. Lovell in a March 14, 1970, news conference. He added, "We've already had quite an odyssey just getting trained for this flight as a matter of fact" ("Lovell Dubs Moon Landing Ship Aquarius," *Chicago Tribune*, March 15, 1970, p. 3).

OFO. Orbiting Frog Otolith *(satellite)*. An experiment designed to study the adaptability of a frog's otolith (inner-ear balance mechanism) to sustained *weightlessness*, to provide information for human space-flight. The Frog Otolith Experiment (FOE), as it was originally called, was developed by Dr. Torquato Gualtierotti of the University of Milan, Italy, while assigned to the *Ames Research Center* as a resident Research Associate under the sponsorship of the National Academy of Sciences. Originally planned in 1966 to be included on an early *Apollo mission*, the experiment was deferred when that mission was canceled. In late 1967, authorization was given to *orbit* the FOE when a supporting *spacecraft* could be designed. The project, part of NASA's Human Factor Systems program, was officially designated OFO in 1968. After a series of delays, OFO was launched on November 9, 1970.

ETYMOLOGY. The name, derived through common use, was a functional description of the biological experiment carried by the *satellite*.

SOURCES. SP-4402, p. 62; Robert W. Dunning, Office of Manned Space Flight and former OFO Experiment Program Manager, Office of Advanced Research and Technology.

OGO. Orbiting Geophysical Observatory. The name derived from NASA's concept for an observatory-class *satellite*. The concept evolved in late 1959 and early 1960 from that of a larger general-purpose scientific satellite (as opposed to the special-purpose *Explorers*), which would be a standardized *spacecraft* housing a variety of instruments to be flown regularly on standardized trajectories. Orbiting Observatory became the term used for this class of spacecraft, and Orbiting Geophysical

Observatory developed as a functional description for this particular satellite. The names EGO and POGO were also developed during this period to apply to OGO satellites in particular orbital trajectories: highly *eccentric* (Eccentric Geophysical Observatory) and *polar orbit* (Polar Orbiting Geophysical Observatory). Between 1964 and 1969, NASA orbited six OGO satellites, and results from the successful OGO program included the first global survey by satellite of the Earth's *magnetic field.* (SP-4402, p. 62; Jack Posner, Office of Space Science and Applications, NASA, telephone interview, August 10, 1965; U.S. Congress, Senate, Committee on Aeronautical and Space Sciences, *Hearings: NASA Scientific and Technical Programs, February and March 1961* [Washington, DC: GPO, 1961], pp. 236–39.)

Oort cloud. The cloud of small objects that surrounds our solar system beyond Pluto and the Kuiper belt objects. The Oort cloud is believed to contain long-period *comets.* The cloud is named after Dutch astronomer Jan Hendrik Oort, who revived the idea in 1950 after it had been originally proposed in the 1930s by Ernst Opik.

open-loop system. Control system that does not provide feedback to the controller.

Operation Paperclip. Code name for the operation mounted by the U.S. intelligence services and military to extract scientists specializing in rocketry (e.g. *V-1, V-2*) from Germany after the collapse of the Nazi government during World War II. These scientists and their families were secretly brought to the United States, without State Department review and approval.

orbit. (1, n.) The closed path of an object that is moving around a second object. (2, v.) To travel around another object in a single path.

orbital decay. The gradual reduction in size of a *satellite*'s elliptical *orbit,* due to air resistance or drag.

Orbital Maneuvering Vehicle. A device used much like a harbor tug in ship operations, with remotely controlled manipulator arms to handle *spacecraft* and refueling operations with great care (Paine Report, p. 198).

orbital period. Time taken by an orbiting body to make a complete *orbit.*

orbital velocity. The velocity necessary to overcome the gravitational attraction of the Earth and so keep a *satellite* in *orbit*—about 17,450 miles per hour (28,080 kph)—close to the Earth.

Orbital Workshop. See *Skylab.*

orbiter. (1) A *spacecraft* designed to go into *orbit,* especially one that does not subsequently land, as in *Lunar Orbiter.* (2) Shuttle *(STS) spacecraft* that carry the designation *OV,* for orbiting vehicle. *Challenger* was OV-099. The other orbiters were *Discovery* (OV-103*), Endeavour* (OV-105*), Columbia* (OV-102), and *Atlantis (*OV-104).

Orbiting Astronomical Observatory. See *OAO*.

Orbiting Frog Otolith satellite. See *OFO*.

Orbiting Geophysical Observatory. *See OGO*.

Orbiting Solar Observatory. See *OSO*.

the Original Seven. The *astronaut*s who entered the *Mercury* program in April 1957. They were picked from more than a hundred test pilots in a list provided by the Pentagon. After a two-month selection process, NASA Administrator T. Keith Glennan introduced the astronauts at an April 7 press conference in Washington: Lt. Col. John H. Glenn Jr. (from the Marine Corps); Lt. Cdr. Walter M. Schirra Jr., Lt. Cdr. Alan B. Shepard Jr., and Lt. M. Scott Carpenter (Navy); and Capt. L. Gordon Cooper, Capt. Virgil I. "Gus" Grissom, and Capt. Donald K. Slayton (Air Force). The seven became household names.

USAGE. The term has come to be used as a descriptor, as in "'Original Seven' astronaut Gordon Cooper."

O-ring. A component of the *Space Shuttle* solid *rocket booster* located between the booster segments. O-ring failure in cold weather was the cause of the Space Shuttle *Challenger* accident.

Orion. (1) *Call sign* for the *Apollo* 16 *Lunar Module* for the *mission* of April 16–27, 1972, used by John W. Young and Charles M. Duke Jr. to explore the lunar surface with the lunar *rover*. The Command Module (commanded by T. K. Mattingly) was called *Casper*.

ETYMOLOGY (definition 1). Named for the constellation Orion because the crew was dependent on star sightings to navigate *cislunar* space. (http://history.nasa.gov/SP-4029/SP-4029.htm.)

(2) The *Crew Exploration Vehicle* that NASA's Constellation Program is developing to carry a new generation of explorers back to the Moon and later to Mars. (See *Project Constellation*.) Orion will succeed the *Space Shuttle* as NASA's primary vehicle for human space exploration. Orion's first flight to the *International Space Station* is planned for no later than 2014. Its first flight to the Moon is planned for no later than 2020. Orion will be capable of transporting cargo and up to six crew members to and from the International Space Station. It can carry four crew members for lunar missions. Later, it can support crew transfers for Mars missions.

Orion borrows its shape from *space capsule*s of the past such as Apollo (see *Apollo on steroids*) but takes advantage of the latest technology in computers, electronics, life-support, propulsion, and heat protection systems. The capsule's conical shape is the safest and most reliable for reentering the Earth's *atmosphere,* especially at the velocities required for a direct return from the Moon. Orion will be 16.5 feet (5 m) in diameter and have a mass of about 25 tons (22,000 kg). Inside, it will have more than 2.5 times the volume of an Apollo capsule.

Mark Carreau, who covers space for the *Houston Chronicle,* wrote that

Orion is "a label that could carry as much meaning for space exploration as Apollo once did" (Carreau, "NASA Lets Its Secret Slip Out: Moon Ship to Be Called Orion," *Houston Chronicle,* August 23, 2006, p. 1).

ETYMOLOGY (definition 2). Named after the figure from ancient Greek mythology, continuing the practice started in the 1960s with the Apollo lunar-exploration spacecraft. Orion was a mortal who was accidentally killed by Artemis, Apollo's huntress sister, who placed him in the heavens as a constellation. The Orion spacecraft is thus also named for one of the brightest, most familiar, and most easily identifiable constellations. "Many of its stars have been used for navigation and guided explorers to new worlds for centuries," said Orion Project Manager Skip Hatfield. "Our team, and all of NASA—and, I believe, our country—grows more excited with every step forward this program takes. The future for space exploration is coming quickly." (NASA News Release 06-299, "NASA Names New Crew Exploration Vehicle Orion.") NASA press materials identified Orion as the "god of the hunt," whereas in all standard references on mythology he is identified as a mortal, not a god.

FIRST USE. The name was released after an advance message being taped by International Space Station crew member Jeff Williams was inadvertently broadcast over an open radio link, according to the Associated Press and various other news accounts. "We've been calling it the Crew Exploration Vehicle for several years, but today it has a name, Orion," Williams said in a statement that was supposed to be released a week later, on August 31, 2006, the day the prime contractor for the new vehicle was to be announced ("Orbiting Astronaut Spills Secret from NASA," AP piece in the *Record* [Bergen County, NJ], August 24, 2006, p. A20). The name Orion had actually first appeared on July 20, 2006, on a website for collectors of space memorabilia called Collectspace.com but without official confirmation by NASA.

OSCAR. Orbiting Satellite Carrying Amateur Radio. A series of satellites built by amateur radio operators for amateur radio communication.

OSO. Orbiting Solar Observatory. OSO evolved from the NASA concept for larger, general-purpose *spacecraft* for scientific experiments. (See *OGO.*) The name was a functional description of the *satellite,* indicating that it was of the orbiting-observatory class of satellites whose purpose was to measure phenomena of the Sun. OSO 1, *launch*ed March 7, 1962, was the first satellite in the Orbiting Observatory series to be placed in *orbit.* OSO 7 was launched on September 29, 1971. The OSO satellites were designed to provide observations of the Sun during most of its 11-year cycle. Results included the first full-disc photograph of the solar corona, the first x-ray observations from a spacecraft of a beginning solar flare and of solar streamers (structures in the coronas), and the first observations of the corona in white light and extreme ultraviolet. (SP-4402, p. 63–64; Jack Posner, Office of Space Science and Applications, NASA, telephone interview, August 10, 1965; U.S. Congress, Senate, Committee on Aeronautical and Space Sciences, *Hearings: NASA Scientific and Technical Programs, February and March 1961* [Washington, DC: GPO, 1961], pp. 240–42.)

outer space. The area beyond the Earth's *atmosphere*. The first *Sputnik* is said to have orbited in outer space, but its *orbit* was within the outer reaches of the Earth's atmosphere. The term still lacks an internationally agreed upon meaning. Between 1958 and 1966 the United Nations worked on and completed an outer space treaty without ever coming to a legal definition. The loose working UN definition was the realm above the range of normal nonexperimental aircraft.

out of family. NASA jargon used to characterize an anomalous event that was not known or expected. See *family*.

outpost. An initial location to provide shelter for a few people on the Moon or Mars. It would not necessarily be permanently occupied. See also *base* and *settlement*. (Paine Report, p. 198.)

OV. Orbiting vehicle.

Ozma. See *Project Ozma*.

ozone depletion. The reduction of the protective layer of ozone in the *upper atmosphere* by chemical pollution. The *Nimbus* 7 Environmental Research Satellite launched on October 24, 1978, provided the global evidence of Antarctic ozone depletion in the 1980s using its Total Ozone Mapping Spectrometer (TOMS).

P

Pacific. Series of Intelsat communications *satellites* (Intelsat II-B, Intelsat II-D, Intelsat-II F-). See *Intelsat*.

Pacific Missile Range (PMR). Former name of USAF Western Test Range, now *Western Space and Missile Center*.

pad. See *launch pad*.

PAD. Project Approval Documents. The package that gives the green light to NASA *mission*s and projects.

pad abort. Stopping the *launch*ing of a *space vehicle* while still on the *launch pad*.

PAGEOS. Passive Geodetic Earth Orbiting Satellite. Inflatable Mylar sphere with no instrumentation on board. It was the second NASA *satellite,* following *GEOS* 1, in the National Geodetic Satellite program. In August 1964 NASA approved *Langley Research Center's* proposal for the PAGEOS project. PAGEOS I, a balloon 98 feet (30 m) in diameter (similar to the *Echo* balloon satellite), achieved *orbit* and inflated June 23, 1966. The passive (uninstrumented) satellite reflected sunlight and, photographed by ground stations around the world, provided a means of precision mapping the Earth's surface. (SP-4402, p. 64; NASA News Release 66-150; Jack Posner, Office of Space Science and Applications, NASA, telephone interview, August 10, 1965.)

    ETYMOLOGY. The term came into use among project officials and found its way into documents through common use. PAGEOS paralleled the name GEOS that designated the active (instrumented) geodetic satellites in the *Explorer* series.

Paine Report. Popular name for the 1986 report of the National Commission on Space, chaired by former NASA Administrator Thomas O. Paine. The report on the U.S. civil *space program,* entitled *Pioneering the Space Frontier: An Exciting Vision of Our Next Fifty Years in Space,* advocated an aggressive space effort oriented toward the exploration and eventual colonization of the Moon and the planets of the solar system.

PAO. *Public affairs officer.*

PARD. *Pilotless Aircraft Research Division.*

parking orbit. An intermediate *satellite orbit* (SP-6001, p. 70).

Passive Geodetic Earth Orbiting Satellite See *PAGEOS.*

Pathfinder. See *Mars Pathfinder.*

payload. The useful load carried by an aircraft or *rocket* over and above what is necessary for the operation of the vehicle during its flight (SP-6001, p. 71).

    ETYMOLOGY. The *OED* suggests that the term is a 20th-century conflation of paying + load.

    FIRST USE. "Research into the question of producing a comfortable machine rather than one showing the highest pay load" (*OED,* citing *London Times,* February 9, 1927, p. 10). Defined in terms of space exploration in 1945 in the lexicon attached to G. Edward Pendray's *The Coming Age of Rocket Power.*

payload bay. The portion of a *spacecraft* reserved for useful loads—experiments, instruments, etc. Most often used today in connection with the *Space Shuttle.*

payload specialist. A member of an *STS* crew primarily concerned with the overall management of *payloads.* The specialist may be an *astronaut* or scientist.

PEACESAT. Pan-Pacific Education and Communications Experiment by Satellite.

Pearl River Test Site. Early designation considered for the Mississippi Test Facility, later *Stennis Space Center.*

Pegasus. *Meteoroid satellite.* The outstanding feature of the Pegasus satellites was their huge winglike panels, 315 feet (96 m) from tip to tip, sweeping through space to determine the rate of meteoroid penetrations. Three Pegasus satellites were placed in *orbit* in 1965, all by *Saturn I* launch vehicles: Pegasus 1 on February 16, Pegasus 2 on May 25, and Pegasus 3 on July 30.

ETYMOLOGY. The Pegasus satellite program office said when choosing from proposed names that the *spacecraft,* to be the heaviest yet orbited, would be "somewhat of a 'horse' as far as payloads are concerned" and there could be "only one name for a horse with wings": Pegasus, the flying horse of ancient Greek mythology. The NASA Project Designation Committee originally agreed on Project Pegasus as the name for the experiments before launch. The satellites were to be supplanted with an Explorer designation in orbit. The original suggestion for the name had come from an employee of the spacecraft contractor, Fairchild Stratos Corporation. The contractor, with the concurrence of the NASA Office of Space Vehicle Research and Technology and *Marshall Space Flight Center,* had held an *in-house* competition in 1963 to select a name for the project. From more than 100 suggestions submitted by Fairchild Stratos employees, the NASA program office recommended the name Pegasus to the Project Designation Committee. The committee approved the selection in July 1964, and NASA announced the name in August. (Julian W. Scheer, Assistant Administrator for Public Affairs, NASA, memorandum for Raymond L. Bisplinghoff, Associate Administrator for Advanced Research and Technology, with concurrence of Robert C. Seamans Jr., Associate Administrator, July 6, 1964.)

SOURCES. SP-4402; Raymond L. Bisplinghoff, Associate Administrator for Advanced Research and Technology, NASA, memorandum to Julian Scheer, Chairman, Project Designation Committee (and Assistant Administrator for Public Affairs), NASA, December 23, 1963; Milton B. Ames Jr., Director, Space Vehicle Division, Office of Advanced Research and Technology, NASA, letter to Edward G. Uhl, President, Fairchild Stratos Corp., July 21, 1964; NASA News Release 64–203.

perigee. Point at which a body in Earth *orbit* is nearest the Earth. See also *apogee.* Strictly speaking, apogee and perigee refer only to Earth orbit, but the terms are often applied to orbits around other celestial bodies.

perihelion. Point at which a body in solar *orbit* is nearest the Sun. See also *aphelion.*

Personal Preference Kit. Sock-sized pouch that *Apollo* astronauts were allowed to fill with their prize possessions.

Phantom Torso. A 95-pound (43-kg) mockup of the human upper body used on the *International Space Station* to study the effects of *radiation* on the body. Nicknamed Fred, the torso contains hundreds of monitoring devices. (NASA Fact Sheet, 2001-02-42-MSFC, February 28, 2001.)

Phobos. Soviet *missions* to Mars, 1988–89. Phobos 1 was *launch*ed on July 7, 1988. It was lost, owing to a command error that occurred on September 2, 1988, while en route to Mars. Phobos 2 was launched on July 12, 1988. It arrived at Mars and was inserted into *orbit* on January 30, 1989. The spacecraft failed after returning 38 high-resolution photos of the Martian moon Phobos.

ETYMOLOGY. Both moon and mission were named for the son of Ares and Aphrodite, of ancient Greek mythology. Phobos means panic and is the basis for the eponym phobia.

Phoenix Mars Lander. *Probe* sent to explore new territory in the northern plains of Mars (analogous to the permafrost regions on Earth). Launched in August 2007, it landed in the northern polar region of the red planet on May 25, 2008, and immediately began examining a site chosen for its likelihood of having frozen water within reach of the lander's robotic arm. On July 31, 2008, it was officially confirmed that water had been found in a soil sample. "We have water," said William Boynton of the University of Arizona (http://phoenix.lpl.arizona.edu/07_31_pr.php), lead scientist for the Thermal and Evolved-Gas Analyzer, or TEGA. "We've seen evidence for this water ice before in observations by the Mars Odyssey orbiter and in disappearing chunks observed by Phoenix last month, but this is the first time Martian water has been touched and tasted."

ETYMOLOGY. Name chosen to reflect the rebound from failed Mars missions: "Like its namesake, Phoenix rises from ashes, carrying the legacies of two earlier attempts to explore Mars. Many of the scientific instruments for Phoenix were built or designed for that mission or flew on the unsuccessful Mars Polar Lander in 1999." (Press Release 05-141, June 2, 2005, "NASA's Phoenix Mars Mission Begins Launch Preparations.")

piggyback. Term describing an experiment that rides along with the primary experiment on a space-available basis, without interfering with the *mission* of the primary experiment (NASA, *Glossary/ Congressional Budget Submission,* p. 31).

piloted spaceflight. Term used to distinguish a *mission* with human operators. Like *crewed spaceflight* or *human spaceflight,* a preferred synonym for *manned.*

Pilotless Aircraft Research Division (PARD). Division of the *NACA's* Langley Aeronautical Laboratory (later NASA's *Langley Research Center*).

Pioneer (space probe). Name chosen for the first U.S. space probe, Pioneer 1, *launch*ed October 11, 1958, as well as for the following series of lunar and *deep space* probes. The Pioneer series had been initiated for the *International Geophysical Year* by the Department of Defense's *Advanced Research Projects Agency* (ARPA), which assigned execution variously to the Air Force Ballistic Missile Division (AFBMD) and to the Army Ballistic Missile Agency *(ABMA)*. Upon its formation in October 1958, NASA inherited responsibility for (and the name of) the probes.

The first series of Pioneer *spacecraft* was flown between 1958 and 1960. Pioneer 1, 2, and 5 were developed by Space Technology Laboratories, Inc., and were launched for NASA by AFBMD. Pioneer 3 and 4 were developed by the *Jet Propulsion Laboratory* and launched for NASA by *ABMA*. In 1960 Pioneer transmitted the first solar flare data and established a communications distance record of 22.5 million miles (36.2 million km).

With the launch of Pioneer 6 (Pioneer A in the new series) in December 1965, NASA resumed the probes to complement inter-planetary data acquired by *Mariner* probes. Pioneer 7, 8, and 9, second-generation *spacecraft* launched between 1966 and 1968, continued the investigation of the interplanetary medium.

Between 1965 and 1967 NASA had been studying the concept for a space probe known as the Galactic Jupiter Probe, or Advanced Planetary Probe, that would investigate solar, interplanetary, and *galactic* phenomena in the outer region of the solar system. By 1968 NASA had included the probe in the Pioneer series, designating two such probes Pioneer F and G.

Pioneer 10 (Pioneer F), launched in March 1972, became the first spacecraft to cross the asteroid belt. It flew by Jupiter in December 1973, returning more than 300 close-up photos of the planet and its inner moons as well as data on its complex *magnetic field* and its *atmosphere*. Accelerated by Jupiter's gravity, the probe headed on a course out of the solar system. In 2001, signals were received from Pioneer 10; it was over 7 billion miles (11.3 million km) from Earth heading toward the star Aldebaran (in the constellation Taurus). Pio-neers 10 and 11 were fitted with a plaque that served as a message to any extraterrestrial intelligence that might encounter it.

Pioneer 11 (Pioneer G), launched in April 1973, crossed the asteroid belt, skimmed by Jupiter three times closer to the planet than Pioneer 10 had, and was thrown by Jupiter's gravity toward Saturn. The space-craft sent back the first photos of Jupiter's poles and information on the atmosphere, the equatorial regions, and the moon Callisto. On the night of December 2, 1974, when Pioneer 11 set its new course for

Saturn, NASA renamed the probe Pioneer Saturn 5. It passed close by Saturn on September 1, 1979. It went on to explore the outer regions of our solar system, studying the *solar wind* and cosmic rays entering our portion of the *Milky Way.* The spacecraft has operated on a backup transmitter since launch. Instrument power sharing began in February 1985 due to declining generator power output. The last signal from Pioneer 11 was received in November 1995 and that from Pioneer 10 on January 23, 2003.

Pioneer Venus 1 was launched May 20, 1978. The Pioneer Venus orbiter carried 17 experiments and studied Venus for a decade. Pioneer Venus 2, also known as Pioneer Venus Multiprobe, was launched August 8, 1978. It was essentially a bus carrying four atmospheric probes, which entered the Venusian atmosphere on December 9, 1978.

ETYMOLOGY. Credit for naming the first probe has been attributed to Stephen A. Saliga, who had been assigned to the Air Force Orientation Group, Wright-Patterson AFB, as chief designer of Air Force exhibits. While he was at a briefing, the *spacecraft* was described to him as a "lunar-orbiting vehicle with an infrared scanning device." Saliga thought the title too long and lacked a theme for an exhibit design. He suggested Pioneer as the name of the probe since "the Army had already launched and orbited the Explorer satellite and their Public Information Office was identifying the Army as 'Pioneers in Space.'" Saliga added that by adopting the name, the Air Force would "make a 'quantum jump' as to who really [were] the 'Pioneers in space.'"

SOURCES. SP-4402, pp. 88–90; David S. Akens, *Historical Origins of the George C. Marshall Space Flight Center,* MHM-1 (Huntsville, AL: MSFC, 1960), p. 51 n. 28; Emme, *Aeronautics and Astronautics,* pp. 102–3; Emme, "Names of Launchings," enclosure to letter, May 12, 1960; Maj. Gen. Reginald M. Cram, Adjutant General, State of Vermont, letter to Stephen A. Saliga, Visual Aids Chief, NASA, February 6, 1970 (Cram was Commander of the Air Force Orientation Group at the time); Saliga, memorandum to Eugene M. Emme, Historian, NASA, April 13, 1972; John F. Clark, Director, GSFC, "Galactic/Jupiter Probes," address at the Fifth Goddard Memorial Symposium (AAS meeting), March 14, 1967; U.S. Congress, House of Representatives, Committee on Science and Astronautics, *Hearings: 1969 NASA Authorization, February 3, 1968* (Washington, DC: GPO, 1968), pp. 207–8, 239–43; George M. Low, Deputy Administrator, NASA, "Letter from Washington," *NASA Activities 5* (December 15, 1974): 3; Peter W. Waller, Public Information Officer, ARC, telephone interview, February 27, 1975; NASA News Release 75-19.

Pioneering the Space Frontier. Report of the National Commission on Space, 1986. See *Paine Report.*

Pioneer Saturn. See *Pioneer*.

Pioneer Venus. See *Pioneer*.

planetary quarantine. Sterilization and decontamination studies directed to the prevention of contamination of planets by *terrestrial* organisms so that the search for *extraterrestrial life* may have validity. Planetary protection also aims to prevent back-contamination to Earth. (Pitts, *The Human Factor*, SP-4213, p. 264.)

planned disequilibrium. Term used by NASA Administrator James E. Webb (1961–68) to describe his management style, which was to keep his staff somewhat off-balance. The term was invoked at the time of his death in all major obituaries (e.g., "James E. Webb Dies at 85, Was NASA Chief in 1960s," *Washington Post*, March 29, 1992, p. B7).

plug repair. Name for the on-orbit repairs to be made on the reinforced carbon-carbon (RCC) panels on *Space Shuttle* wings' leading edge. It can be used to repair holes during *EVA* up to 4 inches (10 cm) in diameter.

Plum Brook Station. Component of the *Glenn Research Center* (GRC).

plus count. During the *launch* of a *rocket*, a count in seconds (plus 1, plus 2, etc.) that immediately follows *T-time*, which has been preceded by the *countdown*. It is used to check on the sequence of events after the action of the countdown has ended. (SP-7.)

PMR. *Pacific Missile Range*.

POES. Polar-orbiting Operational Environmental Satellite (program).

pogo. Vertical vibration of a *launch vehicle* that if continued could destroy it (Kranz, *Failure Is Not an Option*, p. 395). Also used in connection with the *Space Shuttle* main *engine* turbopumps.

ETYMOLOGY. Presumably derived from analogy to a pogo stick, a stilt-like pole with a spring-loaded base on which one jumps about.

FIRST USE. "The first stage of the three-stage rocket went into 'pogo-stick' oscillations" (*New Scientist*, December 19, 1968, pp. 653–54).

POGO. Polar Orbiting Geophysical Observatory. See *OGO*.

polar orbit. Satellite *orbit* passing over both poles of the Earth. During a 12-hour day, a *satellite* in such an orbit can observe all points on Earth.

Polar Orbiting Geophysical Observatory (POGO). See *OGO*.

Post-Orbital Remorse. The feeling that, once you have ridden into space atop a *rocket*, nothing else you ever do will live up to the experience.

ETYMOLOGY. This was the collective title Wolfe gave to his four-part magazine piece on the astronauts in 1973. Doesn't quite have the same ring as "The Brotherhood of The Right Stuff," which was the subtitle. Both this term and "The Right Stuff" are discussed in "What Is 'The Right Stuff'? Name for Space Heroism Has Become Part of American Lexicon" *Daily Press* (Newport News, VA), May 10, 2003, p. D1.

postsatellite era. Also, post-Sputnik era. Term used by some writers—
Isaac Asimov among them—to describe the years since the *launch* of
Sputnik.

potlatch. Metaphoric term used by critics of the *Space Race* to describe
the process by which two superpowers spend vast amounts of money
on space exploration so that a winner can be declared. For instance:
"A cultural anthropologist of the twenty-first century, viewing the hot
fires that consume million-dollar—and million-ruble—rockets in this
decade of the twentieth century, might, understandably enough, be
reminded of the potlatch ceremony" (Edwin Diamond, *The Rise and Fall
of the Space Age* [New York: Doubleday, 1964], p. 1).

ETYMOLOGY. From a Northwest American Indian tribal ritual involving an
extravagant giving away or burning of possessions to enhance one's
prestige or establish one's position.

power landing The landing of a spacecraft on a body in space in which
the thrust of its motion is used as a brake (SP-6001, p. 73).

PPK. *Personal Preference Kit.*

PPS. Precise Positioning System. More accurate version of *GPS*.

probe. Uncrewed instrumented *spacecraft* that obtain scientific information about the Moon, other planets, and the space environment.
Probes are differentiated from *sounding rockets* in that they attain
altitudes of at least 4,000 miles (6,400 km). When a probe is *launch*ed
on an escape *trajectory* attaining sufficient velocity to travel beyond
the Earth's gravitational field, it becomes in effect a *satellite* of the Sun.
The *Lunar Orbiter* probes, however, were sent into *orbit* around the
Earth's natural satellite, the Moon.

　　The first serious consideration of the concept of a *space probe* can
be attributed to Dr. Robert H. Goddard, American *rocket* pioneer. As
early as 1916, Goddard's calculations of his theoretical rocket and his
experiments with flash powders led him to conclude that a rocket-
borne *payload* exploding on the Moon could be detected from Earth.
On September 20, 1952, a paper entitled "The Martian Probe,"
presented by E. Burgess and C. A. Cross to the British Interplanetary
Society, gave the term probe to the language.

　　In May 1960, at the suggestion of Edgar M. Cortright, Assistant
Director of Lunar and Planetary Programs, NASA adopted a system of
naming its space probes. Names of lunar probes were patterned after
land exploration activities (the name *Pioneer,* designating the early
series of lunar and related space probes, was already in use). The
names of planetary *mission* probes were patterned after nautical
terms, to convey "the impression of travel to great distances and
remote lands." Isolated missions to investigate the space environment

were "assigned the name of the mission group of which they are most nearly a part." This 1960 decision was the basis for naming the *Mariner, Ranger, Surveyor,* and *Viking* probes.

SOURCES. SP-4402, p. 83; Milton Lehman, *This High Man: The Life of Robert H. Goddard* (New York: Farrar, Straus and Co., 1963), pp. 81–82; William R. Corliss, *Space Probes and Planetary Exploration* (Princeton: Van Nostrand, 1965), p. 10; Edgar M. Cortright, Assistant Director of Lunar and Planetary Programs, NASA, memorandum to NASA Ad Hoc Committee to Name Space Projects and Objects, May 17, 1960; NASA, Ad Hoc Committee to Name Space Projects and Objects, minutes of meeting, May 19, 1960.

Project Able. See *Able.*

Project Adam. Army proposal as finally developed in April 1958 at the Army Ballistic Missile Agency *(ABMA),* Alabama, to carry a *crewed* instrumented *spacecraft* to a range of approximately 150 statute miles (241 km); to perform psychophysiological experiments during the acceleration phase and the subsequent six minutes of *weightlessness;* and to effect a safe *reentry* and *recovery* of the spacecraft from the sea. The proposal urged that Project Adam be approved as the next significant step toward the development of a U.S. capability for the transportation of troops by ballistic missile, and that funds be provided immediately. The plan stopped there along with the U.S. Air Force *Man in Space Soonest* program; it would not be funded by *ARPA.* ("Development Proposal for Project Adam," report no. D-TR-1-58.)

Project Advent. The military's first effort to employ *repeater* communications satellites. The effort was canceled in May 1962.

Project Ares (pronounced air-eez or ah-rays). Formal name for the *launch vehicles* used in a series of missions that will take humans back to the Moon and onto Mars. See *Ares.*

Project Astronaut. Rejected name for the *Mercury* project (SP-4402, p. 107).

Project CAMEO. Chemically Active Materials in Orbit. Experiment released from the *Nimbus 7 satellite's Delta launch vehicle,* for the study of the Earth's electrical and *magnetic fields.* CAMEO has also been used to mean Composition of the Atmosphere from Mid-Earth Orbit, and Combustion Analysis Model and Optimizer.

Project Clementine. Joint NASA-SDIO (Strategic Defense Initiative Organization) project for lunar *orbit* and an asteroid *flyby.*

Project Constellation. Name for NASA's family of *Crew Exploration Vehicles* (CEVs). The name has come to be used less frequently as the terms *Orion* for the CEV and *Ares* for the *launch vehicle* component have come into play.

ETYMOLOGY. On November 1, 2004, NASA Administrator Sean O'Keefe announced the name at a ceremony on board the historic museum frigate USS Constellation, docked at the U.S. Naval Academy, Annapolis, Maryland: "The proud name 'Constellation' represents the best of progress, valor and the American spirit. First given to one of the finest, most modern sailing ships, a craft built to represent and defend America, the name has been carried proudly by newer, modern vessels and aircraft throughout our nation's history. Today, we help continue that tradition by accepting the spirit of the original Constellation and proudly transferring it to the class of space vehicles that will carry humankind back to the moon, Mars and beyond."

Project Cyclops. Study devoted to the design of powerful search engines and radio telescopes for use in the Search for Extraterrestrial Intelligence *(SETI)*. Based on early studies, Project Cyclops became the name for a proposed *terrestrial* system of phased radio telescopes with a highly sophisticated data-processing system. It was to be a program to create a phased array of 1,000 to 2,500 steerable antennas, each larger than a football field, in an "orchard" about 6 miles (10 km) in diameter located in the American Southwest. According to an internal NASA memo of May 20, 1974, it would cost between $5 billion and $10 billion. It was never built, but many of the ideas in the report have been incorporated into SETI searches. ("Project Cyclops: A Design Study of a System for Detecting Extraterrestrial Intelligent Life," NASA CR114445, 1972; memo of September 20, 1973, filed in Record Number 18550 in NASA History Office.)

ETYMOLOGY. In ancient Greek mythology, a race of one-eyed giants who forged thunderbolts for Zeus. Because Cyclops would serve as a gigantic eye on the universe, the allusion was appropriate. According to Charles Seeger, the name was chosen by Bernard M. Oliver, then head of Research and Development at Hewlett-Packard (Seeger interviewed in David W. Smith, *SETI Pioneers: Scientists Talk about Their Search for Extraterrestrial Intelligence* [Tucson: University of Arizona Press, 1990], p. 253). Oliver believed that Cyclops could give NASA a long-term mandate beyond the Shuttle and robotic exploration of the planets. In a letter to NASA's James C. Fletcher, Oliver stated his case: "At the brink of space—*real* space—our missions must stop, but our curiosity will not. The only way I can see of greatly extending NASA's lifespan is to extend the radius of its sphere of influence a million fold: from a milli–light year to a kilo–light year, with a program like Cyclops" (Oliver, letter to Fletcher, September 20, 1973, filed in Record Number 18550 in NASA History Office).

Project Deal. See *Deal Project*. See also *Explorer*.

Project DODGE. Department of Defense (DOD) Gravity-gradient Experiment. Satellite launched July 1, 1967, aboard a *Titan* IIIC *rocket* to test gravity-gradient stabilization for *spacecraft*. Took first color photographs of full Earth from space.

Project Echo. See *Echo*.

Project Farside. Early U.S. Air Force attempt to reach high altitudes using *sounding rockets* from balloons: *rockoons*.

Project Habitat. From 1982 to 1985, the name for the program that became *Mission to Planet Earth*. It was announced by then–NASA Administrator James Beggs as a term to encompass global monitoring by satellites to chart the Earth's environmental ills. The program was immediately seen as controversial: "Some members of the Reagan White House thought Habitat was too much like something President Jimmy Carter and his environmentalists would propose" (W. Henry Lambright, "Administrative Entrepreneurship and Space Technology: The Ups and Downs of '*Mission to Planet Earth*,'" *Public Administration Review,* March–April 1994, p. 99).

Project Horizon. Army plan for lunar landing in April 1965 with a permanent *outpost* there. The Army concluded that "for political and psychological reasons, anything short of being first on the lunar surface would be catastrophic" ("'Dusty' Moon Paper Urged US First," *Christian Science Monitor,* September 17, 1962, p. 12).

Project Orbiter. Short-lived Army project to put a *satellite* into *orbit* in 1956 on a *Redstone missile* system. It was the precursor to the *Explorer* 1 satellite.

Project Ozma. The first search for radio signals from extraterrestrial civilizations, conducted in 1960 by Cornell University astronomer Frank Drake at the National Radio Astronomy Observatory at Green Bank, West Virginia. The object of the experiment was to search for signs of life in distant solar systems through interstellar radio waves. None were found, but Ozma did become a model for future *SETI* projects.

ETYMOLOGY. The project was named after the Queen of the Land of Oz in the books by L. Frank Baum ("Ozma Picks 2 Stars, Waits for Any Nearby Message," *Baltimore Sun,* April 13, 1960, in *NASA Current News,* April 13, 1960, p. 2). *Ozma of Oz,* published in 1907, was the third book in Baum's Oz series; however, it is also the first in which the majority of the action takes place outside the land of Oz.

Project SCORE. Signal Communication by Orbiting Relay Equipment. An experiment designed to test the feasibility of transmitting messages through the *upper atmosphere* from one ground station to one or more other ground stations. The result of the project was unquestionably a major scientific breakthrough proving that active communications *satellite*s could provide a means of transmitting messages of all sorts from one point to any other on Earth.

The satellite *payload* package was designed, and in large measure built, by personnel of the U.S. Army Signal Research and Development Laboratory at Fort Monmouth, New Jersey. Dr. Hans K. Ziegler, writing in 1960 when he was Chief Scientist at the U.S. Army Signal Research and Development Laboratory, characterized SCORE as the first prototype of a communications satellite, and the first test of any satellite for direct practical applications. According to him, the significance of the experiment lay in the fact that it effectively demonstrated the practical feasibility of worldwide communications in delayed- and real-time mode by means of relatively simple active satellite relays. During the 13-day life of its batteries, the satellite was interrogated by Signal Corps ground stations 78 times, using voice and teletype messages for the communications tests, with excellent results. The project provided valuable information for the design of future communications satellites.

propellant. Material that is used to move an object by applying a motive force. Common propellants are gasoline, jet fuel, and *rocket* fuel.

public affairs officer (PAO). NASA's term for public relations man or woman. Historically, PAOs have played an important role in explaining missions to the public and the media. "The PAO really earned his pay when things went wrong," writes Gene Kranz. "He was our first line of defense—and fortunately we had PAOs who were very good, and unflappable when things got a bit dicey during a mission" (Kranz, *Failure Is Not an Option,* p. 143).

pulsar. *Neutron star.* A body that emits regularly spaced pulses of *radiation* or light while spinning at a rate that defies human comprehension—perhaps a million times per second. When pulsars were first discovered in 1967, it was believed that they could be a signal from an intelligent life form.

purple pigeons. NASA insider terminology, ca. 1976 and later, for new projects with high visibility and substantial impact on the agency's budget. In contrast to *gray mice,* which were new projects with low visibility and minimal budgetary impact. See also *wild turkey.* (NASA Names Files, record no. 17540, memo from John E. Naugle of November 12, 1976, suggests that these terms were Naugle's creation.)

push the envelope. To reach beyond the boundaries; to stretch or exceed known limits. Tom Wolfe popularized the term in *The Right Stuff* (1979): "Pushing the outside, probing the outside limits, of the envelope seemed to be the great challenge and satisfaction of the space flight."

ETYMOLOGY / FIRST USE. William Safire (in "Pushing the Envelope," *New York Times* column, November 7, 1999, p. SM42) reported that the first use of the term he could find in print was in *Aviation Week & Space Technology* for July 3, 1978, where it was applied to an aircraft being flown to altitudes higher than it was designed for: "NASA pilots were to push the envelope to 10,000 feet." Safire called Wolfe, who told him that he first heard it in 1972 "among test pilots who later became astronauts." Wolfe added, "They were speaking of the performance capabilities of an airplane as an envelope, as if there were a boundary. Why they chose envelope, I don't know, but if you get outside the envelope you're in trouble." Wolfe estimated that the term may have originated at the Patuxent River Naval Air Station in Maryland in the 1940s.

pyroxferroite. One of three new minerals, unknown on Earth, that were discovered in the Moon rocks: *tranquillityite,* armalcolite, and pyroxferroite.

Q

quarantine. Period during which humans or animals are isolated after returning from space. Defined in a NASA Policy Directive of July 16, 1969, on the subject of "Extraterrestrial Exposure" as "the detention, examination and decontamination of any person, property, animal or other form of life matter whatever that is extraterrestrially exposed, and includes the apprehension or seizure of such person, animal or other form of life matter whatsoever."

quasar. One of a class of very distant (typically, billions of light-years), extremely bright, and relatively small objects associated with the nuclei of active galaxies. The term means quasi-star—that is, something that looks like a star but can't actually be one. A typical quasar produces more light each second than an entire normal *galaxy* of stars, and it does so from a region of space that may be as small as our solar system. The luminosity of quasars is due to the gravitational action of supermassive black holes at the centers of active galaxies, which pull in matter and form accretion disks. See *black hole.* (ASP Glossary.)

R

RADARSAT. Advanced Earth observation *satellite* project developed by Canada to monitor environmental change and to support resource sustainability.

radiation. Energy transmitted through space as waves or particles.

radiation belts. Regions of charged particles in a *magnetosphere*. The belts contain ions and electrons. Magnetized planets, like Earth, are encircled by zones of particle *radiation* known as the Van Allen Belts, in which charged particles spiral to and fro, trapped by the planet's *magnetic field*.

Radio Astronomy Explorer (RAE). See *Explorer*.

RAE. Radio Astronomy Explorer. See *Explorer*.

Raffaello. One of the Italian-built Multi-Purpose Logistics Modules *(MPLMs)* that are carried in the cargo bay of the *Space Shuttle* and ferried to the *International Space Station*. They contain experiments and supplies. (*Reference Guide to the International Space Station*, SP-2006–557.)

RAM. *Research and Applications Module.*

RAND Corporation. Think tank often involved in studies related to space exploration. In the late 1940s the Department of Defense, especially the U.S. Air Force, was studying multiple scenarios of how the country should proceed into the coming age of jet aircraft, missiles, and rockets. The Air Force saw a need for a stable, highly skilled cadre of analysts to help with the evaluation of these alternatives and established the Rand Corporation in Santa Monica, California, as a civilian think tank to which it could turn for independent analysis.

The first RAND report, released on May 2, 1946, was entitled "Preliminary Design of an Experimental World-Circling Spaceship." That study—conducted by a group within the Douglas Aircraft Company known as the RAND group (for Research and Development) that would later become the RAND Corporation—considered a 1951 *launch* date to be feasible with communications, weather, and spy satellites following in the wake of the first demonstration satellite. The authors of the report admitted that though their crystal ball was cloudy, two things seemed clear. First, "a satellite vehicle with appropriate instrumentation can be expected to be one of the most potent scientific tools of the Twentieth Century." Second, the "achievement of a satellite craft by the United States would inflame the imagination of mankind, and would probably produce repercussions in the world comparable to the explosion of the atomic bomb." The cost would be about $150 million.

Then on October 4, 1950, came a stunning second report from RAND that went beyond the question of whether a satellite could be launched but under what conditions it should be launched. The report stressed the importance of the satellite as an instrument of reconnaissance—spying. In his 1985 book *The Heavens and the Earth,* Walter A. McDougall terms this second RAND paper "the birth certificate of American space policy" because it stressed that the United States should not provoke the Soviets by putting U.S. military spy satellites overhead. RAND recommended that the first U.S. satellite be billed as "experimental," and that it be launched on an equatorial *orbit,* avoiding the Soviet Union altogether. RAND held that this would help establish the concept of "the freedom of space." A follow-on study that appeared in April 1951 went into much greater detail on how all of this would work.

Ranger (lunar *probe*). A probe series designed to gather data about the Moon. NASA initiated Project Ranger, then unnamed, in December 1959, when it requested the *Jet Propulsion Laboratory* (JPL) to study *spacecraft* design and a *mission* to "acquire and transmit a number of images of the lunar surface." In February 1960 Dr. William H. Pickering, JPL Director, recommended that NASA Headquarters approve the name JPL was using for the project, Ranger.

ETYMOLOGY. The name had been introduced by the JPL program director, Clifford D. Cummings, who had noticed while on a camping trip that his pickup truck was called Ranger. Cummings liked the name and, because it referred to "land exploration activities," suggested it as a name for the lunar impact probe. By May 1960 it was in common use.

SOURCES. R. Cargill Hall, *Project Ranger: A Chronology,* JPL/HR-2 (Washington, DC: NASA, 1971); SP-4402, pp. 90–91; Edgar M. Cortright, Assistant Director of Lunar and Planetary Programs, NASA, memorandum to NASA Ad Hoc Committee to Name Space Projects and Objects, May 17, 1960; NASA Ad Hoc Committee to Name Space Projects and Objects, minutes of meeting, May 19, 1960; Oran W. Nicks, Director of Lunar and Planetary Programs, NASA, in U.S. Congress, House of Representatives, Committee on Science and Astronautics, Subcommittee on NASA Oversight, *Hearings: Investigation of Project Ranger, April 1964* (Washington, DC: GPO, 1964), p. 56; William H. Pickering, Director, JPL, letter to Abe Silverstein, Director of Space Flight Programs, NASA, May 6, 1960; Muriel M. Hickey, Secretary to JPL Historian, letter to Historical Staff, NASA, July 18, 1967.

reconnaissance satellite. Proper name for what is popularly known as a spy *satellite.*

reconsat. *Reconnaissance satellite.*

recoverable launch vehicle (RLV). Also known as a reusable *launch vehicle.* Vehicle capable of being launched into space more than once, as opposed to an expendable launch system, wherein each vehicle is launched once and then discarded.

recovery. The retrieval of a *satellite* or *spacecraft* after a flight.

red giant. A very large, distended, and relatively cool star that is in the final stages of its life. A typical red giant, if placed where the Sun is in our solar system, might extend past the *orbit* of Mars. The relatively cool temperature of its outer layers (perhaps only 2,000 degrees Celsius as compared with the Sun's 6,000 degrees) would make it look orange or red instead of yellowish-white. Our Sun will become a red giant in about 5 billion years. (ASP Glossary.)

red-line instrumentation. In human spaceflight, instruments that indicate abnormal or emergency conditions (SP-6001, p. 81).

ETYMOLOGY. From analog dials on planes and cars, which include an actual red line on a gauge. In automobiles the redline is the maximum recommended revolutions per minute (rpm) for an *engine.* On a tachometer, the redline is usually indicated by a thin red line, hence the name.

red shift. The lengthening (or "stretching") of light waves coming from a source moving away from us. If a source of light is moving toward us, the opposite effect—called a blue shift—takes place. Light from all galaxies outside the Local Group is red-shifted, indicating that those galaxies are moving away from us (and from each other). This phenomenon is interpreted as indicating the expansion of the *universe.* (ASP Glossary.)

Redstone *(launch vehicle).* Originally a surface-to-surface *rocket* with a range of between 200 and 300 miles (322 and 483 km) developed by the U.S. Army. Named for the Redstone Arsenal at Huntsville, Alabama, it was the outgrowth of the Nazi V-2 rocket (see *V-1, V-2*) and precursor to the *Jupiter* and *Saturn* rockets. Redstone was a battlefield *missile* developed by the Army and adapted for use by NASA as a launch vehicle for suborbital space flights in Project *Mercury.* On May 5, 1961, the Redstone launched the first U.S. *astronaut,* Alan B. Shepard Jr., into suborbital flight on the *Freedom 7.*

ETYMOLOGY. After having been called by various nicknames, including Ursa and Major, the missile was officially named Redstone on April 8, 1952, for the Redstone Arsenal, where it was developed. The name of the Arsenal, in turn, referred to the rock and soil at Huntsville.

SOURCES. SP-4402, p. 16; Wernher von Braun, "The Redstone, Jupiter, and Juno," in *The History of Rocket Technology,* ed. Eugene M. Emme (Detroit: Wayne State University Press, 1964), p. 109; David S. Akens, "Historical Sketch of Marshall Space Flight Center" (ms., n.d.).

redundancy. (1) The duplication of certain critical components in a space system so that a spare or *backup* exists should there be a failure of the primary system. According to John H. Glenn Jr., "The engineers had a word for this insistence on inserting backups into the system. They called it the principle of 'redundancy.' We felt that anything that was really needed to insure our safety on a flight was not exactly 'redundant.' But we agreed to go along as the engineers *meant* to say that their redundancies were imperative." (Quoted in Carpenter et al., *We Seven*, p. 101.) (2) NASA's practice of having two or more components for each major function of the *Space Shuttle* or providing manual override of automatic controls.

re-entry/reentry. The descent of a *spacecraft* or other body that originated on Earth into Earth's *atmosphere* from space. At an altitude of 400,000 feet (122 km), a spacecraft is considered to be reentering the Earth's atmosphere.

USAGE. Although this term seems self-explanatory, it was defined in early space glossaries and thus became an element of space-talk: "Re-entry. Whenever a space craft comes back into the Earth's atmosphere, it is 're-entering'" ("Glossary of Space Talk," *Popular Mechanics,* March 1959, p. 72).

EXTENDED USE. Of all the space terms that made the leap into the mainstream, this may have become the most common (e.g., to make a reentry into the workforce).

Relay (active-repeater communications satellite). A medium-altitude *active-repeater* communications *satellite* formally named Relay in January 1961 at the suggestion of Abe Silverstein, NASA's Director of Space Flight Programs. The name was considered appropriate because it literally described the function of an active-repeater communications satellite: the satellite received a signal, amplified it within the satellite, and then relayed the signal back toward Earth. Relay 1, launched December 13, 1962, and its successor Relay 2, launched January 21, 1964, both demonstrated the feasibility of this kind of communications satellite. After its research role was completed, Relay 2 was turned over to the Department of Defense to assist in military communications over the Pacific. (SP-4402, p. 67; Abe Silverstein, Director, Office of Space Flight Programs, NASA, memorandum to Robert C. Seamans Jr., Associate Administrator, NASA, with approval signature of Dr. Seamans; Robert Warren, Communication and Navigation Programs, Office of Space Science and Applications, NASA, Letter to Historical Staff, NASA, December 11, 1963.)

Remote Manipulator System (RMS). Also known as Space Station Remote Manipulator System (SSRMS) and Canadarm 2. A Canadian-built 55-foot (17-m) long robotic arm used for assembly and maintenance tasks on the *International Space Station*. It is larger and technically more complex than the Canadarm used on the *Space Shuttle*.

remote sensing. The collection and interpretation of information about an object without being in physical contact with the object; data collection from afar.

ETYMOLOGY. Dr. Nicholas M. Short writes in his online remote sensing tutorial (http://rst.gsfc.nasa.gov): "The term 'remote sensing' is itself a relatively new addition to the technical lexicon. It was coined by Ms. Evelyn Pruitt in the mid-1950's when she, a geographer/oceanographer, was with the US Office of Naval Research (ONR) outside Washington, DC. No specific publication or professional meeting is cited in literature consulted by the writer in which the words 'remote sensing' were stated. Those 'in the know' claim that it was used openly by the time of several ONR sponsored symposia in the late '50s at the University of Michigan. The writer believes he first heard this term at a short course on photogeology coordinated by Dr. Robert Reeves at the annual meeting of the Geological Society of America in 1958. As defined above, the term generally implies that the sensor is placed at some considerable distance from the sensed target, in contrast to close-in measurements made by 'proximate sensing' (sometimes given as 'in situ' sensing), which can apply to some of the set-ups used in medical remote sensing. It seems to have been coined by Ms. Pruitt to take into account the new views from space obtained by the early meteorological satellites which were obviously more 'remote' from their targets than the airplanes that up until then provided mainly aerial photos as the medium for recording images of the Earth's surface."

FIRST USE. The *Chicago Tribune* of January 15, 1957 ("Readings Vary 13 Degrees But Are All Right," p. 21), uses "remote sensing device" to describe a thermometer atop a building that is read elsewhere.

rendezvous. A planned meeting of objects (e.g., a *spaceship* with a *space station*) at a given time and place (*Space: The New Frontier,* EP-6, p. 70).

rendock. *Rendezvous* and *docking mission.*

Research and Applications Module (RAM). *Spacelab* forerunner.

retrorocket. A small *rocket engine* on a larger rocket or *spacecraft* that is fired to slow or alter its course (to produce *thrust* opposed to forward motion).

ETYMOLOGY. According to the glossary attached to *Space: The New Frontier,* EP-6, p. 70, the term is from "retroacting."

FIRST USE. AP report describing the third stage of the upcoming *Vanguard launch:* "A delay fuse is started for third stage rocket engine and with this last 'thinking' act, the 31 foot second stage is separated and slowed slightly by retrorockets which act as brakes." ("First Ten Minutes of Flight Crucial in Launching of Moon," *Chicago Tribune,* October 26, 1957, p. 12).

Ride Report. Informal name for the NASA report *Leadership and America's Future in Space: Report to the Administrator* by Dr. Sally K. Ride (August 1987). The report examined four "bold initiatives" for the future of America in space: (1) *Mission to Planet Earth*: a program that would use the perspective afforded from space to study and characterize our home planet on a global scale. (2) Exploration of the solar system: a program to retain U.S. leadership in exploration of the outer solar system, and regain U.S. leadership in exploration of *comets*, *asteroids*, and Mars. (3) Outpost on the Moon: a program that would build on and extend the legacy of the *Apollo* program, returning Americans to the Moon to continue exploration, to establish a permanent scientific *outpost*, and to begin prospecting the Moon's resources. (4) Humans to Mars: a program to send astronauts on a series of round trips to land on the surface of Mars, leading to the eventual establishment of a permanent *base*. (Ride Report, p. 21.)

the Right Stuff. The character, courage, and savvy of test pilots and *astronaut*s. From the title of the 1979 bestseller by Tom Wolfe and the 1983 movie it inspired. Wolfe's definition included a political sense in the case of Gordon Cooper, who persuaded his estranged wife to end their separation to serve the cause of his career as an astronaut (Wolfe, *Right Stuff*, pp. 146–47).

USAGE. This may be the primary honorific of the *Space Age*—invoked sparingly in its original sense (as when applied to the crew killed in the *Challenger* and *Columbia* disasters) but more commonly as a public relations buzzword, as in a NASA press release of September 16, 2002: "The Right Stuff for Super Spaceships: Tomorrow's Spacecraft Will Be Built Using Advanced Materials with Mind-Boggling Properties."

ETYMOLOGY / FIRST USE. The earliest use of the term in the modern sense appears in a biographical sketch of Edward Preble by James Fenimore Cooper ("Sketches of Naval Men: Edward Preble," *Graham's American Monthly Magazine of Literature, Art and Fashion,* May 1845, p. 205). In attesting to Preble's "courage, determination and high temper," Cooper relates an anecdote in which Preble throws stones at a boating party that included his father, General Jedediah Preble, who had promised to board him but then later refused. "It seems the old general decided that the boy had the 'Right Stuff' in him, and overlooked the gross impropriety of the assault, on account of its justice and spirit." The term has also been used in this sense to describe racehorses with the ability to run. A horse named Miracle Sub is described in the *Chicago Defender* as having "Right Stuff" (Bettor's Edge tip sheet, *Chicago Defender,* July 9, 1975, p. 27). As resurrected by Tom Wolfe to describe the mystique of flying associated with test pilots and the first men in space, the phrase is now inextricably linked with the word astronaut. Wolfe first used it as the subtitle for a series of four 1973 articles in *Rolling Stone* magazine: "Post-Orbital Remorse—The Brotherhood of The Right Stuff." For more information, see "What Is 'The Right Stuff?' Name for Space Heroism Has Become Part of American Lexicon," *Daily Press* (Newport News, VA), May 10, 2003, p. D1.

Robonaut. Humanoid robot under long-term design by the Robot Systems Technology Branch at NASA's *Johnson Space Center* in a collaborative effort with *DARPA*. The Robonaut project seeks to develop and demonstrate an agile tool-using robotic system that can function with the dexterity to work as an *EVA astronaut* equivalent. The initial goal would be a machine capable of lifting some of the *spacewalk* burden from residents of the *International Space Station*. (http://robonaut.jsc.nasa.gov/.)

robotic. Term used to distinguish a *mission* without human operators. Like *uncrewed* or *unpiloted,* a preferred synonym for *unmanned*.

robotic explorer. Vehicle used to collect and return data from space or the surface of a celestial body.

Rockair. A high-altitude sounding system consisting of a small solid-*propellant* research *rocket* carried aloft by an aircraft. The rocket is fired while the aircraft is in vertical ascent. (SP-7.)

rocket. (1) A jet-propulsion device powered by solid or liquid *propellants* that provide the fuel and oxidizer required for combustion. (2) A vehicle that can operate outside the Earth's *atmosphere*.
USAGE. In the days when the Army and Air Force were vying with one another for control of space, the Army used the term *rocket* for the same thing the Air Force called a *missile*.

rocket engine. Also, rocket motor. A reaction *engine* that contains both fuel and oxidizer and can therefore be operated in the absence of air.

Rocket Ranch. *Cape Kennedy,* especially during the *Apollo* years. Headline in the *Roanoke Times and World News* (April 29, 1997, p. C4): "Apollo 13 Astronaut Lectures at Virginia Tech–Haise Talked about the 'Rocket Ranch.'"

rocket scientist. Any scientist or engineer working to put objects into space.
FIRST USE. The term saw post-*Sputnik* use as early as October 10, 1957, in a *New York Times* article referring to the Army team in Huntsville. Later, Wernher von Braun was often identified as America's leading rocket scientist. Prior to Sputnik the term was used as a descriptor for science fiction figures in white lab coats brandishing slide rules.
USAGE. Seldom used in its literal sense, the term is commonly invoked as a measure of intelligence (or lack thereof), as in the construction, "You don't have to be a rocket scientist to . . ." or, by extension, "It ain't rocket

science." This usage probably comes from an oft-quoted full-page ad run in the *New York Times* in 1957, which quoted an October 15 editorial from *Missiles and Rockets* addressed to President Eisenhower. The editorial ends, "You don't have to be a scientist, Mr. President, to solve this problem. You must be a leader." The word rocket is not used, but rockets are the subject, and space lore recalls that this is where the popular usage originated.

Many space historians object to the term as a misnomer. Michael J. Neufeld, Curator at the Smithsonian Air and Space Museum, had this to say in an e-mail to the author on October 29, 2005:

"I think 'rocket science' rests on a fallacious public understanding of the roles of engineering and science, in large part because the public and the media conflate the two under the label 'science.' Journalists are among the worst offenders in continuing to keep this alive. (Witness the constant references to 'NASA scientists' even when often every single person they are referring to is an engineer.) Although the French historian of science and technology Bruno Latour has argued that the two have merged into 'technoscience,' I still find it useful to think about it as range or a spectrum. On the one end is 'basic science' (once known by the loaded term 'pure science'), which is about the search for knowledge about the universe without any regard for practical application; on the other end is engineering or technology, which is about building practical things for service to humans in this world. In the real world, a lot of scientific and engineering work is in between.

"For decades the so-called 'linear model' of science has continued to flourish among scientists and engineers and in the public and news media, namely that science creates knowledge, and applied scientists and engineers convert it to technology. Research has shown this to be grossly simplistic and often just plain wrong. Often scientific understanding comes after the technology has already been invented, or there is a two-way exchange. In rocketry, the development of the practical technology has been done almost entirely by engineers, sometimes with a little scientific theory as a starting point, but often not. In the case of propulsion in particular, the science usually follows later after an empirical phase of invention. The terms 'rocket science' and 'rocket scientists' are therefore wrong and help to perpetuate public misunderstanding of 'rocket engineering' and 'rocket engineers' and engineering and science generally.

"A final unrelated comment is that what you said about 'rocket science' being disparaging is I think completely wrong. It is used to disparage others and other occupations, on the fundamental assumption that 'rocket science' is something really difficult, making those people the smartest people around. Yet it has also been adopted as positive term by practitioners themselves and by the media, although again, most are actually engineers."

rocket sled. A sled that runs on a rail or rails and is accelerated to high velocities by a *rocket engine*. In the early days of human spaceflight such sleds were used in determining *g-tolerance*s and for developing crash survival techniques. (SP-7.)

rockoon. A high-altitude sounding system consisting of a small solid-*propellant* research *rocket* carried aloft by a large plastic balloon. The *rocket* is triggered from the balloon 10–15 miles (16–24 km) above the Earth. Rockoons were used extensively for research during the *International Geophysical Year*. On September 6, 1957, University of Iowa scientists led by Dr. James Van Allen announced that a rockoon had achieved a record altitude of 80 miles (129 km) on August 10, the first known rocket flight through the "visible aurora" ("Rockoon Sets Mark," *New York Times,* September 7, 1957, p. 25).

ETYMOLOGY. Rocket + balloon. Homer Newell attributes the invention and the name to James A. Van Allen. On September 9, 1954, "Van Allen reported that Rockoon flights in the Atlantic had established the existence of a soft radiation in the auroral zone at about 30 miles (50 kilometers) height, which proved to be one of the milestones along the investigative track that ultimately led to the discovery of the Earth's radiation belt." (Newell, *Beyond the Atmosphere,* SP-4211, p. 39.)

FIRST USE. "July 29, 1952. First Rockoon (balloon–launch rocket) launched from icebreaker Eastwind off Greenland by Office of Naval Research group under James A. Van Allen. Rockoon low-cost technique was conceived during Aerobee firing cruise of the Norton Sound in March 1949, and was later used by ONR and University of Iowa research groups in 1953-55 and 1957, from ships at sea between Boston and Thule, Greenland." (Emme, *Aeronautics and Astronautics,* p. 63.)

Rogers Commission. See *Challenger Commission*.

ROSAT. Roentgen satellite. German x-ray *satellite* observatory. NASA is a junior partner.

Rosetta Mission. A *European Space Agency* spacecraft launched in 2004. In 2014 it will orbit *Comet* 67P/Churymov-Gerasimenko, and a small *lander* will attempt a landing.

Rossi X-ray Timing Explorer (RXTE). A *satellite* that observes the fast-moving, high-energy worlds of *black hole*s, *neutron star*s, x-ray *pulsar*s, and bursts of x-rays that light up the sky and then disappear forever.

ROTV. Reusable orbital transfer vehicle. Low *orbit* to *geosynchronous* orbit.

rover. Informal name for the piloted *Lunar Roving Vehicle* (LRV) as well as other robotic vehicles that explore extraterrestrial surfaces.

RSA. Russian Space Agency.

S

safing. Process by which a *space vehicle* is rendered safe after it has returned to Earth, a process that often involves the removal of residual *propellant*s.

Salyut (Russian for salute or fireworks). Series of *space stations* launched by the Soviet Union in the 1970s and 1980s. The Salyuts were all relatively simple structures consisting of a single main *module* placed into *orbit* in a single *launch*. The program was originally designated the DOS 7-K program, with each Salyut station receiving a designation.

San Marco. U.S.-Italian *satellite* project. San Marco, the Italian *space program*, was conceived in 1960 by Professor Luigi Broglio, Professor Carlo Buongiorno, and Dr. Franco Fiario. By 1962 they and their colleagues had decided that an ocean platform in nonterritorial waters should serve as the base for launching their satellite *booster*. ENI, Italy's state-owned oil industry, made available a suitable platform, which happened to be named San Marco.

The San Marco project was a cooperative effort of NASA and the Italian Space Commission, with NASA providing *launch vehicles*, use of its facilities, and training of Italian personnel. On December 15, 1964, the San Marco *Scout* 1 booster, carrying the Italian-designed and built San Marco 1 satellite, was launched from Wallops Station by an Italian crew. The launch was the first satellite launch in NASA's international cooperation program conducted by non-U.S. personnel, and the first Western European satellite launch. San Marco 2 was launched into equatorial *orbit* on April 26, 1967 from the San Marco platform in the Indian Ocean. San Marco 3, launched April 24, 1971, was the third satellite launched from the platform (the second had been NASA's *Explorer* 42, launched December 12, 1970). San Marco 4 was launched from the platform February 18, 1974.

The San Marco satellites were scientific satellites designed to conduct air-density experiments using a variety of instruments. In addition, San Marco 1 and 2 measured ionospheric characteristics related to long-range radio transmission.

ETYMOLOGY. The name grew into the designation for the entire cooperative space project including preparatory phases not associated directly with the sea-based launch site. Professor Broglio was particularly pleased to adopt the name for the project because Saint Mark was the patron saint of Venice, his birthplace. Saint Mark was also the patron saint of all who sailed the sea.

SAO. *Smithsonian Astrophysical Observatory.*

SAROS. Satellite de Radiodiffusion pour Orbit Stationnaire. See *Symphonie.*

SARSAT. Search and Rescue Satellite. System inaugurated in 1982 that detects and locates transmissions from emergency beacons carried by ships, aircraft, and individuals.

SAS. (1) Small Astronomy Satellite. See *Explorer.* (2) Space adaptation syndrome. See *space sickness.*

-sat. Suffix meaning *satellite* used in constructions like *comsat* (communications satellite), *Landsat* (a satellite that looks at the Earth), and spysat.

satellite. (1) A natural or an artificial moon; an object *orbit*ing around another, larger one. Smaller bodies orbiting around planets are called those planets' satellites. (2) Since 1957, any man-made object in orbit around the Earth. Because they orbit the Earth, these objects are Earth satellites. By this definition, orbiting piloted *spacecraft* are also satellites of the Earth. Other satellites, in the strict sense of the word, are the spent *rocket* stages and uninstrumented pieces of hardware (popularly called space junk) placed in orbit incidentally. (3) Any man-made instrumented object placed intentionally in Earth orbit to perform specific functions associated with the space exploration program. **USAGE.** A glossary created by the *Jet Propulsion Laboratory* makes this distinction: "Earth-orbiting spacecraft are called satellites. While deep-space vehicles are technically satellites of the Sun or of another planet, or of the galactic center, they are generally called spacecraft instead of satellites. NASA *unpiloted* satellites have traditionally divided into two categories: scientific satellites (which obtain scientific information about the space environment) and applications satellites (which perform experiments that will have everyday usefulness for humans)." (SP-4402, p. 29; ASP Glossary.)

Satellite de Radiodiffusion pour Orbit Stationnaire (SAROS). See *Symphonie.*

Satellite Italiano Ricerche Orientate. See *SIRIO.*

satelloid. Short-lived (ca. 1955–59) term for a piloted vehicle (half airplane and half *satellite*) designed to *orbit* and then return to Earth. The term was used in the context of the *Dyna-Soar* project (and would describe the *Space Shuttle,* based on its definition). The term had been used earlier to mean "resembling a satellite, as the particles composing the rings of Saturn" (*Knowledge,* January 1906, p. 230).

satelloon. Satellite balloon. The term was coined for Project Echo (1960), which was an aluminum satellite balloon 100 feet (30.5 m) in diameter.

Because of its size and reflectivity, it was seen by many and sometimes called "the most beautiful object ever to be put into space." This short-lived word had enough temporary momentum to be listed in the *1961 Britannica Book of the Year* (p. 753) as one of the significant words to come into the language in 1960.

sattellorb. Satellite-simulating observation and research balloon. Sattellorbs were proposed in 1957 by Dr. David G. Simons, head of the Air Force Office of Space Biology, as a means of training astronauts to go into space ("Training by Balloon in Space Prepared," *New York Times,* October 8, 1957, p 11).

Saturn *(launch vehicle).* Generic name for large U.S. launch vehicles related to the Apollo program. "America's hopes [rest] on the Saturn," wrote science editor William Hines in 1962 (though without distinguishing which Saturn). "The NASA clings to the Saturn for everything bigger than the current series of weapons-derived booster rockets. It is as if General Motors were to designate everything bigger than a Chevrolet by the trade name 'Cadillac.'" (Hines, *Washington Evening Star,* August 23, 1962.) See *Saturn I, Saturn IB,* and *Saturn V.* For more information, see Roger Bilstein, *Stages to Saturn: A Technological History of the Apollo/Saturn Launch Vehicles* (Washington, DC: GPO, 1980); Launius and Jenkins, eds., *To Reach the High Frontier.*

Saturn I, Saturn IB. *Launch vehicles* developed for the *Apollo* program. The Saturn IB launched the first piloted Apollo spacecraft, Apollo 7, on a successful flight on October 11, 1968, and after the completion of the Apollo program it launched three missions to man the *Skylab* Orbital Workshop in 1973. On July 15, 1975, it was used to launch the American crew in the U.S.-Soviet *Apollo-Soyuz Test Project docking* mission.

The nomenclature for the Saturn family of launch vehicles had one of the most complex evolutions of all NASA-associated names. On August 15, 1958, the Department of Defense's *Advanced Research Projects Agency* (ARPA) approved initial work on a multistage launch vehicle with clustered engines in a first stage developing 1.5 million pounds (6.7 million newtons) of *thrust.* Conceived by designers at the Army Ballistic Missile Agency *(ABMA),* the vehicle was unofficially known as *Juno* V. (Juno III and Juno IV were concepts for *space vehicles* to follow Juno II but were not built.) In October 1958 Dr. Wernher von Braun, Director of ABMA's Development Operations Division, proposed that the Juno V be renamed Saturn, and on February 3, 1959, ARPA officially approved the name change. The name Saturn was significant for three reasons: the planet Saturn appeared brighter than a first-magnitude star, so the association of this name with such a powerful new *booster* seemed appropriate; Saturn was the next planet after Jupiter, so the progression was analogous to ABMA's progression from missile and space systems called Jupiter; and Saturn was the name of an ancient Roman god, so the name was in keeping with the U.S. military's custom of naming missiles after mythological gods and heroes.

Throughout the second half of 1959, studies were made of possible *upper stage*s for the new Saturn vehicle. The interagency Saturn Vehicle Evaluation Committee considered many combinations, narrowing the choice to design concepts labeled Saturn A, Saturn B, and Saturn C. In December 1959, following the recommendation of this committee, NASA authorized building 10 research and development models of the first C version, or Saturn C-1 design proposal. For the time being the booster was called Saturn C-1.

In the meantime, Saturn became a NASA project and also had become an important link with the nation's lunar landing program, Project Apollo. In 1962 NASA decided that a more powerful version of the Saturn C-1 would be needed to launch Apollo lunar spacecraft into Earth orbit, to prepare and train for human flights to the Moon later in the 1960s. NASA called this launch vehicle Saturn C-1B. In February 1963 NASA renamed these vehicles. At the suggestion of the NASA Project Designation Committee, Saturn C-1 became simply Saturn I, and the Saturn C-1B became Saturn IB. The Saturn IB was composed of the S-IB first stage, a modified version of the S-I first stage that could develop 1.6 million pounds (7.1 million newtons) of *thrust* by 1973, and the S-IVB second stage an uprated version of the S-IV stage that could develop 230,000 pounds (1 million newtons) of thrust.

On June 9, 1966, NASA changed the name of the Saturn IB to Uprated Saturn I. The redesignation was suggested to the Project Designation Committee by Dr. George E. Mueller, NASA Associate Administrator for Manned Space Flight. The committee agreed with Dr. Mueller, but in December 1967 NASA decided to return to the use of the simpler term, Saturn IB. The proposal was made by the Office of Manned Space Flight and approved by Administrator James E. Webb.

SOURCES. SP-4402, pp. 17–19; Wernher von Braun, "The Redstone, Jupiter, and Juno," in *The History of Rocket Technology,* ed. Eugene M. Emme (Detroit: Wayne State University Press, 1964), p. 119; Abe Silverstein et al., "Report to the Administrator," NASA, on Saturn Development Plan by Saturn Vehicle Team, December 15, 1959; MSFC, Historical Office, *Saturn Illustrated Chronology,* MHR-4 (Huntsville, AL: MSFC, 1965), pp. 8–9, 56, 69; MSFC, Historical Office, *History of the George C. Marshall Space Flight Center:*

January 1–June 30, 1962, MHM-5, vol. 1 (Huntsville, AL: MSFC, 1962), 28; George L. Simpson Jr., Assistant Administrator for Public Affairs, NASA, memorandum for the Associate Administrator, NASA, January 7, 1963.

Saturn V. The *launch vehicle* that took Americans to the Moon from 1969 to 1972. In January 1962 NASA initiated development of the large launch vehicle for Project *Apollo* lunar flight. The vehicle selected was the Saturn C-5, chosen after six months of studying the relative merits of the Saturn C-3, C-4, and C-5 designs. These designs were all based on a large clustered-engine first stage but with various combinations of *upper stage*s. The numerical designation followed the sequence established with the Saturn C-1.

Alternately referred to in 1962 as Advanced Saturn, the Saturn C-5 was renamed early the following year. Nominations were submitted to the NASA Project Designation Committee as well as proposed by the committee members themselves. For a while the leading contender was Kronos. The committee suggested, through Assistant Administrator for Public Affairs George L. Simpson Jr., to NASA Associate Administrator, Dr. Robert C. Seamans Jr., that the new name be Saturn V. The recommendation was approved and the new name adopted early in February 1963.

The final configuration of the Saturn V comprised the S-IC first stage, with 7.7 million pounds (34 million newtons) of *thrust;* the S-II second stage, with 1.2 million pounds (5.1 million newtons) of thrust; and the S-IVB stage of the Saturn IB.

On December 21, 1968, the Saturn V launched Apollo 8, the first *crewed* Apollo spacecraft to escape the Earth's gravitational field, into flight around the Moon. Saturn V launches through Apollo 17 in December 1972 put 27 men into lunar orbit, 12 of them landing on the Moon to explore its surface. On May 14, 1973, the Saturn V launched the first U.S. experimental *space station,* the *Skylab* 1 Orbital Workshop, which was piloted by three successive three-man crews during the year.

SOURCES. SP-4402, pp. 19–20; MSFC, Historical Office, *History of the George C. Marshall Space Flight Center: July 1–December 31, 1961,* MHM-4, vol. 1 (Huntsville, AL: MSFC, 1962), p. 33; George L. Simpson Jr., Assistant Administrator for Public Affairs, NASA, memorandum for the Associate Administrator, NASA, January 7, 1963; MSFC, Historical Office, *Saturn Illustrated Chronology,* MHR-4 (Huntsville, AL: MSFC, 1965), p. 69; NASA News Release 72-220K.

SCORE. See *Project SCORE.*

Scott. Deep Space 2 microprobe named for English explorer Robert Falcon Scott. See *Mars Polar Lander.*

Scout *(launch vehicle)*. Smallest of the basic launch vehicles. Scout was designed at NASA's Langley Research Center as a reliable, relatively inexpensive launch vehicle for high-altitude *probe*s, *reentry* experiments, and small-*satellite* missions. Among the satellites it *launch*ed were scientific satellites such as *Explorer*s and international satellites such as the *San Marco* series. It was the only U.S. satellite launch vehicle to use solid *propellant*s exclusively. The stages for Scout had grown out of the technology developed in the Polaris and Minuteman programs. The Air Force, which used Scout to launch Department of Defense *spacecraft,* called its version *Blue Scout.*

Scout usually consisted of four stages and could put 410 pounds (186 kg) into a 345-mile (555-km) *orbit.* The first stage, *Algol,* was named for a star in the constellation Perseus; the second stage, *Castor,* for the "tamer of the horses" in the constellation Gemini; the third stage, *Antares,* for the brightest star in the constellation Scorpius; and the fourth stage, *Altair,* for a star in the constellation Aquila. In June 1974 a new Scout E, incorporating a solid-fueled *rocket* motor in a fifth stage and adaptable for highly *eccentric* orbits, launched the Hawkeye 1 Explorer satellite.

ETYMOLOGY. The Scout launch vehicle was named in mid-1958 by William E. Stoney Jr., prominent in development of the vehicle at *NACA* Langley Aeronautical Laboratory (later NASA *Langley Research Center*). He thought of the name as a parallel to *Explorer,* the name then being given to a series of spacecraft. Scout seemed appropriate for a vehicle with payloads performing similar tasks while "scouting the frontiers of space environment and paving the way for future space exploration."

SOURCES. SP-4402, pp. 20–21; NASA News Releases 75-19, 74-138; NASA, Wallops Station, Open House Program, September 29–30, 1963; Paul E. Goozh, Scout Program Manager, NASA, telephone interview, June 17, 1974.

SCRAMJET/scramjet. Supersonic Combustion Ramjet. A supersonic combustion ramjet *engine* that can operate in the hypersonic region of space. It is a variation of a ramjet in which combustion of the fuel-air mixture occurs at supersonic speeds. This allows the scramjet to achieve greater speeds than a conventional ramjet, which slows the incoming air to *subsonic* speeds before entering the combustion chamber. Projections for the top speed of a scramjet engine vary between Mach 12 and Mach 24. (Paine Report, p. 199.)

scrub. To cancel a *mission* before or during *countdown*. To *de-scrub* is to reverse the scrub decision.

ETYMOLOGY. Use of the term in the sense of to cancel is attested from 1828 (popularized during World War II with reference to flights), probably from the notion of "to rub out, erase." The *OED* cites use of the word in the modern sense: "1945 *Spectator* May 25, 478/1. The author can possibly justify the inclusion of the term 'scrub', meaning 'to cancel', in a collection of R.A.F. slang. The expression is in common use in the Royal Navy and has been for many generations. It derives from the days when all signals and orders were written on a slate. When the signals were cancelled or orders executed, the words on the slate were 'scrubbed out' or, equally correctly, 'washed out.'" (NASA Names File, record no. 17542.)

scrub club. A program beset with many failures ("Space Age Slang," *Time* magazine, August 10, 1962, p. 12).

séance. Term used by the *Original Seven* astronauts to describe meetings at which they would lock themselves in a room in order to come to a consensus on an issue. Walter M. Schirra Jr. explained: "We called a session like this a 'séance'—because some people thought we were acting like swamis in there, I suppose, and were pulling answers out from under the table. But once we had a séance, we would really bear down, no matter who was against us—even if it happened to be one of our top bosses." (Schirra, "Some Séances in the Room," in Carpenter et al., *We Seven*, p. 90.)

Search for Extra-Terrestrial Intelligence. See *SETI*.

SEASAT. Specialized Experimental Applications Satellite. In 1969 a conference of representatives from the National Oceanic and Atmospheric Administration, the Department of Defense, NASA, other government agencies, universities, and scientific institutions met at Williams College in Williamstown, Massachusetts, to review activities needed in the Earth and ocean physics fields. The conference identified a number of needs, including *satellite* projects. SEASAT and *LAGEOS* were among them, the names growing out of the thinking of a number of the participants and fitting the tasks of the satellites within NASA's Earth and Ocean Physics Applications Program (EOPAP).

After studies and definition of requirements in cooperation with numerous government agencies and private institutions, through the SEASAT User Working Group, NASA introduced SEASAT as a "new start" in its fiscal year 1975 program. The new satellite was launched on June 28, 1978. SEASAT was the first Earth-orbiting satellite designed for *remote sensing* of the Earth's oceans and had on board the first *spaceborne* synthetic aperture radar (SAR).

SOURCES. SP-4402, p. 43; Francis L. Williams, Director of Special Programs, Office of Applications, NASA, telephone interview, June 9, 1975; U.S. Congress, House of Representatives, Committee on Science and Astronautics, Subcommittee on Space Science and Applications, *Hearings: 1975 NASA Authorization, Pt. 3, February and March 1974* (Washington, DC: GPO, 1974), pp. 3-4, 270–71; NASA News Release 75–1; http://southport.jpl.nasa.gov/scienceapps/seasat.html.

SEI. *Space Exploration Initiative.*

self-destruct. To cause a *rocket* or *missile* to destroy itself under a pre-defined set of circumstances. Typically, self-destruct signals activate small explosive devices designed to deliberately destroy an errant rocket before it can jeopardize public safety.

SEOS. Synchronous Earth Observatory Satellite. An advanced study under way in 1974 of a synchronous satellite for experimental meteorological and Earth resources observations using a large telescope with improved resolution and an infrared atmospheric sounder. The *geosynchronous orbit* would provide the short intervals needed to detect and warn of natural disasters such as hurricanes, tornadoes, forest fires, foods, and insect damage to crops. (SP-4402, p. 45; U.S. Congress, House of Representatives, Committee on Science and Astronautics, Subcommittee on Space Science and Applications, *Hearings: 1975 NASA Authorization, Pt. 3, February and March 1974* [Washington, DC: GPO, 1974], p. 50.)

separation. In multistage *launch vehicles*, the action, time, or point in space at which a burned-out stage is discarded and the remaining *missile* continues on its course.

Sergeant (*launch vehicle* first stage). See *Thor.*

Service Module (SM). See *Command and Service Module.*

SETI. Search for Extra-Terrestrial Intelligence. A series of projects usually using radio telescopes to search for signals from extraterrestrial intelligence. The first such project was Frank Drake's *Project Ozma* at the Green Bank Observatory in 1960. NASA had its own SETI project, which received small amounts of funding beginning in the 1970s and moderate funding through the 1980s and early 1990s, but it was terminated by Congress in 1993 less than one year after it became operational. SETI projects are continued at several locations around the world, including those run by the SETI Institute in Mountain View, California. See also *astrobiology, CETI, exobiology,* and *Project Ozma.*
ETYMOLOGY. According to Charles Seeger, "We did not invent the name SETI until 1975 in the Morrison workshops, when we realized we needed an acronym. We choose SETI instead of the older CETI (Communications with ETI) to emphasize that we did not intend to transmit to ETI, but just to detect any radio signals they might transmit." (Seeger interviewed in David W. Smith, *SETI Pioneers: Scientists Talk about Their Search for Extraterrestrial Intelligence* [Tucson: University of Arizona Press, 1990], p. 286.) The Morrison workshops were meetings on SETI under the chairmanship of Philip Morrison. The discussions and results were published in P. Morrison,

J. Billingham, and J. Wolfe, eds., *The Search for Extraterrestrial Intelligence,* SP-419 (Washington, DC: GPO, 1977).

FIRST USE. "It is significant that the National Aeronautics And Space Administration has evolved from CETI to SETI. The former acronym stood for 'communication with extraterrestrial intelligence,' whereas the title now refers to the 'search' for such intelligence" (*New York Times,* November 4, 1976, p. 46). For more information about SETI, see Steven J. Dick, *The Biological Universe: The Twentieth Century Extraterrestrial Life Debate and the Limits of Science* (New York: Cambridge University Press, 1996).

settlement. A permanent community of humans in space, or on the surface of the Moon or Mars with life support, living quarters, and work facilities. Settlements would evolve from *bases*. See also *outpost.* (Paine Report, p. 199.)

shot. A *rocket* flight. Informal and popular term for *launch.*

FIRST USE. Defined in 1945 in the lexicon attached to G. Edward Pendray's *The Coming Age of Rocket Power.*

Shotput *(launch vehicle).* A special-purpose composite *rocket* designed to test balloon-*satellite* ejection and inflation in space. Shotput was used in five launches from Wallops Station in 1959 and 1960 in tests of the *Echo* 1 satellite *payload.* It was also used to test the Italian *San Marco* satellite in suborbital flights. The solid-*propellant* Shotput vehicle consisted of a first-stage *Sergeant* rocket boosted by two Recruit rockets and a second-stage Allegany Ballistics Laboratory X-248 rocket originally designed for the *Vanguard* and *Thor-Able* vehicles and later used as the third stage of the *Delta* launch vehicle. Shotput launched the balloon payload to an altitude of 250 miles (400 km), where the packaged sphere was ejected from the vehicle's nose and inflated above the *atmosphere.* (SP-4402, p. 21; NASA News Releases 60-158, 60-186; William J. O'Sullivan Jr., Head of the Space Vehicle Group, LaRC, letter to Don Murray, September 29, 1960.)

ETYMOLOGY. Shotput was so named because it "tossed" the Echo sphere up above the Earth's atmosphere in a vertical *trajectory.*

the Shuttle. Common name for the *Space Shuttle.*

Shuttle-Mir. Collaborative *space program* involving Russia and the United States that involved the American *Space Shuttle* visiting the Russian *space station Mir.* It was announced in 1993 with the first *mission* occurring in 1995. Despite safety concerns from the Americans, the program continued until its scheduled completion in 1998. The program was an important precursor to the current *International Space Station.*

Sigma 7. *Mercury mission* of October 3, 1962, on which Walter M. Schirra Jr. flew six orbits.

ETYMOLOGY. Named by Schirra, who explained, "Sigma means 'sum of'...
a mathematical term. I wanted to get off the 'Gee Whiz' names and use a
technical/test pilot term as well as acknowledge the original 7. I also toyed
for a while with the names Phoenix and Pioneer, but settled on Sigma
because the flight was the sum of the efforts and energies of a lot of
people." (Quoted in Lattimer, *All We Did Was Fly to the Moon*, p. 15.)

Silverstein Committee. Popular name for the *Saturn* Vehicle Evaluation
Committee chaired by Abe Silverstein.

SIRIO. Satellite Italiano Ricerche Orientate (Italian Research-Oriented
Satellite). The first *satellite* planned for *launch* under a 1970 agreement between the United States and Italy. In March 1970 NASA and
the Italian National Research Commission signed a memorandum
of understanding providing for the reimbursable NASA launch of
Italian scientific *spacecraft*. (SP-4402, pp. 68–69; NASA News Release
70-42.)

sitting fat. *A-OK.* Early colloquialism for a piloted vehicle in *orbit* with all
systems functioning.

ETYMOLOGY. Possibly derived from "fat city," period slang for success,
wealth.

FIRST USE. Defined in "Space Age Slang," *Time* magazine, August 10, 1962,
p. 12. In context it next shows up in "Daddy Lipscomb Wants to Be Next
Astronaut" (*Chicago Daily Defender*, August 16, 1962, p. 24), an article
about a very large tackle for the Baltimore Colts wanting to go into space.
The article mentions that Lipscomb might be cramped for space and adds,
"In a chair he would be 'sitting fat,' space age slang for being successful in
orbit."

skyhook. (1) A term applied in 1966 by four American scientists to the
idea of building a space tower that reached from the Earth's surface to
geosynchronous orbit. A practical idea for a skyhook came in 1973
when Mario Grossi suggested flying a very long wire unreeled from the
Space Shuttle. The wire would act as an antenna that could emit
extremely low frequency radio waves. That notion led to the idea of.
flying a small *satellite* at the end the wire. With this small satellite the
Shuttle could troll the *upper atmosphere* (where the Shuttle couldn't
fly) and collect data. (2) An unpiloted balloon used by the U.S. Navy in
the 1940s and 1960s.

Skylab. Orbital Workshop program; Project Skylab. America's first *space
station*, the 165,000-pound (75-metric-ton) Skylab, was in Earth *orbit*
from 1973 to 1979 and was visited by crews three times in 1973
and 1974.

Planning for post-*Apollo* human spaceflight missions evolved
directly from the capability produced by the Apollo and *Saturn*

technologies, and Project Skylab resulted from the combination of selected program objectives. In 1964, design and feasibility studies had been initiated for missions that could use modified Apollo hardware for a number of possible lunar and Earth-orbital scientific and applications missions. The study concepts were variously known as Extended Apollo (Apollo X) and the Apollo Extension System (AES). In 1965 the program was coordinated under the name Apollo Applications Program (AAP) and by 1966 had narrowed in scope to primarily an Earth-orbital concept.

Projected AAP missions included the use of the Apollo Telescope Mount (ATM). In one plan it was to be launched separately and docked with an orbiting workshop in the "wet" workshop configuration. The wet configuration, using the spent S-IV B stage of the Saturn I *launch vehicle* as a workshop after purging it in orbit of excess fuel, was later dropped in favor of the "dry" configuration using the Saturn V launch vehicle. The extra fuel carried by the S-IV B when used as a third stage on the Saturn V, for Moon launches, would not be required for the Skylab *mission,* and the stage could be completely outfitted as a workshop before *launch,* including the ATM.

Skylab 1 (SL-1), the Orbital Workshop with its Apollo Telescope Mount, was put into orbit on May 14, 1973. Dynamic forces ripped off the *meteoroid* shield and one solar array wing during launch, endangering the entire program, but the three astronauts launched on Skylab 2 (SL-2), the first piloted *mission* to crew the Workshop, were able to repair the *spacecraft* and completed 28 days living and working in space before their safe return. They were followed by two more three-man crews during 1973. The Skylab 3 crew spent 59 days in space, and Skylab 4 spent 84. Each successive Skylab mission became the longest-duration piloted space flight to that date, also setting distance-in-orbit and *extravehicular activity* records. Skylab 4, the final mission (November 16, 1973–February 8, 1974), recorded the longest in-orbit *EVA* (7 hours 1 minute), the longest cumulative orbital EVA time for one mission (22 hours 21 minutes in four EVAs), and the longest distance in orbit for a piloted mission (35 million miles, or 55.5 million km).

The Skylab missions proved that human beings could live and work in space for extended periods. The missions also expanded solar astronomy beyond Earth-based observations, collecting new data that would revise our understanding of the Sun and its effects on the Earth, and they returned much information from surveys of Earth resources with new techniques. The Skylab came back to Earth in 1979 in pieces, creating a great media event.

ETYMOLOGY. The name Skylab, a contraction connoting laboratory in the sky, was suggested by Lt. Col. Donald L. Steelman (USAF) while assigned to NASA. He later received a token reward for his suggestion. Although the name was proposed in mid-1968, NASA decided to postpone renaming the program because of budgetary considerations. The name was later referred to the NASA Project Designation Committee and was approved February 17, 1970.

SOURCES. SP-4402, pp. 109–11; NASA, Historical Staff, *Astronautics and Aeronautics, 1964: Chronology on Science, Technology, and Policy,* SP-4005 (Washington, DC: NASA, 1965), pp. 145, 363, and *Astronautics and Aeronautics, 1965,* SP-4006 (1966), pp. 174, 418, 429; U.S. Senate, Committee on Aeronautical and Space Sciences, *Hearings: NASA Authorization for Fiscal Year 1967, February and March 1966* (Washington, DC: GPO, 1966), pp. 163–66, 238–39.

SL-1. Better known as the R-7. The world's first intercontinental ballistic missile (ICBM) and the vehicle that launched *Sputnik.* A single-stage *rocket,* it used four strap-on *boosters* around the central core *engine.*

SLC-6. Space *Launch Complex* 6, located at Vandenberg Air Force Base in California. Originally built for the Manned Orbiting Laboratory in 1966, it was abandoned after three years and brought back on line a decade later when it was modified to fit the needs of the *Space Shuttle* program. Persistent site technical problems, however, and a joint decision by the Air Force and NASA to consolidate Shuttle operations at Cape Canaveral in Florida, following the *Challenger* tragedy in 1986, resulted in the official termination of the Shuttle program at Vandenberg on December 26, 1989. Today SLC-6 is used by commercial space launch firms and for launches of the new Delta IV EELV (Evolved Expendable *Launch Vehicle*) booster.

slingshot/slingshot effect. (1) Effect produced by lunar gravity on a *spacecraft* passing behind the Moon in its orbital path, which whips it clear of the Moon back toward Earth. "The three men on Apollo 11 had to decide whether to allow themselves to be 'captured' by lunar gravity, take the slingshot and come home" (Armstrong et. al., *First on Moon,* p. viii). (2) Any space flight that uses the gravitational pull of a celestial body in order to accelerate sharply and change course.

Small Astronomy Satellite (SAS). See *Explorer.*

smallsat. Mini-*satellite.* When NASA Administrator Daniel S. Goldin announced in 1994 two Earth-scanning satellites no bigger than a television set, costing less than $60 million apiece, they were dubbed smallsats ("For NASA 'Smallsats,' a Commercial Role," *Washington Post,* June 9, 1994, p. A7).

SMART 1. *European Space Agency* orbiter *mission* to the Moon, 2003.

Smithsonian Astrophysical Observatory (SAO). A "research institute" of the Smithsonian Institution located in Cambridge, Massachusetts. It is joined with the Harvard College Observatory (HCO) to form the Harvard-Smithsonian Center for Astrophysics.

SMS. Synchronous Meteorological Satellite. An operational *satellite* system that could provide continuous observation of weather conditions from a fixed position above the Earth had been under study since the first weather satellites were launched in the early 1960s. Studies of the requirements for a stationary weather satellite were begun in early 1960, and the proposed project was named for Aeros, ancient Greek god of the air. Conceived as the third phase of a program consisting of the Tiros (see *TIROS, TOS, ITOS*) and planned *Nimbus* satellites, Aeros would be a *synchronous* satellite in equatorial *orbit* that could track major storms as well as *relay* cloud-cover photographs of a large portion of the Earth.

By late 1962 the name Aeros had been dropped in favor of the more functional designation Synchronous Meteorological Satellite. Meanwhile, studies were being made of a Tiros *spacecraft* (Tiros-K) that could be modified for a near-synchronous *orbit* to determine the capability of an SMS. Tiros-K was subsequently canceled in 1965 as development plans for the *ATS* satellites permitted the inclusion of experiments to test the proposed instrumentation for the SMS.

After the successful photographic results of ATS 1 and 3, two experimental SMS satellites were approved and tentatively planned for launch. SMS-A and SMS-B, funded by NASA, would be prototypes for the later operational satellites funded by the National Oceanic and Atmospheric Administration (NOAA). Following launch and checkout by NASA, both satellites were to be turned over to NOAA for use in the National Operational Meteorological Satellite System (NOMSS). Successive satellites in the series would be designated *GOES*, for *Geostationary* Operational Environmental Satellite, by NOAA. An operational system of two or more SMS satellites and a single ITOS spacecraft could provide the coverage required for accurate long-range weather forecasts.

SMS-A became SMS 1 upon launch into on orbit May 17, 1974, and supported the international Global Atmospheric Research Program's Atlantic Tropical Experiment (GATE) before becoming part of NOAA's operational system late in the year.

SOURCES. SP-4402, pp. 69–71; NASA Ad Hoc Committee to Name Space Projects and Objects, minutes of meeting, May 19, 1960; U.S. Congress, House of Representatives, Committee on Science and Astronautics, *Hearings: National Meteorological Satellite Program, July 1961* (Washington,

DC: GPO, 1961), p. 32; U.S. Congress, Senate, Committee on Aeronautical and Space Sciences, *Hearings: NASA Authorization for Fiscal Year 1964, Pt. 1, April 1963* (Washington, DC: GPO, 1963), pp. 438–39, 441, 447; NASA News Releases 63-18, 74-95, 74-154; NASA program office.

Snoopy. *Call sign* for the *Apollo* 10 *Lunar Module* when separated from the Command Module *(Charlie Brown)*. The crew members were Col. Thomas Patten Stafford (USAF), commander; Cdr. John Watts Young (USN), Command Module pilot; and Cdr. Eugene Andrew "Gene" Cernan (USN), Lunar Module pilot.

    ETYMOLOGY. The beagle in Charles M. Schulz's comic strip *Peanuts.* The name referred to the fact that the Lunar Module would be "snooping" around the lunar surface in low *orbit*.

soft-land (v.). To effect a *soft landing.*

soft landing (n.). Descent of a *spacecraft* onto a hard surface without damaging the *craft* or its cargo.

    EXTENDED USE. This term first showed up in 1958 in the context of *unpiloted* landings on the Moon. It got applied to economics in the 1950s for a slowdown without a recession that succeeds in avoiding inflation. It has generalized from there.

SOHO. Solar and Heliospheric Observatory. NASA and *European Space Agency spacecraft,* launched 1995. Stationed as of this writing about 930,000 miles (1.5 million km) from the Earth, where it constantly watches the Sun, returning data and spectacular pictures of the storms that rage across its surface. SOHO's studies include the Sun's hot interior through its visible surface and stormy *atmosphere*. (www.esa.int/esaSC/120373_index_0_m.html.)

Sojourner. Name of the first *rover* on Mars. This rover, which landed on July 4, 1995, was deployed from the *Mars Pathfinder* and was the first rover sent to another planet. During its 83 Martian days, or *sols*, of operation, it sent 550 photographs to Earth and analyzed the chemical properties of sixteen locations near the *lander.*

    ETYMOLOGY. Named by Valerie Ambrose, then 12, of Bridgeport, Connecticut, in a naming contest involving more than 3,500 entries. She submitted the winning essay about Sojourner Truth, an African-American reformist who lived during the Civil War era. An abolitionist and champion of women's rights, Sojourner Truth made it her mission to "travel up and down the land," advocating the right of all people to be free and the right of women to participate fully in society. The name Sojourner was selected because it means traveler. (NASA News Release 95–112, "NASA Names First Rover to Explore the Surface of Mars.")

sol. One day on Mars. This term was created to distinguish a Martian day from the *terrestrial* day, which is 39 minutes shorter. According to an

AP article, "More abstract notions, like time, also have their own names to help scientists and engineers remember where they're working, 150 million miles away. 'Yestersol'—Martian for yesterday—is one example. On Mars, a day . . . is called 'sol'. . . 'Have a good sol,' 'tosol' and 'morrowsol' are all in varying degrees of use." (Andrew Bridges, "Mars Missions Spawns Its Own Unworldly Lingo," *Chico Enterprise,* February 23, 2004, p. A4.)

Solar and Heliospheric Observatory. See *SOHO.*

solar cycle. The roughly 11-year, quasi-periodic variation in the frequency or number of sunspots, solar flares, and other solar activity.

Solar Explorer. See *Explorer.*

Solar Radiation and Climate Experiment (SORCE). A NASA-sponsored *satellite mission* that will provide state-of-the-art measurements of incoming x-ray, ultraviolet, visible, near-infrared, and total solar *radiation.*

solar wind. The charged particles (plasma), primarily protons and electrons, that are emitted from the Sun and stream outward throughout the solar system at speeds of hundreds of kilometers per second.

solid-fuel rocket. Also solid rocket. *Rocket* with a motor that uses solid *propellant* (fuel/oxidizer). The *Space Shuttle* solid rocket *booster*s are an example.

sonic. Describing speed approximately equal to that of sound in air at sea level.

Sortie Lab. See *Spacelab.*

sounding rocket. A *rocket* that carries instruments into the *upper atmosphere* to investigate its nature and characteristics, gathering data for meteorological measurements at altitudes as low as 20 miles (32 km), and data for ionospheric and cosmic physics at altitudes up to 4,000 miles (6,400 km). Sounding rockets also flight-test instruments to be used in satellites. The term is derived from the analogy to maritime soundings made of the ocean depths.

Sending measurement instruments into the high atmosphere was one of the principal motives for 20th-century rocket development. This was the stated purpose of Dr. Robert H. Goddard's rocket design studies as early as 1914. But it was not until 1945 that the first U.S. government–sponsored sounding rocket was launched—the *Wac Corporal,* a project of the *Jet Propulsion Laboratory* and U.S. Army Ordnance.

Sounding rockets played an important role in the *International Geophysical Year* (IGY), an 18-month period (July 1, 1957–December 31, 1958) coinciding with high solar activity. The IGY was an intensive investigation of the natural environment—the Earth, the oceans, and the atmosphere—by 30,000 participants representing 66 nations. More than 300 instrumented sounding rockets launched from sites around the world made significant discoveries regarding the atmosphere, the ionosphere, cosmic *radiation,* auroras, and geomagnetism.

The *International Years of the Quiet Sun* (January 1, 1964–December 31, 1965), a full-scale follow-up to the *International Geophysical Year,* was an intensive effort to collect geophysical observations in a period of minimum solar activity. Instrumented sounding rockets again played a significant role in the investigation of Earth-Sun interactions. By the end of 1974, some 20 countries had joined NASA in cooperative projects launching more than 1,700 rockets from ranges in the United States and abroad.

Sounding rocket research gave rise to three new branches of astronomy: ultraviolet, x-ray, and gamma ray. Experiments launched on sounding rockets have characterized the main features of the Earth's upper atmosphere and contributed the first recognition of the geocorona (i.e., the solar far-ultraviolet light that is reflected off the cloud of neutral hydrogen atoms that surrounds the Earth), knowledge of ionospheric chemistry, detection of electrical currents in the ionosphere, and description of particle flux in auroras. One of the earliest discoveries was of solar x-rays originating in the solar corona.

Because higher-performance sounding rockets were not economical for low-altitude experiments and lower-performance rockets were not useful for high-altitude experiments, NASA used a number of rockets of varying capabilities, including *Aerobee* and Astrobee, Arcas, Argo D-4 *(Javelin), Nike-Apache,* Nike-*Cajun,* Nike-*Hawk,* Nike-*Malemute,* Nike-*Tomahawk, Terrier-Malemute,* and *Black Brant.* A high-performance *rocket,* the Aries, was under development in 1974. Vehicles could economically place 11–2,000 pounds (5–900 kg) at altitudes up to 1,400 miles (2,200 km). Highly accurate *payload* pointing and also payload *recovery* were possible when needed.

SOURCES. Homer E. Newell, ed., *Sounding Rockets* (New York: McGraw-Hill, 1959), pp. 1–2, 28; *Space: The New Frontier,* EP-6, pp. 37–41; *Meteorological Satellites and Sounding Rockets,* EP-27 (Washington, DC: NASA, 1965), p. 17; Robert H. Goddard, *A Method of Reaching Extreme Altitudes,* Smithsonian Miscellaneous Collections 71, no. 2 (Washington, DC: Smithsonian Institution, 1919); Emme, *Aeronautics and Astronautics,* p. 51; Wallace W. Atwood Jr., *The International Geophysical Year in Retrospect,* Dept. of State Publication 6850 (Washington, DC: Dept. of State, 1959), from *Department of State Bulletin,* May 11, 1959; GSFC, *United States Sounding Rocket Program,* p. 1; NASA News Release 75-19.

Soyuz (Russian for union). (1) Soviet *spacecraft*. Longest-lived, most adaptable, and most successful *spacecraft* design in the history of *crewed* spaceflight. In production for over 30 years, more than 220 Soyuz spacecraft have been built and flown on a wide range of missions. The design will remain in use well into the 21st century with the activation of the *International Space Station*. A Soyuz spacecraft is always on standby at the International Space Station to be used as an emergency escape vehicle should the need arise. (http://aerospace scholars.jsc.nasa.gov/HAS/cirr/glossary.) (2) Soviet three-stage *launch vehicle* used for launching the Soyuz spacecraft and other missions. Introduced in 1966 and based on the Vostok launcher.

space. (1) All of the *universe* lying outside the limits of the Earth's *atmosphere*. (2) More generally, the volume in which all celestial bodies, including the Earth, move. (SP-7.)

space adaptation syndrome (SAS). See *space sickness*.

Space Age. (1, n.) The period in human history that began on October 4, 1957, with the *launch* of *Sputnik*. The term has served as a marker for progress in assertions such as this from Dr. I. M. Levitt: "We have learned more about the sun since the advent of the Space Age than in all the years leading up to 1957" (*Philadelphia Inquirer*, August 20, 1967, p. 8). Space historian Roger D. Launius has noted, "The phrase that soon replaced earlier definitions of the time was 'Space Age.' With the launch of Sputnik 1, the Space Age had been born and the world would be different ever after."

FIRST USE. One of the early important uses of the term was in the title to Harry Harper's *Dawn of the Space Age* published just after World War II (London: Sampson Low, Marston and Co., 1946). The gist of the book was that when the V-2 weapon (see *V-1, V-2*) began to be equipped with instruments rather than explosives, space exploration could begin. The London *Daily Express* may have been the first to actually proclaim it in a headline of October 5, 1957: "The Space Age Is Here."

(2, adj.) Anything using state-of-the-art technology. For example: "I have an arsenal of space-age lures" (Erwin A. Bauer, "Hooked on Bass," *National Wildlife*, June–July 1982, p. 13).

space agency. Familiar name for NASA or other national agencies involved in space operations and space exploration. For example: "The Space Shuttle Challenger lost power in one of its three main engines minutes after blasting off Monday, but the ship safely reached a lower-than-expected orbit and the space agency said it hoped to carry out a full weekend, long science mission" (*Chicago Tribune*, July 30, 1985, p. 1).

Referring to NASA as the space agency may be an allusion to the first name suggested for NASA by President Eisenhower in his letter sent to Congress on April 2, 1958: the "national aeronautics and space agency."

space-ager. One living in the *Space Age,* a rare Space Age term that is included in the *OED.*

space-agey. Characteristic of the *Space Age.*

space barrel (space program + pork barrel). Term describing efforts to bring space dollars to a particular state or congressional district. The barrel alluded to here is the federal Treasury.

space biology. See *bioastronautics.*

space blanket. Aluminum-coated *Mylar* sheets sold commercially as emergency protection from the cold, such as those sold to protect hikers from exposure (hypothermia). Functioned as insulation barriers in *astronauts' Moon suits* and as *radiation* barriers in virtually all *spacecraft* dating back to the *Apollo* project. See *Mylar.*

FIRST USE. Although the space blanket and its variants have appeared in many advertisements (including the Space Rescue Blanket, Space Sportsman's Blanket, and Space Stadium Blanket), the first appearance in a news context was in a 1957 AP dispatch from Juneau, Alaska. The story described the rescue of an Alaskan educator and his 9-year-old daughter lost in the Yukon wilds after having survived a plane crash. "After staying with the plane for a day and a half [the pair] set out . . . taking with them about 50 pounds of survival gear, including food, matches and 'space blankets.' The blankets are silver on one side and red on the other for reflection and signaling. They used the blanket to attract the attention of the search plane that spotted them." (*New York Times,* "Rescued Pair Tell of 3 Days in Yukon," June 19, 1957, p. 19.)

space-borne. (1) Carried through space. (2) Carried out in space or by means of instruments in space.

space-bound. Bound or limited by the properties of space.

space bus. (1) Synonym for *space tug.* (2) An architectural design upon which satellites are based.

space cabin. A chamber designed to support human life in space.

space cadet. According to the *OED,* "a trainee spaceman . . . esp. a (young) enthusiast for space travel." Nickname for space enthusiasts, initially used by the U.S. Army's *missile* development group to describe themselves. In the post-Apollo era, the term tended to be used by those who advocate a return of humans to the Moon and onward to Mars.

ETYMOLOGY. The term comes from Robert A. Heinlein's 1949 novel of that title, a cautionary tale about things nuclear. The first uses of the term in major American newspapers are in connection with that book (e.g., "'*Space*

Cadet' Inter-Global Rocket Travel," *Chicago Tribune,* November 14, 1948, p. F14). The term was reinforced by the television show *Tom Corbett—Space Cadet,* which premiered in 1950. It was set in the year 2355 and aired three times a week in 15-minute segments on the CBS network. It was the prototypical "space opera" in that it used the same serial formula of the soap operas of the time. An element of the show that created attention was Corbett's colorful verbal expressions: "By the clouds of Venus," "By the wings of Saturn" "By the moons of Jupiter," etc. In his syndicated television column John Crosby wrote, "Space cadets don't blow their tops, they blow their jets. They don't kick up a rumpus; they blow up some Meteor dust" ("The Kids Really Go for Space Shows," *Washington Post,* October 28, 1950, p. B13).

FIRST USE. An article in the *St. Petersburg Times* contains one of the earliest uses of the phrase with reference to the U.S. *space program:* "There is one John Bruce Medaris. He is tough, hard-headed, a two-star general with two feet on the ground. Then there is John Bruce Medaris, space cadet" (*St. Petersburg Times,* February 2, 1958). On October 16, 1958, the *Minneapolis Star* carried a feature entitled "Army Missile Men Are 'Space Cadets.'"

USAGE. The term quickly acquired negative connotations in many circles. "Fear of being labeled 'space cadets' is inhibiting many reputable scientists from working on space research," wrote Walter Sullivan in 1959, alluding to a speech given by Dr. Lloyd V. Berkner, Chairman of the Space Science Board of the National Academy of Sciences ("Scientists Called Shy on Satellites: Fear of 'Space Cadet' Label Found Keeping Many out of Profitable Research," *New York Times,* January 21, 1959, p. 13). According to Homer Newell, NASA's first Administrator, T. Keith Glennan, stated that he was "no space cadet," which in Newell's opinion allowed him to achieve "just the right balance of conservatism and interest in space to make him congenial to President Eisenhower and acceptable to the Congress" (Newell, *Beyond the Atmosphere,* SP-4211, pp. 111–12).

EXTENDED USE. A person out of touch with reality. The term is common in baseball: "The land-locked Dodgers found themselves being strafed by a space cadet named Joaquin Andujar, who is one weird Dominican" (Mike Downey, "No One Including Andujar Wants to Talk about Andujar," *Los Angeles Times,* October 11, 1985, p. C15). The term is also used of a person who is experiencing the effects of drugs. This use is certainly influenced by the terms spaced, spaced out, and spacey, which came out of the drug culture of the late 1960s.

space capsule. A container used for carrying out an experiment or operation in space (SP-7).

space caucus. Informal congressional group that surfaced during the cost-cutting of the Reagan presidency to protect NASA interests.

space colony. Human group of people imagined as living and working in a *space station* or on another planet.

spacecraft. Traditionally defined as a vehicle designed to be placed in *orbit* about the Earth or into a *trajectory* toward another celestial body. The term was given a more specific definition in a memo written by John E. Naugle and Tom Chappelle: "A self-sufficient device, manned or unmanned, capable of traveling in space outside Earth's atmosphere to perform a useful function. Generally, a spacecraft performs one or more functions in space either continuously over an appreciable period or repeatedly; is capable of communicating, actively or passively, with Earth or other spacecraft; and is not, in effect, an expendable portion of a spacecraft, or a portion that essentially only extends the ability of the parent spacecraft to perform its total mission" ("Definition of Spacecraft," March 16, 1979, in NASA Names Files, record no. 177491).

spacecraft housekeeping. See *housekeeping.*

Space Day. Since 1997, a day set aside to increase public interest in and understanding of space staged by a coalition of space interests led by the Lockheed Martin Corporation. It is held annually on the Thursday that falls closest to the anniversary of the May 25, 1961, speech in which John F. Kennedy committed the United States to a lunar landing. The first national Space Day occurred on May 22, 1997. (*Launchspace* magazine, March 1997, p. 18.)

space density. The expected frequency of objects in space measured in number of objects per square minutes of arc on the sky.

space elevator. A physical connection from the surface of the Earth to a *geostationary orbit* above the Earth. Also termed an orbital tower. Popularized by writers such as Arthur C. Clarke, the concept has been the subject of various serious studies.

Space Exploration Initiative (SEI). An effort, proposed by President George H. W. Bush on July 20, 1989, to put a permanent human base on the Moon, and then go to Mars. The initiative ended by the early 1990s, but many of the same ideas are found in George W. Bush's *Vision for Space Exploration,* announced January 14, 2004.

Space Exploration Technologies Corporation. See *SpaceX.*

space-faring (adj.). Relating to or involved in outer-space activity.
ETYMOLOGY. Directly modeled on seafaring, the term was used to describe the U.S. *space program* as venturesome. Once the decision was made to land humans on the Moon, America began calling itself a "space-faring nation." The first public use of the term may have come on September 12, 1962, when President John F. Kennedy told Congress that he wanted to lead the nation to the Moon and make America "the world's leading space-faring nation."

space fever. Strong public support of space, such as occurred when the Manned Spaceflight Center was located in Houston in the early to mid-1960s. According to the web history of the *Johnson Space Center,* "The people of the Houston area welcomed MSC personnel with open arms and offered complete cooperation in all facets of the operation. The city was ecstatic. Space fever promptly swept the town. The baseball team was named the Astros, and the basketball team was called the Rockets. The Astrodome, Astroworld and countless businesses with 'space city' somewhere in the title blossomed over the years." (www.nasa.gov/centers/johnson/about/history/jsc40/jsc40_pg4.html.) **FIRST USE.** The term was first invoked in the early 1950s when a group of television shows including *Captain Video, Tom Corbett—Space Cadet, Rod Brown of the Rocket Rangers,* and *Space Patrol* created an alternative to Westerns for many young people (Meyer Berger, "'Space Fever' Hits the Small Fry," *New York Times* magazine, March 10, 1952, p. 17).

spaceflight. A voyage outside the Earth's atmosphere.

Spaceflight Tracking and Data Network (STDN). Formerly Space Tracking and Data Acquisition Network (STADAN). Early network established by NASA for space-to-ground communications with objects in near-Earth orbit. Its role was later taken over by the Tracking and Data Relay Satellite System *(TDRSS).* The *Deep Space Network* (DSN) is used for deep space communications.

space food. Foods prepared for humans wearing pressurized suits or occupying pressurized *spacecraft.* Originally applied to food for pilots flying jets at high altitudes. (2) Name for various commercial products that approximate or imitate food consumed by humans in space. Pillsbury's Space Food Sticks were a prime example. They were marketed in 1970 with ads proclaiming, "Space Food Now Available As a High-Energy Snack" (*Washington Evening Star,* April 8, 1970, sec. F, back page). **FIRST USE.** "Air Force Testing New Space Foods in Tubes," *Los Angeles Times,* November 6, 1957, p. 15.

Spaceguard. Term referring to various efforts to discover and study *near-Earth objects* (NEOs). Arthur C. Clarke coined the term in his novel *Rendezvous with Rama* (1973), with Spaceguard being the name of an early warning system created following a catastrophic asteroid impact on Italy. This name was later adopted by a number of real-life efforts to discover and study near-Earth objects. In Clarke's fictional account, Project Spaceguard is a system of radars, orbiting telescopes, and computers developed to track asteroids so that preventive action could be taken against future meteorites. (Robin Anne Reid, *Arthur C. Clarke: A Critical Companion* [Westport, CT: Greenwood, 1997], p. 41.)

space habitat. Any artificial environment designed to sustain life in space.

space industry. Those companies and institutions involved in the exploration or utilization of space. The term came into widespread use in 1958 as more than 300 companies had contracted to work on the *X-15* piloted aircraft and other projects such as *Dyna-Soar*. The term was given a boost in 1962 when the book *The Space Industry: America's Newest Giant,* by the editors of *Fortune* magazine, was published (Englewood Cliffs, NJ: Prentice-Hall). Also termed the *aerospace* industry.

space junk. Debris left in Earth *orbit* as a consequence of the human push into space; orbital trash. Space junk is a hazard to active *spacecraft* and multiplies as pieces collide and fragment into smaller pieces. By 1979 there were an estimated 4,600 pieces of space junk in orbit ("Space Junk Falls All the Time," *Washington Post,* July 10, 1979, p. A1).

ETYMOLOGY. The term was used during the early 1950s in television shows like *Tom Corbett—Space Cadet* and *Rod Brown of the Rocket Rangers.* It was defined in an early TV space glossary as "small pieces of rock or small meteors that float aimlessly in space and are a hazard." (NASA Names Files, record no. 15137; this file contains an unattributed "Galaxy Glossary" of terms from early TV science fiction shows.)

FIRST USE. An AP story of late December 1960 with wide distribution on the destruction of Discoverer XVII ended with this note: "Destruction of this satellite leaves a total of 34 objects in orbit around the Earth. These include rocket bodies and space junk—pieces of rockets after they break up." ("Discoverer Breaks up in Atmosphere," *Chicago Tribune,* December 31, 1960, p. 4.)

Spacelab. A reusable *space laboratory* in which scientists and engineers could work in Earth *orbit* without spacesuits or extensive training aboard the *Space Shuttle.* The program drew the United States and Europe into closer cooperation in space efforts.

The name finally chosen for the space laboratory was the one used by the European developers. It followed several earlier names used as NASA's program developed toward its 1980s operational goal. In 1971 NASA awarded a contract for preliminary design of *Research and Applications Module*s (RAMs) to fly on the Space Shuttle. A family of human-piloted *payload* carriers, the RAMs were to provide versatile laboratory facilities for research and applications work in Earth orbit. Later modules were expected to be attached to *space station*s, in addition to the earlier versions operating inside the Shuttle payload bay. The simplest RAM mode was called a Sortie Can at *Marshall Space Flight Center.* It was a low-cost, simplified pressurized laboratory to be carried on the Shuttle orbiter for short sortie missions into space. In June 1971 the NASA Project Designation Committee redesignated the Sortie Can the Sortie Lab, as a more fitting name.

In 1969, when the President's *Space Task Group* had originally recommended development of the Space Shuttle, it had also recommended broad international participation in the *space program*, and greater international cooperation was one of President Richard M. Nixon's Space Policy Statement goals announced in March 1970. NASA Administrator Thomas O. Paine visited European capitals in October 1969 to explain Shuttle plans and invite European interest, and 43 European representatives attended a Shuttle Conference in Washington. One area of consideration for European effort was development of the Sortie Lab.

On December 20, 1972, a European Space Council ministerial meeting formally endorsed European Space Research Organization *(ESRO)* development of Sortie Lab. An intergovernmental agreement was signed August 10, 1973, and ESRO and NASA initialed a memorandum of understanding. The memorandum was signed September 24, 1973. Ten nations—Austria, Belgium, Denmark, France, West Germany, Italy, the Netherlands, Spain, Switzerland, and the United Kingdom—would develop and manufacture the units.

In its planning and studies, ESRO called the laboratory Spacelab. And when NASA and ESRO signed the September 1973 memorandum on cooperation, NASA Administrator James C. Fletcher announced that NASA's Sortie Lab program was officially renamed Spacelab.

Spacelab was designed as a low-cost laboratory to be quickly available to users for a wide variety of orbital research and other applications. It consisted of two elements, carried together or separately in the Shuttle orbiter: a pressurized laboratory, where scientists and engineers with only brief flight training could work in a normal environment; and an instrument platform, or "pallet," to support telescopes, antennas, and other equipment exposed to space. A tunnel is used to gain access to the module; there is also an instrument pointing subsystem. Spacelab is not deployed free of the orbiter.

Reusable for 50 flights, the laboratory remained in the Shuttle hold, or cargo bay, while in orbit, with the bay doors held open for experiments and observations in space. NASA astronauts called *mission* specialists, as well as noncareer astronauts called *payload* specialists, flew aboard Spacelab to operate experiments. Payload specialists were nominated by the scientists sponsoring the experiments aboard Spacelab, and were accepted, trained, and certified for flight by NASA. Their training included familiarization with experiments and payloads as well as information and procedures to fly aboard the Space Shuttle. From one to four payload specialists could be accommodated aboard a Spacelab flight. These specialists rode into space and returned to Earth in the orbiter crew compartment cabin, but they worked with Spacelab on orbit. Because Spacelab missions, once on orbit, may operate on a 24-hour basis, the flight crew was usually divided into two teams.

At the end of each flight, the orbiter would make a runway landing, and the laboratory would be removed and prepared for its next flight. Racks of experiments would be prepared in the home laboratories on the ground, ready for installation in Spacelab for flight and then removal on return. Twenty-five Spacelab missions were flown between 1983 and 1999.

SOURCES. NASA News Releases 71-6, 71-67, 73-191, 74-198, 75-28; MSFC News Releases 71-34, 72-41; U.S. Congress, House of Representatives, Committee on Science and Astronautics, Subcommittee on Manned Space Flight, *Hearings: 1973 NASA Authorization, Pt. 2, February and March 1972* (Washington, DC: GPO, 1972), pp. 238–45; NASA, *Astronautics and Aeronautics: Chronology on Science, Technology, and Policy for 1969, 1970, 1971, and 1972*, SP-4014–SP-4017 (Washington, DC: NASA, 1970, 1972, and 1974); JSC, *Space Shuttle* (Washington, DC: GPO, February 1975); James C. Fletcher, Administrator, NASA, memorandum to Administrators, Associate Administrators, Assistant Administrators, and Directors of Field Centers, September 24, 1973; Office of Manned Space Flight, NASA, Spacelab/CVT Program Approval Document, December 4, 1974; ESRO News Release, June 5, 1974; September 4, 1974; February 21, 1975; ESRO, *Europe in Space* (Paris: ESRO, March 1974), pp. 42-47; Fletcher, prepared text for address before the National Space Club, Washington, DC, February 14, 1974; Douglas R. Lord, Director, Spacelab Program, NASA, prepared testimony for *Hearings, Fiscal Year 1976 NASA Authorization, before House Committee on Science and Technology, Subcommittee on Space Science and Applications, February 20, 1975.*

space laboratory. A laboratory in space; in particular, a *spacecraft* equipped as a laboratory.

space launcher. A *rocket* used to lift a *spacecraft* into space.

space law. Area of the law that encompasses national and international law governing activities in outer space.

spaceman. Term long used in science fiction to describe a space-farer, employed as well in the early days of NASA before the name *astronaut* had been adopted: "US Picking 1st Space Men" (*Chicago Tribune* headline, January 28, 1959, p. 1).

space medicine. A branch of *aerospace medicine* specifically concerned with the health of humans who make or expect to make flights into space beyond the part of the atmosphere that offers resistance to a body passing through it.

spacenik. Post-*Sputnik* space enthusiast.

space observatory. An astronomical observatory in space.

space physics. The physics of extraterrestrial phenomena and bodies, especially within the solar system.

space pilot. Term used by the military for a pilot who flew to a height of 50 miles (80 km) or higher; the military equivalent of *astronaut*. Neil A. Armstrong noted in a 2005 interview, "The military somehow felt that they ought to have some comparable sort of thing, but they shied away from the term astronaut . . . however they thought that it would be a good idea to have some emblem on their wings that indicated that they were space pilots." (NASA Oral History Interviews, 31248, "Oral History Interview of Neil A. Armstrong," Office of History Division, pp. 6–7.)

Space Plasma High Voltage Interaction Experiment satellite. See *SPHINX*.

spaceport/space-port. (1) A transportation center in space that acts like an airport on Earth. It provides a transport node where passengers or cargo can switch from one *spaceship* to another, and a facility where spaceships can be berthed, serviced, and repaired. (Paine Report, p. 198.) (2) Term sometimes used to refer to Earth-based launch sites.

  FIRST USE. "Spaceport—Sensational City of the Future," exhibit featuring a spaceship replica designed by Dr. Willy Ley, was set up in the New York department store Gimbels in March 1951. (NASA Names Files, record no. 15137, file contains information on the spaceport.)

Spaceport USA. Former name for the *Kennedy Space Center* Visitor Complex. See *spaceport*.

space probe. Any *unpiloted spacecraft* for research or reconnaissance. See *probe*.

space program. A nation's plan for the exploration of space and the development of space technology.

Space Race. The Cold War contest between the United States and the Soviet Union that began before the *launch* of *Sputnik* in 1957 and ended with exploration of the Moon, beginning in July 1969, by twelve U.S. astronauts.

  FIRST USE. Although there were incidental uses of the term earlier, it gets early exposure in an article by Robert C. Cowen, "U.S. and U.S.S.R. in Space Race" (*Christian Science Monitor*, March 25, 1955, p. 7). An AP headline that ran weeks before *Sputnik* announced, "Russia Termed Ahead in Space Race" (*Los Angeles Times* June 23, 1957, p. 10).

USAGE. Although it was often argued by U.S. officials that there was no official race, the term gained fast public acceptance. It was used as the title of the Smithsonian National Air and Space Museum's May 1997 exhibit sponsored by the Perot Foundation.

space science. Scientific investigations made possible or significantly aided by *rockets*, *satellites*, and *space probes*.

ETYMOLOGY / FIRST USE. According to Homer E. Newell, the first Director of Space Science for NASA and widely regarded as the father of space science, "The first formal use of the phrase that I recall was in the pamphlet Introduction to Outer Space prepared by members of the President's Science Advisory Committee and issued on March 26, 1958 by President Eisenhower . . . A few months later the phrase appeared in the title of the Space Science Board which the National Academy of Sciences established in June 1958." (Newell, *Beyond the Atmosphere,* SP-4211, pp. 11–15.)

spaceship. Piloted *space vehicle.*

Spaceship Earth. A term of awareness that the human race is aboard a *craft* in *orbit* around the Sun. Employed widely to underscore the notion that humans are not taking care of their planet. The term was popularized by R. Buckminster Fuller (author of *A Manual for Spaceship Earth*), who said in a 1981 interview, "We are on a Spaceship Earth, which is traveling around the Sun at the speed of 60,000 miles per hour. Humans are 60 percent water. Think of it, water freezes and water boils. In 85 years, I have consumed 1000 tons of food and 1000 tons of water and the food I've eaten and the water I've drunk isn't the real me. You are looking at a transreceiver. The real me is not the physical me. We are metaphysical, you and I." ("'Bucky' Fuller—Courage Born of Crisis," *Boston Globe,* April 10, 1981.)

FIRST USE. "We are all 'helionauts' aboard Spaceship Earth in *orbit* around the sun" (Irving S. Bengelsdorf in the *Los Angeles Times* of July 11, 1966, p. A6). An editorial in the *Los Angeles Times* for March 18, 1967, p. J7, is entitled "Let's Clean Up Spaceship Earth."

SpaceShipOne. Vehicle that completed the first privately funded human spaceflight on June 21, 2004, and repeated the feat on October 4, 2004, thereby winning the $10 million *Ansari X Prize,* which requires reaching 62 miles (100 km) in altitude twice in a two-week period carrying the weight-equivalent of three people, with no more than 10 percent of the nonfuel weight of the *spacecraft* replaced between flights.

Space Shuttle. (1) Any space transport vehicle used to deliver personnel or cargo from Earth or another celestial body to another destination and then return for additional loads. (2) Reusable U.S. *spacecraft*: the orbiters in the *STS* program. (3) Since 1990, the official name for the program formerly known as the National Space Transportation System.

Since its establishment in 1958, NASA had studied aspects of reusable launch vehicles and *spacecraft* that could return to the Earth. *Marshall Space Flight Center* (MSFC) sponsored studies of recovery and reuse of the *Saturn V* launch vehicle. In 1962, MSFC Director of Future Projects Heinz H. Koelle projected a "commercial space line to Earth orbit and the moon," for cargo transportation by 1980 or 1990. The following year Leonard M. Tinnan of MSFC published a description of a winged fly-back Saturn V. Other studies of "logistics spacecraft systems," "orbital carrier vehicles," and "reusable orbital transports" followed throughout the 1960s in NASA, the Department of Defense, and industry.

As the *Apollo* program neared its goal, NASA's space program objectives widened, and the need for a fully reusable, economical space transportation system became more urgent. In 1966 the NASA budget briefing outlined a fiscal year 1967 program including advanced studies of "ferry and logistics vehicles." In February 1967 the President's Science Advisory Committee recommended studies of more economical ferry systems with total recovery and rescue possibilities. Industry studies under NASA contracts during 1969–71 led to definition of a reusable Space Shuttle system and to a 1972 decision to develop the Shuttle. On January 5, 1972, President Nixon announced that the United States would develop the Space Shuttle.

The Space Shuttle as it was developed over the next decade consists of four main components: the *orbiter,* in which the astronauts and payloads reside; the external tank, carrying more than half a million gallons of liquid hydrogen and liquid oxygen; the Space Shuttle main engines, fed by the fuels from the external tank; and two solid rocket boosters that provide 3.3 million pounds (15 million newtons) of *thrust* at liftoff. All components except the external tank are reusable, making the Space Shuttle the first reusable (as opposed to expendable) *launch vehicle.*

The individual orbiters were all named after heroic sea vessels. The original plan was to call them Constitution, *Columbia, Challenger,* and *Atlantis.* The first Shuttle was renamed *Enterprise* (used for approach and landing tests only and now in the National Air and Space Museum Udvar-Hazy Center). An additional orbiter was named *Discovery.* The Shuttle *Endeavour* was built in the wake of the Challenger disaster. The first Space Shuttle flight was Columbia, launched April 12, 1981. Since then the Space Shuttle program has had more than 100 successful flights, as well as two disasters, Challenger in 1986 and Columbia in 2003, resulting in a total loss of crew and vehicle.

While many people associate the Shuttle era with the searing images of the Challenger accident in 1986 and the Columbia accident in 2003, those events should not obscure the accomplishments of the only functional winged *space vehicle* ever to fly. While the Shuttle program's goals of low cost and routine frequent access to space were not met, its numerous flights did transport payloads of up to 55,000 pounds (25,000 kg) into low Earth orbit— the only reusable system capable of doing so. During 117 flights in its first 25 years (1981–2006), the Space Shuttle's accomplishments include 24 commercial satellites deployed prior to the Challenger accident; major scientific missions including *Galileo, Magellan, Chandra,* and the launching and servicing missions of the *Hubble Space Telescope;* the *Spacelab* and Spacehab missions with their material, *microgravity,* and life sciences experiments; deployment of the Tracking and Data Relay Satellite System *(TDRSS)* network; and numerous flights in support of *Mir* and the *International Space Station.* The Shuttle's primary purpose until it is retired in 2010 will be to complete assembly of the International Space Station.

ETYMOLOGY. Application of the word shuttle to anything that moves quickly back and forth (from shuttlecock to shuttle train) arose from the name of the weaving instrument that passed or "shot" the thread of the woof from one edge of the cloth to the other. The English word came from the Anglo-Saxon *scytel* (missile), related to the Danish *skyttel* (shuttle), the Old Norwegian *skutill* (harpoon), and the English "shoot." (*Webster's International Dictionary,* 2nd ed., unabridged.)

The term shuttle crept into forecasts of space transportation at least as early as 1952. In a *Collier's* article published in that year, Dr. Wernher von Braun, then Director of the U.S. Army Ordnance Guided Missiles Development Group, envisioned *space station*s supplied by *rocket* ships that would enter *orbit* and return to Earth to land "like a normal airplane," with small rocket-powered "shuttle-craft" or *"space taxis"* to ferry men and materials between rocket ship and space station.

In October 1959, Lockheed Aircraft Corporation and Hughes Aircraft Company reported plans for a space ferry or "commuter express" for "shuttling" men and materials between Earth and outer space. In December, *Christian Science Monitor* Correspondent Courtney Sheldon wrote of the future possibility of a "man-carrying space shuttle to the nearest planets."

The term reappeared occasionally in studies through the early 1960s. A 1963 NASA contract to Douglas Aircraft Company was to produce a conceptual design for Philip Bono's Reusable Orbital Module Booster and Utility Shuttle (ROMBUS), to orbit and return to touchdown with legs like the lunar landing module's. Jettison of eight strap-on hydrogen tanks for *recovery* and reuse was part of the concept. The pressing accounts of European discussions of Space Transporter proposals and in articles on the Aerospaceplane, NASA contract studies, U.S. Air Force START *reentry* studies, and the joint lifting-body flights referred to "shuttle service," "reusable orbital shuttle transport," and "space shuttle" forerunners. The *Defense/Space Business Daily* newsletter was persistent in referring to USAF and NASA reentry and lifting-body tests as "Space Shuttle" tests. Editor-in-Chief Norman L. Baker said the newsletter had first tried to reduce the name Aerospaceplane to Spaceplane for that project, and had shifted from that to Space Shuttle for reusable, back-and-forth space transport concepts, as early as 1963. The name was suggested to him by the Washington (DC)–to–New York airline shuttle flights. (Telephone interview, April 22, 1975.)

The name Space Shuttle thus evolved from descriptive references in the press, aerospace industry, and government, and gradually came into use as concepts of reusable space transportation developed.

FIRST USE. The name came into official use as early NASA advanced studies grew into a full program. In 1965 Dr. Walter R. Dornberger, Vice President for Research of Textron Corporation's Bell Aerosystems Company, published "Space Shuttle of the Future: The Aerospaceplane," in Bell's periodical *Rendezvous.* In July Dr. Dornberger gave the main address in a University of Tennessee Space Institute short course: "The Recoverable, Reusable Space Shuttle." In 1968, NASA officially used the term shuttle for its reusable transportation concept officially. Associate Administrator for Manned Space Flight George E. Mueller briefed the British Interplanetary Society in London in August with charts and drawings of "space shuttle" operations and concepts. In November, addressing the National Space Club in Washington, DC, Dr. Mueller declared that the next major thrust in space should be the space shuttle. By 1969, Space Shuttle was the standard NASA designation, although some efforts were made to find another name as studies were pursued.

SOURCES. SP-4402, pp. 111–14; "Commercial Moon Flights Predicted within 30 Years," *Birmingham Post-Herald,* November 9, 1962 (report of Koelle's October paper for the American Rocket Society); George S. James, "New Space Transportation System—An AIAA Assessment," unpublished draft paper December 1972; NASA News Release, "Background Material and NASA Fiscal Year 1967 Budget Briefing," January 22, 1966; *The Space Program in the Post-Apollo Period,* report of President's Science Advisory Committee (Washington, DC: GPO, February 1967), p. 37; Wernher von Braun, "Crossing the Last Frontier," *Collier's* 129, no. 12 (March 22, 1952): 24–29, 72–74; "Space Technology Highlights," *Astronautical Sciences Review,* October–December 1959, pp. 6–8, 29; Courtney Sheldon, "Shuttle to

Planets Awaits Development," *Christian Science Monitor,* December 8, 1959; Walter J. Dornberger, "Space Shuttle for the Future: The Aerospaceplane," *Rendezvous* 4, no. 1 (1965): 2–5; Dornberger, "The Recoverable, Reusable Space Shuttle," *Astronautics & Aeronautics,* November 1965, pp. 88–94; *The Post-Apollo Space Program: Directions for the Future,* Space Task Group report to the President (Washington, DC: GOP, September 1969); NASA News Release 74-211; U.S. Congress, House of Representatives, Committee on Science and Astronautics, Subcommittee on Manned Space Flight, *Hearings: Fiscal Year 1975, NASA Authorization, Pt. 2, February and March 1974* (Washington, DC: GPO, 1974), p. 9; JSC, *Space Shuttle* (Washington, DC: GPO, February 1975); memo to Dr. Lovelace, "Orbiter Names," June 10, 1979, NASA Names Files, record no. 17547.

space sick / space-sick. Adjective describing *space sickness.*

space sickness. Illness due to prolonged weightless, analogous to car sickness, air sickness, and overall motion sickness. First used in a diagnostic sense in discussing Gherman Titov's 25-hour 17-minute *orbit* on August 6, 1961. In a report delivered to the International Astronautical Congress in October of that year, the condition was likened to sea sickness or a merry-go-round effect that became stronger when the *cosmonaut* turned his head sharply or was observing quickly moving objects (Howard Simons, "Titov Was Ill in Orbit, Red Scientists Reveal," *Washington Post,* October 5, 1961, p. A6).

FIRST USE. The term was introduced to the public at the first annual Symposium on Space Travel at New York's Hayden Planetarium in 1951: "'Space sickness' was suggested by Dr. Heinz Haber of the Air Force's department of space medicine at Randolph Field, Texas." Haber went on to say that "it is the nervous system that needs watching" ("Platforms in Sky Now Seem Feasible," *New York Times,* October 13, 1951, p. 14).

space simulator. A device that simulates the conditions of space, or of the interior of a *spacecraft.*

space sovereignty. The issue of how far into space nationality extends and whether it can extend to celestial bodies, including mineral rights. The issue was thrust into the fore in the wake of *Sputnik* 1, when one issue at hand was whether the *satellite* had trespassed on the airspace of nations over which it passed.

space-speak. The jargon of space technologists, considered as a corruption of standard English *(OED).* Also known as *NASAese* or *NASA-speak.*

space station. A large orbiting structure designed for humans to live on in outer space. A space station is distinguished from other *spacecraft* by its lack of major propulsion or landing facilities; other vehicles are used as transport to and from the station. Space stations are designed for medium-term living in *orbit,* for periods of weeks, months, or even years.

The Soviet Union launched the world's first dedicated space station, *Salyut* 1, in 1971—a decade after launching the first human into space. The United States sent its first space station, the larger *Skylab,* into orbit in 1973; it hosted three crews before it was abandoned in 1974. Russia continued to focus on long-duration space missions and in 1986 launched the first modules of the *Mir* space station. The 16-nation *International Space Station,* led by the United States, is the largest space station project to date. (http://spaceflight.nasa.gov/history/station/index.html.)

ETYMOLOGY / FIRST USE. The first proposal for a *crewed* station was published in 1869, when the American novelist Edward Everett Hale described a fanciful "Brick Moon" propelled into Earth *orbit* to help ships navigate at sea. However, the castle in space remained a topic for fantasy and idle speculation until the Russian "Father of Cosmonautics," Konstantin Tsiolkovsky, first mapped the science that would sustain life in artificial orbital stations. In his labyrinthine treatise-novel *Vne Zemli* ("Beyond Earth"), written and revised between 1896 and 1920, Tsiolkovsky outlined how human beings could feasibly build self-sustaining habitats in space. The term space station was first used by the Romanian physicist Hermann Oberth (in his dissertation *The Rocket into Planetary Space*) to describe a wheel-like facility that would serve as a jumping-off place for human journeys to the Moon and Mars. In 1952 Dr. Wernher von Braun published his concept of a space station in *Collier's* magazine. He envisioned a facility that would have a diameter of 250 feet (76 m), orbiting more than 1,000 miles (1,600 km) above the Earth and spinning to provide artificial gravity through centrifugal force.

Space Station Alpha. See *International Space Station.*

Space Station Freedom. See *International Space Station.*

space suit. A pressure suit for wearing in space or at very low ambient pressures within the *atmosphere,* designed to permit the wearer to leave the protection of a pressurized cabin.

Space Task Group. (1) Organization created on November 5, 1958, at Langley Research Center to, in the words of NASA Administrator T. Keith Glennan, " implement the manned satellite project," Project *Mercury.* The group, headed by Robert Gilruth, drew up specifications for the Mercury *capsule* and participated in the selection of the *Original Seven* astronauts. It became the Manned Spacecraft Center and then the *Johnson Space Center.* (2) Presidentially appointed panel that issued its report on a post-*Apollo space program* on September 15,

1969. Under the chairmanship of Vice President Spiro T. Agnew, this group met throughout the spring and summer of 1969 to plot a course for the space program. The politics of this effort was intense. NASA lobbied hard with the group and especially its chair for a far-reaching post-Apollo space program that included development of a *space station,* a reusable *Space Shuttle,* a Moon base, and a human expedition to Mars. The NASA position was well reflected in the group's final report, but Nixon did not act on its recommendations. He remained silent on the future of the U.S. space program until a March 1970 statement in which he said, "We must also recognize that many critical problems here on this planet make high priority demands on our attention and our resources."

space taxi. (1) Term used by Wernher von Braun to describe a reusable *Space Shuttle* in his series for *Collier's* magazine. (2) Nickname for the *Lunar Excursion Module.*

space tether. A long cable used to couple *spacecraft* to one another or to other masses.

spacetime. A system of looking at the *universe* as one in which the three spatial dimensions are unified with the time dimension.

space tourism. Travel into space by individuals for the purpose of personal pleasure. At the moment, space tourism is affordable only to exceptionally wealthy individuals and corporations, with the Russian *space program* providing transport.

space tourist. One who pays a fee to be transported into space for reasons of personal gratification and curiosity.

ETYMOLOGY. Over the objections of the majority of the *International Space Station* partners, Californian Dennis A. Tito became the first space tourist, paying $20 million for a trip aboard a Russian *Soyuz capsule* and the *International Space Station,* April 29–May 7, 2001.

Space Tracking and Data Acquisition Network (STADAN). See *Spaceflight Tracking and Data Network.*

Space Transportation System. Formal name of the *Space Shuttle* Program. See *STS.*

space tug. Term used frequently in studies and proposals through the years to describe any of several proposed reusable vehicles that would service *space stations* and large *orbiters.* During the early days of the *STS* development a space tug was envisioned that would fit into the cargo bay to deploy and retrieve payloads beyond the orbiter's reach and to achieve Earth-escape speeds for deep-space exploration.

ETYMOLOGY. Joseph E. McGolrick of the NASA Office of Launch Vehicles used the term in a 1961 memorandum suggesting that as capabilities and business in space increased, a need might arise for "a space tug, a space vehicle capable of orbital rendezvous and of imparting velocities to other bodies in space." He foresaw a number of uses for such a vehicle and suggested that it be considered with other concepts for the period after 1970. (SP-4402, p. 113.)

space vehicle. *Booster* plus *spacecraft* (NASA, *Glossary/Congressional Budget Submission*, p. 39). *The OED* defines the term as a "spacecraft, esp. a large one."

space velocity. The velocity in space of a star relative to the Sun, equal to the vector sum of its proper motion and its radial velocity *(OED)*.

spacewalk. Human physical activity undertaken in space outside a *spacecraft*. Popular name for *extravehicular activity* (EVA).

FIRST USE. The first attributed use is from a statement by Kenneth Garland, Vice President of the British Interplanetary Society, in a Reuters dispatch that appeared in the *New York Times* (March 19, 1965, p. 24) under the title "Russian 'Spacewalk' Hailed by Briton." The article alluded to the extravehicular activity of Alexei Leonov.

space warp. An imaginary distortion of space-time that is conceived as enabling space travelers to make journeys that would otherwise be contrary to the known laws of nature *(OED)*.

space weather. The conditions and processes occurring in space that have the potential to affect the near-Earth environment. Space weather processes can include changes in the interplanetary *magnetic field,* coronal mass ejections from the Sun, and disturbances in the Earth's magnetic field. The effects can range from damage to satellites to disruption of power grids on Earth.

SpaceX. Popular name for the Space Exploration Technologies Corporation headquartered in El Segundo, California. SpaceX is developing a family of *launch vehicle*s intended to ultimately reduce the cost and increase the reliability of access to space by a factor of ten. *Falcon,* a two-stage liquid-fueled orbital launch vehicle, is the company's first product. The entire vehicle, including main and *upper-stage* engines, primary structure, *avionics*, and guidance control, is being developed internally at SpaceX.

SpaceX was founded in June 2002 by Elon Musk, co-founder of PayPal, the world's leading electronic payment system. He created SpaceX to help make humanity a *space-faring* civilization. In an online report of July 6, 2006, Musk reported that the company has ten launches on its schedule and that it was on the verge of becoming "cash flow positive," but this goal proved to be elusive as the first three SpaceX launches failed. The third failure, August 2, 2008, had three payloads—a U.S. military satellite and two from NASA—which were all lost. Musk immediately responded, "The most important message

I'd like to send right now is that SpaceX will not skip a beat in execution going forward" (www.spacex.com/).

spam in a can. Derisive phrase used by and about *Mercury astronauts* willing to orbit the Earth in a small ballistic *capsule.* It also alluded to the idea that these test pilots would do little if any piloting: "These men were the first ones to undertake such a dangerous mission as sitting atop a rocket booster in a tiny capsule, where they felt like 'Spam in a can'" (Swenson, Grimwood, and Alexander, *This New Ocean,* SP-4201).

The term was central to the debate about early human spaceflight. According to James Hansen, "Ironically, the greatest skepticism about the Mercury concept existed inside the family of test pilots. Path breaking NACA/NASA test pilots like A. Scott Crossfield, Joseph A. Walker, and even the young Neil Armstrong, who in 10 years was to become the first man to walk on the Moon, were at first not in favor of Project Mercury. Their attitude was that the astronaut inside the ballistic spacecraft was no more than 'spam-in-a-can.' Charles E. 'Chuck' Yeager, the air force test pilot who broke 'the sound barrier' in 1947 in the X-1, expressed this prejudice: 'Who wanted to climb into a cockpit full of monkey crap?' (This was a reference to the primates who flew in the Mercury spacecraft prior to the astronauts)." (Hansen, *Spaceflight Revolution: NASA Langley Research Center from Sputnik to Apollo* [Washington, DC: GPO, 1995], SP-4308, p. 41.)

The term was rendered appropriate as astronauts proved their skill as pilots. Concerning Gordon Cooper's skilled return from *orbit* in *Faith 7,* Leon Wagener wrote, "Cooper was greeted as an American hero with a ticker tape parade in New York City, and, just as important to him, was considered a champion to his colleagues for proving, for once and for all, that astronauts in a space craft were not 'Spam in a can,' but highly skilled pilots." (Wagener, *One Giant Leap: Neil Armstrong's Stellar American Journey* [New York: Forge, 2004], p. 141.)

The term has also been invoked to mark changes in the position of the astronaut. Joan Johnson-Freese and Roger Handberg described the "dramatic pendulum swing in the role of the astronaut from 'Spam in a Can' passengers in the early Mercury capsules to installation of astronauts in senior management positions and ultimately leading to Astronaut/Admiral Richard Truly being named as NASA Administrator during the Bush Administration." (Johnson-Freese and Handberg, *Space, the Dormant Frontier: Changing the Paradigm for the 21st Century* [Westport, CT: Praeger, 1997], p. 19.)

ETYMOLOGY. Spam (spiced + ham) is a canned pork product introduced by the Hormel Foods Corporation in 1937. During World War II the North

American P-51 Mustang fighter was given several nicknames, including Spam Can, which in the postwar world became the name for any propeller-driven aircraft ("A Prang in a Spam Can," *New York Times,* March 19, 1954, p. 22).

Specialized Experimental Applications Satellite. See *SEASAT.*

SPHINX. Space Plasma High Voltage Interaction Experiment (satellite). Planned as one of NASA's smallest scientific *satellites,* the 250-pound (113-kg) SPHINX was to be launched *piggyback* on the proof flight of the newly combined *Titan III–Centaur launch vehicle,* along with a dynamic model of the *Viking spacecraft.* The planned year-long *mission* was to measure effects of charged particles in space on high-voltage solar cells, insulators, and conductors. The data would help determine if future spacecraft could use high-voltage solar cells, instead of the existing low-voltage cells, to operate at higher power levels without added weight or cost. The Centaur stage failed during launch on February 11, 1974, however, and the satellite was destroyed. (SP-4402, p. 71; NASA News Release 74–25.)

Spider. *Call sign* for the *Apollo 9 Lunar Module,* piloted by James A. McDivitt, David R. Scott, and Russell L. Schweickart. Conducted March 3–13, 1969, this was the first human flight tested in Earth *orbit* with all the lunar hardware required for a lunar landing.

ETYMOLOGY. Apollo 9 was the first *mission* since the early *Gemini* flights in which the crew was able to name its craft, namely its Command Module and Lunar Module. The Lunar Module was named Spider because it looked like a spider, just as the Command Module was dubbed *Gumdrop* because its blue shroud suggested a candy wrapper. (Lattimer, *All We Did Was Fly to the Moon,* p. 57.)

spillover. Synonym for *spinoff* used before that term became dominant. "The topic of spillover from space research was a matter of lively discussion several years ago and has been frequently regarded as a joke since" (Frederick Seitz, "Science and the Space Program," *Science,* June 24, 1966, p. 1720).

spinoff/spin-off. Term used to describe the secondary applications of space research. The 1958 Space Act that created the NASA stipulated that the agency's vast body of scientific and technical knowledge should also "benefit mankind." These benefits are commonly referred to as "NASA spinoffs." Unlike mainline programs in such areas as Earth imaging, communications, and aeronautics that are aimed directly at Earth applications, spinoffs are taken from their original applications and put to other uses by corporations, universities, and individual entrepreneurs. NASA refers to this practice as "technology twice used." Each year NASA documents these secondary applications in a slick annual report entitled *Spinoff.* One example from a recent *Spinoff* report—ingestible, foamless toothpaste, developed for astronauts in a zero-gravity environment where spitting and frothing present a host of *housekeeping* problems—shows how down-to-Earth these things can be at the end of the spinoff cycle. The toothpaste has been marketed commercially as NASAdent. It is being marketed for total-care nursing patients (who may choke on air bubbles), hospital patients and others who are not always near a basin, and young children (who often swallow toothpaste).

Other spinoffs include inorganic paints (which help coastal bridges resist corrosion), collapsible towers (for applications ranging from portable radar to rock concert acoustics), air tank breathing systems for firefighters (based on breathing systems that were used on the Moon), watch batteries, food sticks, reservations systems, police radios, robotic systems, measuring instruments, insulation material, heart rate monitors, high-temperature lubricants, ceramic powders, solar panels, poison detectors, heated ski goggles, scratch resistant sunglass lenses, Retin A to combat acne and skin wrinkling, water filters that attach to faucets, the Jarvik artificial heart, cordless tools, an insulin infusion pump, the liquid crystal wristwatch, freeze-dried food (first developed for John Glenn's 1962 Mercury orbits), the graphite of tennis rackets, golf clubs, and fishing poles, and much more. (The NASA Spinoff database, www.sti.nasa.gov/spinoff/database, contains abstracts of articles featured in the publication *Spinoff* since 1976.)

ETYMOLOGY. A distribution of stock in a new company to shareholders of a parent company; a company so created. Also, in journalism a story that is generated from an earlier one, or a new television show based on a minor character from the original.

USAGE. Early uses of the term in the period 1961–68 occasionally appeared in a negative context. For example, a headline in *Science* (September 1, 1967, p. 157) proclaimed, "NASA Study Finds Little 'Spin-off'" (an allusion to a Denver Research Institute report that could find no large-scale technology transfer). A letter published in *Science* magazine from David McNeill (May 13, 1966, p. 878) decried spinoff as a "spurious" word coined by newspapermen. NASA nevertheless embraced the term as a way of emphasizing that the money being spent on space had a beneficial effect on Earthbound processes and products.

Spitzer Space Telescope. Formerly the Space Infrared Telescope Facility (SIRTF). One of NASA's Great Observatories, along with the *Hubble Space Telescope,* Compton Gamma Ray Observatory, and *Chandra.* The Spitzer Space Telescope was launched into space by a *Delta* rocket from *Cape Canaveral,* Florida, on August 25, 2003. In the years since it went into operation it has made numerous significant astronomical findings, including hundreds of *black holes* hidden deep inside dusty galaxies billions of light-years away.

ETYMOLOGY. Named for astrophysicist Lyman Spitzer Jr. (1914–97), the first to propose the idea of placing a large telescope in space and the driving force behind the development of the Hubble Space Telescope. (From news releases at www.spitzer.caltech.edu/spitzer/index.shtml.)

splashdown. The landing of a *spacecraft* by parachute in the sea at the end of a space flight. This was the method used by American *crewed* spacecraft prior to the *Space Shuttle.*

FIRST USE. The term first comes into play during planning for the *Mercury* flight of Alan B. Shepard, which was to be the first human splashdown. It made its public debut in the *Washington Post:* "The Lake Champlain, an aircraft carrier and 6 destroyers have continued steaming in the area of Grand Bahamas Island since they took up their stations Tuesday morning. They are strung out about 60 miles (97 km) apart, and their *mission* is to retrieve Shepard after 'splashdown.'" (Edward T. Folliard, "Astronaut Flies Today If Weather Permits," *Washington Post,* May 5, 1961, p. B2.)

SPOT. Système Probatoire pour l'Observation de la Terre. French *satellite* that takes high-resolution pictures of the Earth.

Sputnik (pronounced spoot-neek). (1) The first Earth-circling *satellite.* Sputnik 1, launched on October 4, 1957, was designed to send radio signals to Earth and determine the density of the *upper atmosphere.* However, it transmitted signals to Earth for only 21 days after *launch.* Its *orbit decay*ed, and it fell to Earth on January 4, 1958. (2) Any Earth-circling *satellite.* (3) Any of the satellites in the first Soviet satellite program consisting of three Sputnik satellites. (The Sputnik launched between Sputniks 2 and 3 failed to reach orbit.) Sputnik 2 was launched on November 3, 1957, and carried aboard it a dog, *Laika.* Biological data was returned for approximately one week, the first data of its kind. *Laika* died shortly after going into orbit. The satellite itself remained in orbit 162 days. Sputnik 3 was launched on May 15, 1958. It was originally intended as the first launch in the Sputnik program, but because of delays in construction it orbited later. It was designed to be a geophysical laboratory, performing experiments on the Earth's *magnetic field, radiation* belt, and ionosphere. It orbited Earth and transmitted data until April 6, 1960, when its orbit decayed.

ETYMOLOGY. The Russian name for the satellite was Sputnik Zemlyi (traveling companion of the world). Almost at once the name was shortened to Sputnik, and in that form it entered the languages of the world. The first syllable of the word was hardly ever pronounced as it would have been in Russian (spoot). Indeed, it became an American word.

"Some new words take years to get into the language," said lexicographer Clarence L. Barnhart at the time. "She's a record breaking word." So sure was he that 24 hours after the launch of Sputnik 1, Barnhart called his printer to dictate a definition, and it was included in the next Thorndike-Barnhart *Comprehensive Desk Dictionary* (1958). Barnhart and others defined the word as meaning *Earth satellite*. (John G. Rogers, "Sputnik Soars into Type in a U.S. Dictionary," *New York Herald Tribune*, December 18, 1957, p. 17.)

USAGE. Sputnik is an important word, one of those rarities that claimed its place in the English language—and other languages, for that matter—the minute it was heard. It is, of course, only conjecture to say that the look and sound of the word added to its impact, but the word did seem to carry its own shock value—completely the opposite of other Russian missions such as *Salyut, Soyuz,* or the positively placid *Mir*. The shock value wore off as Americans became familiar with the term and the headline writers and wags began playing with it, but it opened strong.

EXTENDED USE. While the early Sputniks were still in orbit, all sorts of things were done with the word. It was clipped by headline writers, as in "Sput 2: No Surprise," and was turned into a verb: for instance, the sudden popularity of space toys for Christmas had people talking about their homes being Sputnikked for the holidays. It was turned into a slang expression, as in "to go Sputnik" (to go into orbit), and was even used as a marketing buzzword: "Try our oranges—they are sputnik (Russian for 'Out of this world')," claimed a January 1958 magazine ad. It was combined with other words (Sputnik + sorcery yielded sputnickery, which made its debut in the *New York Times* in 1958) and was given odd suffixes (as in the case of sputnikitis, an unhealthy obsession with the satellite). It attracted puckish definitions that purported to explain "what the word Sputnik really means in Russian," such as "no tax cut for the Americans next year." It was even used as a term of deprecation, as when Governor Theodore R. McKeldin of Maryland described segregationist governor Orval E. Faubus "that sputtering sputnik from the Ozarks."

The second syllable became the all-purpose suffix de jour, attracting all sorts of odd and playful coinages. With the dog Laika aboard, Sputnik 2 was Muttnik, Poochnik, Whuffnik, and Dognik. The inability of America to get its own satellite aloft was known as Bottlenik. The failed American *Vanguard* was Flopnik, Dudnik, Puffnik, Pfftnik, Sputternik, and Oopsnik. *Time* magazine (December 16, 1957) reported of the Vanguard disaster, "Samnik is kaputnik." (One series of spacecraft with a -nik ending that had nothing do with Sputnik were the Canadian *Anik* communications satellites. The name was chosen in a nationwide contest and is the Inuit word for brother.)

While the event itself was still reverberating, the shock value of the word soon evaporated and it became, in the light of continued Soviet success and

U.S. failure, a term of self-deprecation in jokes having the United States as the "buttnik." Sample: "Did you hear that America's scientists have invented a new Sputnik cocktail? It's two parts vodka and one part sour grapes."

Most of these terms were ephemeral, and there is doubt as to whether some were ever actually used. For instance, the unlikely word smoochnik was presented in the *American Weekly* (February 9, 1958) as a "kissing date" in the teenage vernacular. On the other hand, a golfer's Sputnik—an extremely high drive off a tee—had a long life on the links.

Other usages stuck. Sputnik was promptly hyphenated, and remains so. NASA historian Roger D. Launius has noted, "Almost immediately, two phrases entered the American lexicon to define time": pre-Sputnik and post-Sputnik. The word was immediately used as a modifier for in terms like Sputnik crisis, Sputnik diplomacy, Sputnik shock, and Sputnik debacle that are still used. Perhaps the strongest survival is in the word beatnik, created by *San Francisco Chronicle* columnist Herb Caen in his column of April 2, 1958. Caen was later quoted, "I coined the word 'beatnik' simply because Russia's Sputnik was aloft at the time and the word popped out." In his book on American youth slang, *Flappers 2 Rappers* Springfield, MA: Merriam-Webster, 1996), Tom Dalzell says, "Beatnik must be considered one of the most successful intentionally coined slang terms in the realm of 20th century American English." Sputnik/beatnik led to a host of variations including neatnik (someone who is compulsively well dressed and well groomed), Vietnik (someone opposed to the war in Vietnam), and peacenik (for individuals who were anti-war).

The original English meaning of the word—any uncrewed Earth satellite—began to lose ground about the time that America's named satellites began to succeed, and the name Sputnik began to get wide application in nicknames, sobriquets, slang, allusions, and more. A singer who made a rapid rise to fame, for example, was dubbed the "Sputnik tenor," and the British humor magazine *Punch* called Frank Sinatra, Sammy Davis Jr., and company "Hollywood sputniks," presumably because they were always in orbit fueled with alcohol. Bill Gold of the *Washington Post* noted that bureaucrats were sometimes called sputniks because "they ran around in circles." The nickname of National Basketball Association point guard Anthony "Spud" Webb was a shortened version of Sputnik, and Lenny Moore of the Baltimore Colts was nicknamed Sputnik, so that to this day there are old-time Baltimoreans who think football whenever the word Sputnik is uttered. Speaking of sports, Sputnik is alive and well as a simile. A 1995 *Washington Times* feature on Reggie Roby, a National Football League punter who routinely put the ball into orbit, reported that "Roby's sputnik-like shots have inspired awe among fans, teammates and opponents alike."

Sputnik was used to name all sorts of things. A Milwaukee bakery advertised doughnuts as sputnuts; the bun on the sputnikburger was topped with a pickle in orbit on the end of a toothpick; and Pednik, a revolving toy spacecraft powered by a foot pedal, was one of the toys rushed to the market for Christmas 1957. The Sputnik hairdo was born at a hairdressers' convention in Lansing, Michigan, on October 7, 1957. The subject's hair was wrapped around her head with an upward flair, covering her ears, and was topped with a four-inch plastic model *satellite* complete with antennas. The Equadors' "Sputnik Dance" was a minor hit record in 1958, and an entrepreneur with an eye to the sky produced the Sputnik Fly Killer—two little pesticide-laden sputniks that hung from the ceiling and were guaranteed to kill flies. Later, the name would be given to a Scandinavian rock group that performed in tacky space suits, a Norwegian folk singer, a ballet, a Ukrainian travel agency, a major Swedish horse race, a Russian forestry holding company, and a take-out double in the game of bridge.

The name shows up as a modifier in some unlikely places. "The Sputnik Christ" is the popular name for a mass-produced print from a late-1950s painting by Warner E. Sallman depicting a monumental Jesus standing atop the Earth and extending into outer space, with planets, meteors, and satellites whizzing by. Sputnik-weed is the popular name of an invasive one-celled algae that attaches itself to the back of oysters and whose single cell may extend for a mile. It got its name when it showed up in Long Island in 1958.

The word also has found new life in the world of gangs and punk rock bands, underscoring the fact that the word still retains a little of its old shock value. The English group Sigue Sigue Sputnik got its name from an article in the *Paris Herald Tribune* about a Russian street gang of that name. In the words of one of the band's members in a rambling web-page explanation, "Somehow it captures the essence of the band . . . the idea of this Fagin like group of money launderers in Moscow is so right . . . the name makes a great story every time its printed . . . like all great names . . . so different . . . even the Sputnik connection is great . . . mans first object in space . . . like some kind of man made god!" The punk rocker is thrilled to find that "Sputnik is [also] the name of a Manila motorcycle gang who when they kill people cut a notch in their arm."

Sputnik was also the generic name for and basis of a hot style of design that has become newly popular among collectors of 1950s modern furniture and accessories, a genre that includes lava lamps and Eames-designed furnishings. Sputnik-inspired design became common in the late 1950s, creating such outrageous items as a hanging Sputnik lamp that had a round chrome center and 18 long arms terminating in star-shaped bulbs (the bulbs for these lamps were and are still known as Sputnik bulbs). On the eBay Internet auction and elsewhere there is a market for Sputnik design as exemplified by spiked Christmas ornaments (originally used for the futuristic aluminum Christmas trees of the early *Space Age*), Sputnik-inspired radio antennas, and lighting fixtures like the aforementioned item (including not only originals but replicas and variants such as the twisted Sputnik lamp, a table version, and a floor lamp that sports six illuminated spheres sprouting from a tulip pedestal).

Sputnik eventually became, for some, both a crude unit of size and a measure of time—even into the 1990s. *New York Times* style editor Amy M. Spindler said, "Karl Lagerfeld taught the public to expect the unexpected. Collections swung from long to short, shifted with the popular music of the moment, were covered with exaggerated chains or pearls and topped by Sputnik-size hats." (*New York Times*, October 1, 1995.) For a while in the early 1990s, one way for sportswriters to convey that a team was getting old was to say, "Most of these guys are older than Sputnik," meaning they were born before the first launch (just as, back in 1957, something could be "as new as Sputnik"). Upon the 125th anniversary of Frank Lloyd Wright's birth it was noted that "Wright was born in 1867, two years after the end of the Civil War, and died April 9, 1959, two years after the launching of Sputnik" (*Arizona Republic*, May 17, 1992).

But perhaps the longest lasting use to which the term Sputnik has been put in America is as a metaphor for a national challenge. "It is time for American policy-makers to meet the challenge posed by the new global information economy much the way an earlier generation responded when the Russians launched Sputnik"—so spoke Senator Larry Pressler, Chairman of the Commerce Committee, addressing a 1995 plan to overhaul the nation's communications laws. This generic use dates back at least to the *Newsweek* of February 10, 1958, when America was told, "We may find ourselves confronted with a sputnik in the chemical, biological and radiological field as we did with missiles." The metaphor seems to have become more powerful over time. In America, every real or imagined weakness in the fabric of national life calls for a Sputnik-style mobilization of national resources—or, as it is sometimes expressed, "a new Sputnik challenge."

Perhaps the ultimate irony is the extent to which the term has become symbolic, underscoring faded Soviet glory and Russian decline. "There are the open manholes, the broken pavement," said an American Peace Corps volunteer in Moscow to a *Baltimore Sun* reporter in 1994, "in a country that built a sputnik before we did. Yet they don't care enough to do anything for people. It's idiotic and insane." Eventually, the name and the very idea of a Russian Sputnik lost its ability to turn heads. A post-Soviet Sputnik fired in

friendship was of so little import that many, if not most, American
newspapers did not even bother to run this small story released by the
Russians: "A space capsule containing a cut-glass replica of the Statue of
Liberty and greetings from Russian President Boris Yeltsin has splashed
down off the Washington coast on November 23, 1992. The Resource
500 Sputnik, which had circled the Earth 111 times in six days, was billed
as Russia's first private space launch and was targeted at the United States
as a peaceful promotion marking the end of the Cold War."

squib. (1) Any of various small explosive devices. (2) An explosive device
used in the *ignition* of a rocket. Usually called an *igniter*. (SP-7.)

squitch. A *squib*-operated switch (SP-7).

squitter. Random firing, intentional or otherwise, of a transponder
transmitter (SP-7).

STADAN (Space Tracking and Data Acquisition Network). Former
name for *Spaceflight Tracking and Data Network.*

staging. The *separation* of engines of a multistage *rocket* during flight.

Star City. An extensive complex of service and office buildings outside
Moscow, where Russian and foreign *cosmonauts* live and undergo
training. It is also the site of the Yuri Gagarin Cosmonaut Training Center.

state-of-the-art. Level to which technology and scientific knowledge
has developed to a certain time. The term was invoked often in the
early days of the U.S. *human spaceflight program* (e.g., NASA, *Glossary/
Congressional Budget Submission,* p. 40).

steely-eyed missile man. NASA *astronaut* or aerospace engineer who
makes a significant contribution or devises an ingenious solution to a
tough problem. It is a term of high honor as heard in the film *Apollo 13*
and is found in numerous biographies of NASA astronauts and ground
controllers. "There weren't many steely-eyed missile men in the NASA
family. Von Braun was certainly one. Kraft [Christopher C. Kraft, NASA's
first Flight Director] was probably one too." (Lovell and Kluger, *Lost
Moon,* p. 170.)

steely-eyed rocket scientist. Synonym for *steely-eyed missile man.*
USAGE. In the post-*Challenger* era the term took on a negative
connotation in the sense that it implied a certain recklessness and dis-
regard for safety: "On this week in 1986, the steely-eyed rocket scientists
from Von Braun's vaunted Marshall Space Flight Center in Alabama threw
caution to the frigid January wind, pushing hard to light Challenger's
candles, even though they knew the score." (Miles O'Brien, space correspondent, "Fifteen Years after Challenger, NASA Inoculates against 'Go
Fever,'" CNN, January 18, 2001, http://archives.cnn.com/2001/TECH/
space/01/18/downlinks.40/.)

Stennis Space Center (SSC). Formally known as the John C. Stennis Space Center, SSC is NASA's primary center for *rocket* propulsion testing. It has undergone a number of name changes. Its original name, Mississippi Test Operations, was changed to Mississippi Test Facility (MTF) in 1965. In 1974 the facility was named the National Space Technology Laboratories (NSTL), reporting to NASA Headquarters in Washington, DC.

NASA announced on October 25, 1961, that it had selected southwestern Mississippi as the site for a large *booster (Saturn)* test facility under the direction of the *Marshall Space Flight Center* (MSFC). Pending official naming of the site, NASA encouraged use of the name Mississippi Test Facility, which seems to have been already in informal use. On December 18, 1961, the name Mississippi Test Operations was officially adopted, but the site was still widely called the Mississippi Test Facility, particularly by Headquarters and MSFC offices concerned in the installation's development. On July 1, 1965, MSFC announced the official redesignation Mississippi Test Facility. The change was said to better "reflect the mission of the facility."

MTF test stands were put into standby status November 9, 1970, after more than four years and the test-firing of 13 S-IC first stages and 15 S-II second stages of the *Saturn V.* With the close of Saturn production and the approaching end of the *Apollo* program, NASA had established an Earth Resources Laboratory at MTF in September 1970, stressing applications of remote-sensing data from aircraft and *satellites.* A number of other government agencies, at NASA invitation, moved research activities related to resources and the environment to MTF to take advantage of its facilities. And on March 1, 1971, NASA announced that MTF had also been selected for sea-level testing of the *Space Shuttle*'s main *engine.*

On June 14, 1974, the Mississippi Test Facility was renamed the National Space Technology Laboratories and became a permanent NASA field installation reporting directly to NASA Headquarters, "because of the growing importance of the activities at NSTL and of the agencies taking advantage of NSTL capabilities."

A new chapter was added in June 1975 when the Space Shuttle main engine was tested at Stennis for the first time. All the engines used to boost the Shuttle into low-Earth orbit are tested and proven flightworthy at SSC on the same stands used to test-fire all first and second stages of the Saturn V in the Apollo and Skylab programs. The A Test Complex of the Rocket Propulsion Test Complex now supports engine testing for the Space Shuttle program.

Stennis has expanded its rocket test capabilities. The E Test Complex was constructed as a result of several national propulsion development programs in the late 1980s and early 1990s. The versatile three-stand complex includes seven separate test cells capable of testing that involves ultra high-pressure gases and high-pressure, super-cold fluids. These *state-of-the-art* propulsion test facilities are designed for testing developmental components to full-scale engines. Today Stennis also plays a key role in the *Constellation* program. It continues as a collaborative research community, with residents including the U.S. Navy, Lockheed Martin, and NASA researchers.

ETYMOLOGY. In May 1988, the Mississippi Test Facility was renamed the John C. Stennis Space Center in honor of Senator John C. Stennis for his steadfast leadership and staunch support of the nation's *space program*.

SOURCES. SP-4402, pp. 156–57; MSFC News Release 65-167; MSFC, *Marshall Star,* November 10, 1970, pp. 1, 4; NASA News Releases 70-98, 70-147, 71-30, 72-167, 74-159. For more information about Stennis Space Center, see Herring, *Way Station to Space,* SP-4310.

STS. Space Transportation System. This is the formal name of the *Space Shuttle* program. All Shuttle flights are designated with the abbreviation STS and the number.

USAGE. Unlike earlier piloted spaceflight program names *(Mercury, Gemini, Apollo),* STS carries little name recognition among the public: "The first hearing of the Challenger investigating commission was only minutes old when Chairman William P. Rogers posed his first question. 'What,' he asked, 'does STS stand for?' He probably knew as well as any of the others on the 13-member commission that STS is the space agency's way of referring to the 'Space Transportation System'—or, in plain English, the fleet of Space Shuttles. But Rogers thought the group should not only find what caused the shuttle to explode but also should make the meaning of STS clear to everyone. The question marked Rogers, from the very beginning, as the panel member who would cut through engineering fog and aerospace jargon of the witnesses before him and ask the questions ordinary citizens would ask." (Harry F. Rosenthal, "Rogers Cut through Witnesses' Jargon," *Lexington (KY) Herald-Leader,* June 10, 1986, p. A3.)

suborbital. Not attaining *orbit,* i.e., a ballistic space *shot.*

sub-satellite. A secondary object released from a larger parent *satellite* in *orbit,* e.g., an electronic *ferret* released by a *reconnaissance satellite.*

subsonic. Describing a speed that is less than the speed of sound.

SuitSat/Suitsat. An empty spacesuit equipped with a radio transmitter launched from the *International Space Station* (ISS) on February 3, 2006. The suit hurled from the ISS was a Russian Orlan spacesuit with three batteries, a radio transmitter, and internal sensors to measure temperature and battery power. The SuitSat provided recorded greetings in six languages to ham radio operators for about two *orbits* of the Earth before it stopped transmitting, perhaps because its batteries had failed in the cold environment of space, according to amateur radio coordinators affiliated with the station program (International Space Station Status Report, SS06–005, February 4, 2006).

sunspot. A magnetic disturbance on the Sun that appears as a dark blotch on its surface.

Super Explorer. See *High Energy Astronomy Observatory*.

supernova. An explosion that marks the end of a very massive star's life. When it occurs, the exploding star can outshine the totality of all the other stars in the *galaxy* for several days and may leave behind only a crushed core (perhaps a *neutron star* or *black hole*). Astronomers estimate that a supernova explosion takes place about once a century in a galaxy like our *Milky Way*. While most supernovae in our galaxy are probably hidden from our view by interstellar gas and dust, astronomers can detect supernova explosions in other galaxies relatively frequently. (ASP Glossary.)

Surveyor (lunar landing *probe*). Name chosen in May 1960 to designate an advanced *spacecraft* series to explore and analyze the Moon's surface. The designation was in keeping with the policy of naming lunar probes after "land exploration activities" established under the Cortright system of naming space probes. Following the *Ranger* photographic lunar hard *landers* (see *hard landing*), Surveyor probes marked an important advance in space technology: a *soft landing* on the Moon's surface to survey it with television cameras and analyze its characteristics using scientific instruments. Five Surveyor spacecraft—Surveyor 1 in 1966, Surveyor 3, 5, and 6 in 1967, and Surveyor 7 in 1968—soft-landed on the Moon and operated on the lunar surface for a combined total time of approximately 17 months. They transmitted more than 87,000 photographs and made chemical and mechanical analyses of surface and subsurface samples. (SP-4402, pp. 84–85; Edgar M. Cortright, Assistant Director of Lunar and Planetary Programs, NASA, memorandum to NASA Ad Hoc Committee to Name Space Projects and Objects, May 17, 1960; NASA Ad Hoc Committee to Name Space Projects and Objects, minutes of meetings, May 19, 1960; NASA, Office of Technology Utilization, *Surveyor Program Results*, SP-184 [Washington, DC: NASA, 1969], pp. v–vii.)

swizzle stick. A rod 10 inches (25.4 cm) long, with a hook on one end for pulling levers and a stub on the other for pushing buttons. Carried in early human spaceflight to extend one's reach with a fully pressurized suit. "We call it, naturally, a 'swizzle stick'" (John H. Glenn Jr., "Seven

Miles of Wire—And a Swizzle Stick," in Carpenter et al., *We Seven*, p. 107).

Symphonie. Franco-German communications *satellite*. Experimental communications satellites designed and built in Europe for *launch* by NASA with *launch vehicle*s and services paid for by France and West Germany.

In 1967 France had a stationary *(synchronous) orbit* communications satellite, SAROS (Satellite de Radiodiffusion pour Orbit Stationnaire), in the design stage, and West Germany was about to begin designing its Olympia *satellite*. The two nations agreed in June 1967 to combine their programs in a new joint effort. Two satellites were developed by the joint Consortium Industriel Franco-Allemand pour le Satellite Symphonie (CIFAS) under the direction of the French *space agency* Centre National d'Études Spatiales *(CNES)* and the West German space agency Gesellschaft für Weltraumforschung (GfW). The three-axis stabilized satellites were to test equipment for television, radio, telephone, telegraph, and data transmission from *geosynchronous orbit*, 22,300 miles (36,000 km) above the equator. They were planned for launch from French Guiana on the Europa II launch vehicle, but when the European Launcher Development Organization (ELDO) canceled its vehicle project, the countries turned to NASA. The contract for NASA launch services was signed in June 1974. NASA launched Symphonie 1 (Symphonie-A, before launch) into orbit from the Eastern Test Range December 18, 1974. Symphonie-B was launched on August 27, 1975.

ETYMOLOGY. Participants in the 1967 discussions in Bonn (the Federal Republic of Germany's capital on the Rhine River) sought a new name for the joint satellite just before the agreement was signed. The negotiator for France was Foreign Affairs Minister Maurice Schumann, whose name reminded Gerard Dieulot, technical director of the French program, of the German composer Robert Schumann. The new accord in the Rhine Valley, Dieulot suggested, was a "symphony by Schumann." Symphonie, the French spelling of the Latin and Greek *symphonia* (harmony or agreement), was adopted when the agreement was signed in June. (SP-4402, p. 72.)

Syncom. Synchronous communications satellite. A NASA program for communications *satellite*s in *geosynchronous orbit*. Three Syncom satellites were developed and *launch*ed during the 1960s. After a launching success but communications failure with Syncom 1 (February 14, 1963), Syncom 2 was launched on July 26, 1963, into the first synchronous orbit ever achieved, and Syncom 3 (launched August 19, 1964) was put into the first truly stationary orbit. The Department of Defense participated in Syncom research and development, providing ground stations and conducting communications experiments. Early in 1965, after completing the research and development program, NASA transferred use of the two successful Syncom satellites to the Department of Defense. In the 1980s the series was continued as Syncom IV. Four were launched from the Space Shuttle under the name Leasat in 1984 and 1985.

ETYMOLOGY. The name was devised by Alton E. Jones of NASA's *Goddard Space Flight Center* in August 1961, when he was working on the preliminary project development plan. Having decided that a name was required before the plan could go to press the next day, he invented the name Syncom (from the first syllables of the words synchronous and communications). By the end of August, NASA Headquarters had approved the preliminary plan and a press release has been issued using the name.

SOURCES. SP-4402, p. 73; Alton E. Jones, GSFC, letter to Historical Staff, NASA, April 7, 1964; GSFC, "Syncom Preliminary Project Development Plan," August 5, 1961; Robert C. Seamans Jr., Associate Administrator, NASA, memorandum to Director, Office of Space Flight Programs, NASA, August 17, 1961; NASA News Release 61-178.

T

TAD. *Thrust*-Augmented *Delta*. See *Thor*.

taikonaut. Anglicized spelling of the Chinese equivalent of *astronaut/cosmonaut*.

ETYMOLOGY. According to the oddly phrased official Chinese space agency website www.taikonaut.com, "'Taikong' is a Chinese word that means space or cosmos. The resulted [sic] prefix 'taiko-' is similar to 'astro-' and 'cosmo-' that makes three words perfectly symmetric, both in meaning and in form. Removing 'g' from 'taikong' is to make the word short and easy to pronounce. On the other side, its pronunciation is also close to 'taikong ren,' the Chinese words 'space men.'"

FIRST USE. According to the official www.taikonaut.com website, "In May 1998, Mr. Chiew Lee Yih from Malaysia used it first in interest [Internet?] newsgroups. Almost at the same time, Mr. Chen Lan used it and announced it at 'Go Taikonauts' site."

takeoff. The *launch* of a *spacecraft*. Informally, *blastoff*. See also *liftoff*.

Tang. An orange-flavored soft drink packaged in powdered form. A commercial product of General Foods initially developed for the Army for prepackaged field rations, Tang was bought off the supermarket shelf by NASA for the *Apollo* astronauts, who consumed it on the Moon. General Foods made much of this fact in its advertising. Along with *Teflon* and *Velcro*, Tang is at the heart of one of the great myths of the *Space Age:* the idea that Space Age technology brought the world little in return. The glib and all too common way of looking at it is that we spent billions, and all that came of it was "Teflon, Tang, and Velcro." These were actually products of earlier days, and all three were in fact used by NASA.

TAT. Thrust-Augmented Thor. See *Thor*.

taxi crew. Humans who fly to and from the *International Space Station* to deliver goods and equipment.

TD. ESRO Thor-Delta solar astronomy *satellite* project. TD, an abbreviation for the U.S. Thor-Delta *launch vehicle*, was the name given to the satellite project by the European Space Research Organization (*ESRO*). Under a 1966 memorandum of understanding with ESRO, NASA was reimbursed for the *launch*. TD was the second reimbursable launch under this agreement; *HEOS* 1 was the first.

Proposals for the satellite, then unnamed, had been discussed at an astronomy colloquium soon after the formal establishment of ESRO in March 1964. By 1965, ESRO had planned a series of TD satellites and in 1967, after several program delays, signed a contract with NASA for the launch of two satellites, TD-1 and TD-2. In April 1968, however, ESRO announced the cancellation of both satellites because of problems in financing. The project was later reinstated, and a second contract for a single Thor-Delta launch was signed with NASA in June 1970. The satellite was subsequently redesignated TD-1A because it differed from the two earlier configurations and combined the TD-2 design with several experiments originally planned for TD-1. T-1A, a solar astronomy satellite designed to carry a variety of instruments including a large telescope, was launched by NASA on March 11, 1972.

SOURCES. SP-4402, p. 75; George D. Baker, Delta Project Office, GSFC, telephone interview, July 23, 1971; NASA News Release 66-332; R. Lust, "The European Space Research Organisation," *Science* 149 (July 23, 1965): 394–96; "Europeans Reviewing Space Goals through Early 1970's," *Aviation*

Week & Space Technology 82 (June 14, 1965): 200; John L. Hess, "European Communication Satellite Seems Doomed," *New York Times,* April 28, 1968, p. 24.

TDRSS (commonly pronounced tee-drus). Tracking and Data Relay Satellite System. TDRSS refers to the network and TDRS to each satellite in the network. On April 4–9, 1983, the crew of *STS-6,* the *Space Shuttle Challenger,* deployed the first shuttle-launched TDRS (TDRS A) into *geostationary orbit.* TDRS B was destroyed in the Challenger disaster in 1986. TDRS C through G were launched by Shuttles, and H through I by *Atlas* IIA's. In 2007 NASA awarded Boeing contracts for TDRS J and TDRS K, to be deployed in 2012 and 2013 as replacement satellites.

Teflon. Brand name for products made with a fluorine-containing polymer. Along with *Velcro* and the powdered beverage *Tang,* Teflon is one of the products erroneously identified as a byproduct of the U.S. *space program.* Like the other two products, it was used by the *space program* but had origins elsewhere. Teflon was invented by accident as a residue of refrigeration gases in the DuPont chemical company's labs in 1938, by Dr. Roy J. Plunkett, a chemist who saw that the substance had unusual properties. First used only in defense projects, it became a commercial product in 1948. It was mated with the electric frying pan to create the "Happy Pan" in 1961, at about the same time that NASA started using it for a host of applications from *space suits* to *nosecones.* (*New York Times* obituary of Plunkett, May 15, 1994, p. 44.)

telemetry. (1) Measuring an object or phenomenon from a distance. Radio signals from a *spacecraft* are used to encode and transmit data to a ground station. Without telemetry, uncrewed space exploration would not exist and human spaceflight would be severely hampered. (2) The data so treated.

telepresence. The use of real-time video communications coupled with remote control techniques so as to provide an operator on the surface of the Earth or in another location with the capability to carry out complex operations in space or on the surface of a planet or Moon (Paine Report, p. 199).

telepuppet, telerobot. Early names for robotic devices that would operate in space and explore planets. Terms replaced by *rover* and *lander.*

Telesat. Telecommunications Satellite. In early 1969 the Canadian Ministry of Communications proposed plans for a *satellite* system that could be used entirely for domestic communications. The system would be managed and operated by Telesat Canada, a new corporation supported by industry, government, and public investment. The first two satellites in the system, Telesat A and B, would be launched into synchronous equatorial *orbit* and be capable of relaying TV, telephone, and data transmissions throughout Canada. Under an agreement with Telesat, NASA would provide the Thor-Delta *launch vehicle*s and be reimbursed for the satellite launches. (SP-4402, p. 75; "Canadian Satellite," *Washington Post,* April 16, 1969, p. A17; NASA News Release 71-85.)

telescience. Conducting scientific operations in remote locations by teleoperation (Paine Report, p. 199).

Television and Infra-Red Observation Satellite (TIROS). Later, Tiros meteorological *satellite.* See *TIROS, TOS, ITOS.*

Telstar (active-repeater communications satellite). The active communications satellites developed by American Telephone & Telegraph Company. In November 1961, at the request of AT&T's Bell Telephone Laboratories, NASA endorsed the selection of Telstar (telecommunications + star) as a name for the project. NASA was responsible for *launch, tracking,* and data acquisition for the AT&T-built satellites on a cost-reimbursable basis. Telstar 1, the first active-repeater communications satellite, was the first privately funded satellite and relayed the first live transatlantic telecast after its July 10, 1962, launch. It was followed by the equally successful Telstar 2, launched May 7, 1963. (SP-4402, p. 75; David Williamson Jr., Office of Tracking and Data Acquisition, NASA, memorandum to Boyd C. Myers II [Chairman, NASA Project Designation Committee], Director, Program Review and Resources Management, NASA, October 18, 1961; Myers, memorandum to Robert C. Seamans Jr., Associate Administrator, NASA, October 30, 1961, with approval signature of Dr. Seamans, 2 November 1961.)

termites. Term of derogation for Pentagon and civilian overseers during the pre- and immediate post-*Sputnik* days at the Army Ballistic Missile Agency *(ABMA)* in Huntsville.

ETYMOLOGY. On December 14, 1958, Wernher von Braun testified before Senator Lyndon Johnson, Chairman of the Preparedness Subcommittee of the Senate Committee on the Armed Forces, on problems encountered with the progress and funding for the *Jupiter* C. The transcript provided this exchange in an otherwise somber discussion:

"Dr. VON BRAUN. I would say at first things were going very smoothly. We had everything we asked for, and it was only after about a year or so that—how shall I express myself—that the termites got into the system.

"Senator JOHNSON. The what?

"Dr. VON BRAUN. The termites.

"Senator JOHNSON. Termites?

"Dr. VON BRAUN. As General Medaris mentioned before, the money was allocated, but somebody withheld it, and we got only part of it.

"Senator JOHNSON. When did the termites come?

"Dr. VON BRAUN. I would say we had clear sailing for about a year.

"Senator JOHNSON. When did the termites come?"

Finally Von Braun reveals the dates of the infestation.

Terra. The flagship of NASA's Earth Observing System *(EOS)*. Terra was launched on December 18, 1999, from Vandenberg Air Force Base. It is part of an international program to monitor climate and environmental change on Earth over a 15-year period. Part of a series of EOS satellites, Terra will enable new research into the ways that Earth's lands, oceans, air, ice, and life function as a total environmental system.

terrestrial. Of or pertaining to the Earth.

Terrier *(sounding rocket* first stage). A sounding rocket first stage used by NASA. It was developed by Hercules Powder Company as the first stage of the Navy's Terrier antiaircraft *missile,* and NASA inherited the name (originally given to it by the Johns Hopkins Applied Physics Laboratory, where it was created). NASA used Terrier with the *Malemute* second stage, as the Terrier-Malemute. (Edward E. Mayo, Flight Performance Branch, Sounding Rocket Division, GSFC, information sent to Historical Office, NASA, January 30, 1975.)

ETYMOLOGY. Richard Kirschner, who had been in charge of the guidance research effort at the Applied Physics Lab, gave the missile its name because of its tenacious ability to hold to the center of the radar guidance beam. All of the early rockets and missiles developed by the Johns Hopkins Applied Physics Laboratory began with the letter T—Talos, Tarter, Terrier, Transit, Triad, Trident, and Triton—because the Lab was operating under a Section T contract with the U.S. Navy. (Klingaman, *APL—Fifty Years of Service to the Nation,* p. 64.)

Terrier-Malemute *(sounding rocket).* See *Malemute* and *Terrier.*

Tethered Satellite System (TSS). A joint NASA–Italian Space Agency project for a *satellite* connected to another by means of a thin wire. During the first test of the Tethered Satellite System (TSS-1), in 1992, a fault with the reel mechanism allowed only 840 feet (256 m) of the tether to be deployed, although the satellite was recovered. On the second *mission,* TSS-1R in 1996, 12 miles (19.6 km) of tether was deployed before the tether broke and the *satellite* was lost.

Thor *(launch vehicle).* Originally developed as a U.S. Air Force intermediate-range ballistic *missile* by Douglas Aircraft Company and adapted for use as a *launch vehicle* in combination. NASA used Thor as a first stage with both *Agena* and *Delta* upper stages. The Air Force–developed Thrust-Augmented Thor (TAT), with three added solid-*propellant rocket* motors strapped on the base of the Thor, also was used with both Agena and Delta upper stages. When TAT was used with Agena, the configuration was called Thrust-Augmented Thor-Agena. With Delta, the vehicle was known as Thrust-Augmented Delta (TAD) or Thrust-Augmented Thor-Delta (TAT-Delta).

In 1966 the Air Force procured a new version of the Thor first stage, elongated to increase fuel capacity, for heavier payloads—the Long-Tank Thrust-Augmented Thor (LTTAT), sometimes also called Thorad. LTTAT used with an Agena upper stage was called Long-Tank Thrust-Augmented Thor-Agena or Thorad-Agena. With Delta, it was Long-Tank Thrust-Augmented Thor-Delta. NASA began using the long-tank Thor with the improved Delta second stage in 1968, going to six strap-on rockets for extra thrust in 1970 and introducing nine strap-on rockets in 1972. Combinations varied according to the performance needed for the *mission. See Delta.*

ETYMOLOGY. The name, which came into use in 1955, derived from the ancient Norse god of thunder, "the strongest of gods and men." The origin of the missile's name has been traced to Joe Rowland, Director of Public Relations at the Martin Company, who was assigned to suggest names for Martin's new intercontinental ballistic missile in preparation for a meeting at Air Research and Development Command (ARDC) Headquarters. The meeting was to be attended by representatives of other missile contractors, Convair/Astronautics Division of General Dynamics Corporation and Douglas Aircraft Company. Of Rowland's list of proposed names, *Titan* was the one preferred by his colleagues, with Thor as second choice. At the ARDC meeting, the first-choice Titan was accepted as the appropriate name for the Martin Company's project. Through a misunderstanding, Douglas had prepared no name to propose for its missile. Rowland, with Titan now firm for his company's project, offered his alternative to Donald Douglas Jr. Douglas's Vice President of Public Relations agreed it was an attractive name and proposed it to ARDC officials. It was officially adopted.

SOURCES. SP-4402, pp. 12-13; W. C. Cleveland, Director of Public Relations, Douglas Missile and Space Systems Division, letter to Historical Staff, NASA, August 27, 1965; Bulfinch, *Mythology,* p. 243; Mallan, *Peace Is a Three-Edged Sword,* pp. 190–92; NASA News Release 68-84; USAF News Release 205.65.

Thor-Able *(launch vehicle). See Thor.*

Thorad. Long-Tank Thrust-Augmented Thor. *See Thor.*

Thorad-Agena. *See Thor.*

Thor-Agena. *See Thor.*

Thor-Delta. *See Thor.*

thrust. The force a *rocket engine* exerts to overcome inertia and accelerate a vehicle to required velocity. Thrust is expressed in pounds (or, in metric units, newtons) of force.

tickety-boo / tickety boo. *A-OK.* A term that has been used on occasion to signal success in space. On the *Space Shuttle Challenger*'s first flight, *astronaut* Paul J. Weitz said, "Everything is going tickety-boo so far" ("Quote of the Day," *New York Times,* April 5, 1983, p. C51).
FIRST USE. "Churchill Pilots Plane," *Christian Science Monitor,* January 19, 1942, p. 6. The article reported that British Prime Minister Winston Churchill piloted a passenger plane for a portion of his wartime trip to Washington. The British Air Ministry queried the plane with its VIP pilot: "How's that British Airways Boeing getting on?" "All tickety boo," replied the Wing Commander in charge. "Dead right on the course and time." (NASA Names Files, record no. 17503.)

TIROS, TOS, ITOS (meteorological satellites). A series of meteorological *satellites* that provided weather data from high above the Earth's cloud cover, 1960–74. In mid-1958 the Department of Defense's *Advanced Research Projects Agency* (initiator of the project) requested the Radio Corporation of America (contractor for the project) to supply a name for the satellite. RCA personnel concocted the name TIROS, an acronym derived from the descriptive title Television and Infra-Red Observation Satellite. The name eventually came to be written Tiros (first letter only capitalized) because it was used as part of other acronyms.

In April 1959, responsibility for the Tiros research and development program was transferred from the Department of Defense to NASA, and on April 1, 1960, Tiros 1 was launched into *orbit.* Meteorologists were to receive valuable data including more than 5 million usable cloud pictures from 10 Tiros weather satellites. By early 1964 NASA had orbited Tiros 1 through Tiros 8, and the U.S. Weather Bureau (USWB) was making operational use of the meteorological data from them. These satellites were able to photograph about 20 percent of the Earth each day.

On May 28, 1964, NASA and the Weather Bureau announced a plan for an operational meteorological satellite system based on Tiros research and development. They called the system TOS, an acronym for Tiros Operational Satellite. In accordance with the NASA-USWB agreement, Tiros 9 was a NASA-financed modified Tiros satellite, designed to test the new "cartwheel" configuration on which the TOS would be based. Tiros 10 was a USWB-financed Tiros satellite similar to Tiros 9, intended to continue testing the TOS concept. Early in 1966 NASA launched the two operational satellites in the TOS system financed, managed, and operated by the Weather Bureau, by then an agency of the new Environmental Science Services Administration *(ESSA).* Upon their successful orbit, ESSA designated the TOS satellites ESSA 1 and ESSA 2C (ESSA in this case being derived from an acronym for Environmental Survey Satellite). These two satellites provided continuous cloud-cover pictures of the entire sunlit portion of the Earth at least once daily.

In 1966 NASA announced plans for a design study of an improved TOS *spacecraft* that would be twice as large as the previous TOS satellites. This spacecraft would be able to scan the Earth's nighttime cloud cover and would more than double the daily weather coverage obtained in the TOS series of ESSA satellites. The first satellite in the Improved Tiros Operational Satellite (ITOS) series, ITOS 1, launched January 23, 1970, was a joint project of NASA and ESSA. With the exception of ITOS 1, spacecraft in the ITOS series would be funded by ESSA.

On October 3, 1970, ESSA was combined with the major federal programs concerned with the environments of the sea and air; programs from four departments and one agency were consolidated to form the National Oceanic and Atmospheric Administration (NOAA) in the Department of Commerce. The first operational ITOS spacecraft funded by NOAA, designated NOAA 1 in *orbit,* was launched December 11, 1970. NOAA 4 (ITOS-G) was put into orbit December 15, 1974, to join the still orbiting NOAA 2 and 3 (launched October 15, 1972, and November 6, 1973) in obtaining global cloud-cover data day and night and global measurements of the Earth's atmospheric structure for weather prediction.

SOURCES. SP-4402, pp. 48–49, 76–78; Robert F. Garbarini, Office of Space Science and Applications, NASA, letter to Historical Staff, NASA, December 30, 1963; NASA News Release 66-115.

Tiros Operational Satellite (TOS). See *TIROS, TOS, ITOS.*

Titan *(launch vehicle).* Family of launch vehicles used between 1959 and 2005. The Titan II launch vehicle was adapted from the U.S. Air Force intercontinental ballistic *missile* to serve as the *Gemini* launch vehicle in NASA's second *human spaceflight program.* The Titan III, an improved Titan II with two solid-*propellant* strap-on rockets, was developed for use by the Air Force as a standardized launch vehicle that could lift large payloads into Earth *orbit.* NASA contracted for Titan III vehicles for a limited number of missions to begin in the mid-1970s. *ATS* satellites would require the Titan IIIC vehicles and *HEAO* satellites, the Titan IIID configuration. Interplanetary missions requiring high-velocity escape trajectories, such as the *Viking* Mars *probes* and *Helios* solar probes, began using the Titan III–*Centaur* configuration on completion of the Centaur integration program in 1974. A Titan IIIE–Centaur launched Helios 1 into orbit of the Sun on December 10, 1974.

The Titan IIIC (which launched *ATS* 6 on May 30, 1974) could put a 26,000-pound (11,820-kg) *payload* into a 345-mile (555-km) orbit, or

3,300 pounds (1,500 kg) into synchronous orbit. The Titan IIIE–Centaur could launch 11,000 pounds (5,135 kg) into an Earth-escape orbit, or 8,700 pounds (3,960 kg) to Mars or Venus.

Titan IV rockets were developed by the U.S. Air Force in the 1990s. A Titan IV-B–Centaur launched the *Cassini-Huygens probe* to Saturn in 1997. Titan IV rockets last flew in 2005 and were superseded by Atlas V and Delta IV rockets.

ETYMOLOGY. Originating in 1955, the name Titan was proposed by Joe Rowland, Director of Public Relations at the Martin Company (producer of the missile for the Air Force). Rowland was assigned the task of suggesting possible names for the project, requested of Martin by the Air Research and Development Command. Of the list of possible names, Titan was preferred. Rowland took the name from Roman mythology: the Titans were a race of giants who inhabited the Earth before men were created. ARDC approved the nomination, and Titan became the official name. When the missile was developed, the original Titan came to be known as Titan I, and the second improved version was named Titan II. Titan II was chosen as the Gemini launch vehicle because greater *thrust* was required to launch the 7,000-pound (3.5-metric-ton) Gemini spacecraft. Moreover, its storable fuels promised the split-second launch needed for *rendezvous* with the target vehicle.

SOURCES. SP-4402, pp. 24–25; R. L. Tonsing, Director of Public Relations, Aerospace Division, Martin Marietta Corp., letter to Historical Staff, NASA, September 26, 1965; Mallan, *Peace Is a Three-Edged Sword*, pp. 190–92; Bulfinch, *Mythology*, p. 15. For more information about Titan, see Launius and Jenkins, eds., *To Reach the High Frontier*.

Titan-Centaur *(launch vehicle)*. See *Titan*.

Tomahawk *(sounding rocket upper stage)*. The Tomahawk, a sounding rocket upper *stage* used with the *Nike booster* stage, was named by Thiokol Corporation for the Indian weapon, in Thiokol's tradition of giving its motors ethnic, regional, and Indian-related names. (See *Cajun*.) The Nike-Tomahawk could lift 59-pound (27-kg) instrumented payloads to a 305-mile (490-km) operating altitude, or 260 pounds (118 kg) to 130 miles (210 km). (GSFC, *United States Sounding Rocket Program*, pp. 38, 47.)

topside sounder. A satellite designed to determine ion concentration within the ionosphere as measured from above the ionosphere.

TOS. Tiros Operational Satellite (meteorological). See *TIROS, TOS, ITOS*.

touchdown. The moment of landing of a *space vehicle*.

tracking. The science of monitoring satellite locations by means of radio or radar antennas or visually at ground stations or by using other satellite systems in space.

Tracking and Data Relay Satellite System. See *TDRSS*.

tracking station. A station set up to track an object through the *atmosphere* or space, usually by means of radar or radio.

trailing side. For a *satellite* that keeps the same face toward the planet, the hemisphere that faces backwards, away from the direction of motion.

trajectory. Path followed by a vehicle.

tranquillityite. Extraterrestrial mineral discovered at the Sea of Tranquillity, the Moon. See also *pyroxferroite*.

transhab. Concept of an inflatable living quarters proposed as a crew quarters for the *International Space Station*. The concept was tested but not chosen, and NASA is no longer pursuing the concept.

trench. Name for the *Mercury* Control Center at *Cape Canaveral* (1960–65) or *Mission Control Center* at Houston (1965–72). The key personnel in the trench are the retrofire officer (RETRO), the flight dynamics officer (FIDO), and the specialist in navigation and computer software (GUIDO). (Kranz, *Failure Is Not an Option,* pp. 395–96.)

Triana. *Satellite* proposed in 1998 by former Vice President Al Gore for the purpose of Earth observation. It never flew. It was intended to be positioned so as to provide a near-continuous view of the sunlit side of the Earth and to make that live image available via the Internet. Gore hoped not only to advance science with these images but also to raise awareness of the Earth itself, updating the influential "Blue Marble" photograph taken by *Apollo* 17, a copy of which had long hung in his office. In addition to an imaging camera, the satellite would carry a radiometer that would take the first direct measurements of how much sunlight is reflected and emitted from the whole Earth. This data could constitute a barometer for the process of global warming. NASA proposed expanding these scientific goals to include measuring the amount of solar energy reaching Earth, observing cloud patterns and weather systems, monitoring the health of Earth's vegetation, and tracking the amount of ultraviolet light reaching the surface through the ozone layer.

Derided by critics as an unfocused project, the satellite was nicknamed GoreSat and GorCam, and was often referred to as an "overpriced screen saver" by its opponents. The *New York Times* termed the idea boring, compared it to watching grass grow, and sided with an unnamed Republican who described it as "a satellite that in essence allows people to sit in front of the TV set and think about nothing" ("Like Watching Al Gore," editorial, *New York Times,* March 18, 1998, p. A26).

Congress asked the National Academy of Sciences if the project was worthwhile. The resulting report stated that the *mission* was "strong and vital." Faced with political hostility on one side and scientific support on the other, Triana could neither be launched nor be terminated. It was removed from its original *launch* opportunity on *STS*-107. In an attempt to regain support for the project, NASA renamed the satellite *Deep Space Climate Observatory* (DSCOVR). In 2001 the $120 million *spacecraft* was mothballed at the *Goddard Space Flight Center,* where it remains in storage at a cost of $1 million annually.

ETYMOLOGY. The name was bestowed on the project by Gore. Rodrigo de Triana was the lookout who first sighted land from aboard the Pinta on October 12, 1492, during Columbus' first voyage to the New World. Although no documentary evidence has been found, Triana was probably named for him. Two U.S. Navy ships have carried the name.

TRMM. *Tropical Rainfall Measuring Mission.*

Tropical Rainfall Measuring Mission (TRMM). Joint *mission* between NASA and the *Japan Aerospace Exploration Agency* (JAXA) designed to monitor and study tropical rainfall. TRMM is the first *satellite* to measure rainfall over the global tropics, allowing scientists to study the transfer of water and energy among the global *atmosphere* and ocean surface that form the vastest portions of the Earth's climate system. Because TRMM's radar enables it to "see through" clouds, it allows weather researchers to make the equivalent of a CAT scan of hurricanes and helps improve prediction of severe storms. Launched in 1997, TRMM was originally designed as a three-year research mission. Following four years of extending TRMM, NASA and JAXA recently announced a decision to decommission TRMM and proceed with a safe, controlled *deorbit,* but this plan was abandoned to allow the satellite to serve through the 2004 hurricane season. Options for safe *reentry* become increasingly limited the longer TRMM is operated, as it is already more than three years beyond design life. (NASA News Release 04-261a, corrected, "NASA Extends TRMM Operations through 2004 Hurricane Season.")

T-time. Any specific time, minus or plus, as referenced to zero or *launch* time, during a *countdown* sequence that is intended to result in the firing of a *rocket* propulsion unit that launches a *rocket* vehicle or *missile.*

tumble. Describing an oblong *Earth satellite* rotating on its horizontal axis, end over end.

FIRST USE. Heflin, *Interim Glossary Aero-Space Terms,* p. 33, where it is said to have first described *Sputnik* 2.

TV. Test vehicle. The designation of the first *Vanguard* attempt on December 6, 1957, carried the designation TV 3.

UE. *Redstone* numbering system. See *Juno.*

Uhuru. Small Astronomy Satellite A. See *Explorer.*

ullage. The amount that a container, such as a fuel tank, lacks of being full.

Ulysses. A NASA–*European Space Agency* (ESA) *mission* to study the Sun at all latitudes, launched on October 6, 1990, aboard *Space Shuttle Discovery* as part of the *STS*-41 mission. It achieved high latitude over the solar north pole on July 27, 1995, after having traveled 1.86 billion miles (3 billion km). (NASA News Release 95-125, July 27, 1995.)

ETYMOLOGY. The name—the Latin form of Odysseus, the hero of Homer's *Odyssey*—was proposed by ESA and agreed to by NASA. A NASA news release also points to "the Italian poet Dante's description (in the 26th Canto of his 'Inferno') of Ulysses' urge to explore 'an uninhabited world behind the sun'" (News Release 84-127, "Ulysses New Name for International Solar Polar Mission," September 10, 1984).

umbilicals. Lines carrying electrical signals or fluids, between the *gantry* and the upright *rocket* before *launch.*

ETYMOLOGY. From the umbilical cord, the membranous duct connecting the human fetus with the placenta. The term was first adopted to describe the hoses and wires connecting a deep-sea diver to his or her air supply.

FIRST USE. A 1959 display ad for the Martin Company called for engineers with experience with "connectors, umbilicals, wire installations and electrical test procedures" (*Chicago Tribune,* November 29, 1959, p. A10). At this point Martin was a contractor on the *Titan* intercontinental ballistic missile (ICBM), *Dyna-Soar,* and *Saturn.*

umbilical tower. Vertical structure supporting the servicing electrical or fluid lines running into a *rocket* in *launch* position (NASA, *Glossary/Congressional Budget Submission,* p. 45).

uncrewed. Term used to distinguish a *mission* without human operators. Like *robotic* or *unpiloted,* a preferred synonym for *unmanned.*

universe. The vast space that contains all of the matter and energy in existence. Everything in space is part of the universe, though some now talk of a multiverse that contains many universes, including our own. The *Wilkinson Microwave Anisotropy Probe* has shown that our universe was formed 13.7 billion years ago with a big explosion known as the *Big Bang.* When the universe cooled down, huge swirls of dust and gas clung together to form stars and galaxies.

unk. An unknown.

unk unk. An unknown unknown. Especially in engineering, something (such as a problem) that has not been and could not have been imagined or anticipated. Formally, in the field of decision analysis an unk unk is an uncertainty that is unanticipated and, hence, unaccounted for.

unmanned. Term used to distinguish *missions* without human crew.

    USAGE. The terms *uncrewed, robotic,* and *unpiloted* are now preferred as a matter of NASA style, just as *crewed, human,* and *piloted spaceflight* are preferred to *manned spaceflight.*

unmanned spacecraft. Vehicle not carrying an *astronaut* or *cosmonaut.*

unmanned spaceflight. Robotic flight.

unobtanium. Facetious though useful term from the early days of the American *space program* for "a substance having the exact high test properties required for a piece of hardware or other item of use, but not obtainable either because it theoretically cannot exist or because technology is insufficiently advanced to produce it" (Heflin, *Interim Glossary Aero-Space Terms,* p. 33).

    FIRST USE. Defined as above in *Interim Glossary Aero-Space Terms* (March 1958), where it is tagged "humorous or ironical."

unpiloted. Term used to distinguish a *spacecraft* without human operators. Like *robotic* or *uncrewed,* an increasingly preferred synonym for *unmanned.*

untethered. Term describing an *EVA* or *spacewalk* performed without the crew member being attached to the *spacecraft.* The first such spacewalks took place during the *mission* of STS-41B, the *Space Shuttle Challenger,* February 4, 1984. American astronauts Bruce McCandless and Robert Stewart made the first untethered spacewalks wearing a propulsion backpack called a Manned Maneuvering Unit (MMU).

uplink. Signal sent to a *spacecraft.*

upper atmosphere. The region of the Earth's *atmosphere* above the troposphere (which extends to about 12 miles, or 20 km). The regions of the upper atmosphere are the stratosphere, the mesosphere, and the thermosphere.

upper stage. Final *booster*(s) for a *spacecraft.* Dwayne A. Day and Roger Guillemette define the term in an essay on the Centennial of Flight website: "Although there is no strict definition of an 'upper stage,' it usually refers to the third and fourth stages of a rocket, fired at high altitude. Because upper stages are rarely visible from the ground and leave no long firetrails to see, they attract little attention and are the unsung workhorses of the Space Age." (www.centennialofflight.gov/essay/SPACEFLIGHT/upper_stages/SP12.htm.)

V

V-1, V-2. German "Revenge" (Vergeltungswaffe) weapons used during World War II. The V-1 was a *rocket* bomb and the V-2 a liquid-fueled rocket used as a ballistic *missile*.

ETYMOLOGY. Michael J. Neufeld, a Museum Curator at the National Air and Space Museum, Smithsonian Institution, and the author of *The Rocket and the Reich: Peenemünde and the Coming of the Ballistic Missile Era* (1995), provides this explanation: "V2 (the Germans did not use hyphens) was a designation that Goebbels' Propaganda Ministry applied to the A4 missile in fall 1944, following on the V1. The full name was Vergeltungswaffe 1 or 2. Conveniently, 'Vergeltung' can translated as 'Vengeance,' hence 'Vengeance Weapon 2.' But it could also be translated as 'Revenge' but keeping the 'V' is obviously a more convenient translation. The vengeance/revenge was for the Allied destruction of German cities by bombers." (E-mail to the author, October 29, 2005.)

VAB. (1) *Vehicle Assembly Building (Kennedy Space Center).* (2) *Vertical Assembly Building (Michoud Assembly Facility).*

Van Allen radiation belt(s). Also called Van Allen belt(s) or zones. Two doughnut-shaped zones of intense *radiation—magnetic field*s and trapped atomic particles—surrounding the Earth outside the *atmosphere,* concentrated at altitudes of 3,000 and 10,000 miles (4,800 and 16,000 km).

ETYMOLOGY. Named for the astrophysicist James A. Van Allen, who placed Geiger counters on two early U.S. satellites, *Explorer*s 1 and 3, which made the initial discovery as part of the *International Geophysical Year* (IGY).

FIRST USE. Van Allen presented his findings on May 1, 1958, to a joint meeting of the National Academy of Sciences and the Physical Society, and the radiation belt "at once became known as the Van Allen Radiation Belt" (Newell, *Beyond the Atmosphere,* SP-4211, p. 84).

Vanguard. Project of the U.S. Naval Research Laboratory concerned with developing a carrier *rocket* and *satellite*s, the latter to be placed in *orbit* during the *International Geophysical Year* (IGY). The name Vanguard applied to both the first satellite series undertaken by the United States and to the *launch vehicle* developed to put the satellites into orbit.

In the spring of 1955 there was growing scientific interest in putting an artificial *Earth satellite* into orbit for the International Geophysical Year (July 1, 1957–December 31, 1958). Several launch vehicle proposals were developed for placing a U.S. satellite in orbit. The proposal chosen in August 1955 to be the U.S. satellite project for the IGY was the one offered by the Naval Research Laboratory (NRL), based on Milton W. Rosen's concept of a new launch vehicle combining the *Viking* first stage, the *Aerobee* second stage, and a new third stage. Rosen became technical director of the new project at NRL. (The NRL Vanguard project team was transferred to NASA when the space agency was established October 1, 1958.)

Vanguard 1, a 3.3-pound (1.5-kg) scientific satellite, was launched on March 17, 1958. It was followed in 1959 by Vanguard 2 and 3. Scientific results from this series included the first geodetic studies indicating the Earth's slightly "pear" shape, a survey of the Earth's *magnetic field,* the location of the lower edge of the Earth's *radiation belts,* and a count of micrometeorite impacts. Vanguard 1 can be considered one of the most successful *missions* ever. On its second anniversary in orbit it was still transmitting after having put 131,318,211 miles (over 210 million km) on its odometer. Its radio transmitter went dead in 1964. Today it is the oldest man-made object in space, with a current life expectancy that will take it to around the year 3000. Vanguard has its own website, where its orbits and time in space are tracked.

But historical inaccuracy often prevails over the subtleties of what really happened, and the blurred conventional wisdom today sees 1957–58 as a series of explosions during its early launch attempts, including one that was televised live. In fact, Vanguard's objective was accomplished during the International Geophysical Year, as planned, and U.S. military forces achieved a retaliatory power before the Soviet Union could amass the hardware needed for a ballistic-*missile* attack. The lasting impression of the Vanguard failures are worse than the reality, but they are clearly still with us. Vanguard has been plagued with the image of total disaster even in its own home town. In a May 3, 1993, article on the local Martin Marietta Corporation complex, the *Baltimore Sun* devotes two lines to Vanguard: "Baltimore was the home of the ill-fated Vanguard rocket that the government called on in the late 1950s to answer the Soviet Union's launch of Sputnik. Vanguard's first attempt to launch a grapefruit-sized satellite ended in a fiery explosion."

ETYMOLOGY. The name Vanguard was suggested by Rosen's wife, Josephine. Rosen forwarded the name to his NRL superiors, who approved it. The Chief of Naval Research approved the name on September 16, 1956. The word denotes that which is out ahead, in the forefront. (Ironically, though, Vanguard was not the first U.S. satellite; Explorer 1 had been launched into orbit by the Army on January 31, 1958.)

FIRST USE. Even before the first Vanguard launch, the program had generated much public anticipation because of a massive publicity effort by the U.S. Navy, which was building it. By mid-1957 several Vanguard books were already in stores, and hundreds of magazine feature articles were previewing what was presumed to be the first satellite. Few of these articles even mentioned the possibility that the honor might go to a Russian craft. "America's man-made moon" would be the first object in orbit, according to *Life* magazine (June 1957). In May 1957 a new edition of *Discover the Stars,* a popular book for hobbyists, was published with the image of Vanguard on the cover and detailed plans inside for building a model of the satellite. The book claimed that the *Space Age* would begin in early 1958 with a Vanguard launch from Banana River, Florida, also known as *Cape Canaveral. National Geographic* referred to the planned Vanguard spacecraft as "history's first artificial earth-circling satellite" (February 1956) and as the "first true space vehicle" (April 1956). Vanguard had become a household word, Martin Caidin wrote in *Overture to Space,* "Scientists had given speeches and lectures on the miracle we were about to bring to the world. Artificial satellites had become synonymous with American genius, technology, engineering, science, and leadership."

"Everyone knew in 1957 that space exploration was the next item on the scientific and technological agenda, and almost everybody assumed that the United States would lead the way as usual," John Brooks wrote in *The Great Leap,* a look at the years 1939–65. In fact, Americans were so complacent that they weren't even prepared to monitor other satellites, so on "Sputnik night" the Russian satellite twice passed within easy detection range of the United States before anyone in authority knew of its existence. Sputnik was announced by an Associated Press report from London.

SOURCES. SP-4402, pp. 78–79; Milton W. Rosen, Office of Defense Affairs, NASA, letter to R. Cargill Hall, Lockheed Missiles & Space Co., August 1963; Rosen, telephone interview, February 16, 1965; Chief of Naval Research, letter to Director, Naval Research Laboratory, September 16, 1955; Martin Caidin, *Overture to Space* (New York: Dell, Sloan and Pierce, 1963); John Brooks, *The Great Leap: The Past Twenty-five Years in America* (London: Gollancz, 1967).

Vega. Soviet *missions* to Venus and *Comet* Halley, 1984.

Vehicle Assembly Building (VAB). Facility at *Kennedy Space Center* originally built for assembly of *Apollo-Saturn* vehicles. It was later modified to support *Space Shuttle* operations. One of the largest buildings in the world in terms of volume, the VAB sometimes generates its own weather. (*Historic Properties,* KSC Archives, John F. Kennedy Space Center, Florida, August 2005, NP-2005–08-KSC.)

Vela (Spanish for vigil or watch). Name for a group of *satellites* launched by the United States to monitor Soviet compliance with the 1963 Partial Test Ban Treaty.

Velcro (velvet + crochet; French for hook). Trade name for hook-and-pile fastener, generally of nylon, used to replace zippers and other means of securing items. Velcro was invented in 1948 by George de Mestral, a Swiss engineer who got burrs caught in his heavy woolen stockings, saw the principle of tiny hooks and loops at work, and reproduced the effect in woven nylon. NASA has always used it, and each *Space Shuttle* contains about 10,000 square inches (64,500 sq cm) of Astro Velcro, a special variety of the stuff.

Venera. Soviet *missions* to Venus, 1967–83.

VentureStar. Name for the short-lived concept of a *recoverable launch vehicle* (RLV), which was to be tested in a half-scale model in the *X-33* test aircraft. Static tests of the X-33 fuel tank in 1999 brought the technical feasibility of the design into question, and the program was canceled in 2001. (NASA News Release 96-128, "Lockheed Martin Selected to Build X-33," July 2, 1996.)

Venus Express. *European Space Agency mission* to study the *atmosphere* and plasma environment of Venus from *orbit,* 2006 through 2009. The *spacecraft* will primarily be investigating the role played by the greenhouse effect in creation of the atmosphere; the behavior and characteristics of cloud and haze formation at different altitudes; processes at work in atmospheric escape and its interaction with solar winds; and the mechanism behind the super rotation in the *upper atmosphere.* It will also study the weak Venus *magnetic field,* the UV absorption features at 49 miles (80 km) altitude, the high radio wave reflectivity areas on the surface, the atmosphere-surface interaction, and the possibility of volcanic or seismic activity.

Venus Express was successfully launched on November 9, 2005, on a *Soyuz*-Fregat *booster* from the *Baikonur Cosmodrome* in Kazakhstan. The *nominal mission* duration is roughly two Venus sidereal days (486 Earth days). Total budget for the mission is 220 million euros (US $262 million 2005), of which 82.4 million euros is for *satellite* construction and instrument integration and 37 million euros is for the *launch.* (National Space Science Data Center Master Catalog Display: *Spacecraft.*)

Vertical Assembly Building (VAB). Structure at *Michoud Assembly Facility* used to assemble *STS* equipment. Sometimes confused with the *Vehicle Assembly Building* at *Kennedy Space Center,* another large structure with the same initialism.

VHF. Very high frequency (TV channels 2–13).

Viking. (1) *Sounding rocket* built by Glenn L. Martin Co. and Reaction
Motors for the Navy, first launched May 3, 1949. This single-stage
rocket later became the prototype for the first stage of the Vanguard
launch vehicle. (Milton W. Rosen, *The Viking Rocket Story* [London: Faber
and Faber, 1956]; Constance McLaughlin Green and Milton Lomask,
Vanguard: A History, SP-4202 [Washington, DC: NASA, 1970].)

(2) Name of two NASA *missions* sent to look for life on Mars,
1975–76. The successor to Project *Voyager,* which was canceled in
1968, the Viking program was to send two uncrewed *spacecraft,* each
consisting of an *orbiter* and *lander,* to make detailed scientific measurements of the Martian surface and search for indications of past or
present life forms. The two Viking spacecraft, planned for launch in
1975 on *Titan III–Centaur* launch vehicles, were to reach Mars in 1976.
Viking 1 landed on July 20, 1976, on the Chryse Planitia (Golden
Plains). Viking 2 was launched on November 9, 1975 and landed on
Utopia Planitia on September 3, 1976. The Viking project's primary
mission ended on November 15, 1976, 11 days before Mars' superior
conjunction (its passage behind the Sun), although the spacecraft
continued to operate for six years after first reaching Mars, sending
back detailed images of the planet. The last transmission from the
Viking 1 lander reached Earth on November 13, 1982. Controllers at
NASA's *Jet Propulsion Laboratory* tried unsuccessfully for another six
and a half months to regain contact with the lander but finally closed
down the overall mission on May 21, 1983. (SP-4402, pp. 93–94; Walter
Jacobowski, Office of Space Science and Applications, NASA, telephone interview, July 16, 1969; Peter F. Korycinski, Office of the Director,
LaRC, memorandum to Historical Division, NASA, September 4, 1969.)
ETYMOLOGY (definition 2). The name had been suggested by Walter
Jacobowski in the Planetary Programs Office at NASA Headquarters and
discussed at a management review held at *Langley Research Center* in
November 1968. It was the consensus at the meeting that Viking was a
suitable name in that it reflected the spirit of nautical exploration in the
same manner as *Mariner,* in keeping with the Cortright system of naming
space probes. The name was subsequently sent to the NASA Project
Designation Committee and approved.

Vision for Space Exploration. U.S. space policy announced on January
14, 2004, by President George W. Bush. The main points of the policy,
as quoted and abridged from the Fact Sheet distributed by the White
House after the speech, are as follows: "First, America will complete its
work on the International Space Station by 2010, fulfilling our commit-
ment to our 15 partner countries . . . Second, the United States will
begin developing a new manned exploration vehicle to explore

beyond our orbit to other worlds—the first of its kind since the Apollo Command Module. The new spacecraft, the Crew Exploration Vehicle, will be developed and tested by 2008 and will conduct its first manned mission no later than 2014 . . . Third, America will return to the Moon as early as 2015 and no later than 2020 and use it as a stepping stone for more ambitious missions. The experience and knowledge gained on the Moon will serve as a foundation for human missions beyond the Moon, beginning with Mars. NASA will increase the use of robotic exploration to maximize our understanding of the solar system and pave the way for more ambitious manned missions. Probes, landers, and similar uncrewed vehicles will serve as trailblazers and send vast amounts of knowledge back to scientists on Earth." (www.whitehouse .gov/news/releases/2004/01/20040114-1.html.)

vomit comet. Training aircraft for astronauts that flew such high parabolas as to produce *weightlessness*. In recent years a commercial firm, the Zero Gravity Corporation, has offered rides on its version of the vomit comet for $3,500 a ticket. In April 2007 the term received great publicity when astrophysicist Stephen Hawking rode on the vehicle (at no cost) and was able to escape his wheelchair, if only for a short time. The famed professor and astrophysicist has been paralyzed by ALS, also known as Lou Gehrig's disease, and confined to a wheelchair for a good portion of his life.

ETYMOLOGY. Originally used to describe a particularly rough commercial aircraft flight (in the same way that a dirty plane would be called a "roach coach" and a packed flight a "cattle car"), the term was adopted by astronauts for the modified Boeing 707 used in training.

von Braun paradigm. The belief that humans are destined to physically explore the solar system, described in *Collier's* magazine in the early 1950s by Wernher von Braun. For von Braun, humans were the most powerful and flexible exploration tool imaginable.

Voskhod (Russian for dawn). The second series of crewed Soviet spacecraft. Following the triumph of the *Vostok* launchings that had put the first human into space, the Soviets developed the first spacecraft capable of carrying more than one crew member. On October 12, 1964, Voskhod 1 carried three *cosmonaut*s into Earth *orbit*.

Vostok (Russian for east). Series of early Soviet spacecraft designed as both a camera platform and a vehicle to take humans into space. Vostok 1 was the first crewed space mission. Launched on April 12, 1961, Vostok 1 took Soviet *cosmonaut* Yuri Gagarin into space, making him the first human to journey beyond the Earth's atmosphere and the first to go into orbit.

Voyager. Name for two spacecraft launched in 1977 that have explored Jupiter, Uranus, Saturn, and Neptune, along with dozens of their moons. In addition, they have been studying the solar wind, the stream of charged particles spewing from the Sun at nearly a million miles per hour (1.6 million kph). The Voyager *missions* were designed to study the region in space where the Sun's influence ends and the dark recesses of interstellar space begin—in other words, they were created as *spacecraft* destined to leave our solar system.

During the latter 1960s NASA scientists found that once every 176 years both the Earth and all the giant planets of the solar system gather on one side of the Sun. This geometric line-up made possible the close-up observation of all the planets in the outer solar system (with the exception of Pluto) in a single flight—a *Grand Tour.* To take advantage of this opportunity, NASA launched two spacecraft from *Cape Canaveral,* Florida: Voyager 2, lifting off on August 20, 1977, and Voyager 1, entering space on a faster, shorter *trajectory* on September 5, 1977.

Voyager 1, already the most distant human-made object in the cosmos, reached 100 *astronomical units* from the Sun on August 15, 2006. This means that the spacecraft, launched nearly three decades earlier, was 100 times more distant from the Sun than Earth is. (http://voyager.jpl.nasa.gov/.)

Wac Corporal *(sounding rocket).* Name for the first U.S. government sounding rocket—a project co-sponsored by the *Jet Propulsion Laboratory* and U.S. Army Ordnance. It has been suggested but never shown conclusively that this *rocket* was a "little sister" to the larger Corporal, the Army's surface-to-surface guided missile. "Wac" may have been added to the name as a pun on the acronym of the Women's Army Corps of the World War II era.

walkaround. Visual inspection of a *spacecraft.*

walk-around bottle. A personal supply of oxygen for the use of crew members when temporarily disconnected from the *spacecraft'*s system (SP-7).

Wallops Flight Center (WFC). This was *Wallops Station* before April 1974. See *Goddard Space Flight Center.*

Wallops Flight Facility (WFF). This was *Wallops Flight Center* until 1982. See *Goddard Space Flight Center.*

Wallops Station (WS). Became *Wallops Flight Center* in 1974. See *Goddard Space Flight Center.*

Webb's Giant. Also known as the James E. Webb Memorial Rocket. Name given to a postulated Soviet rocket that NASA Administrator James E. Webb invoked beginning in 1964 to generate support for his programs, especially the *Saturn V.* As Donald K. Slayton recalled in his autobiography, Webb claimed that "the Russians were building a big lunar booster. He got so identified with it that it was known for years as 'Webb's Giant.'" Slayton adds, "Webb's Giant was actually called the N-1 and sure enough, it was about the size and capacity of the Saturn V" (Slayton, *Deke,* p. 217). Dwayne A. Day points out that Webb's warnings "closely tracked what the intelligence community—primarily the CIA—was telling him the Soviet Union was doing."

FIRST USE. *TRW Space Log,* Winter 1968–69, used the term "Webb's Soviet giant," and *Aerospace Daily,* April 8, 1969, referred to it as "Webb's postulated giant."

Webb Space Telescope (JWST). *James Webb Space Telescope.*

week in the barrel. *Astronaut* slang for periods of time spent publicizing and marketing the *space program,* especially in the home districts of key members of Congress. Answering the question, "What was the most difficult aspect of your involvement in the space program?" Walter M. Schirra Jr. answered first that it was the funerals but was quick to add, "We were thrown to the press corps, literally, so that might be the second most difficult part . . . We had our routine; we called it 'week in the barrel.' We had to go around the country visiting congressmen, or having Press media. Now I get paid for doing that. But in those days, that was pretty difficult for a guy from a small town of 2200." (Oral History Transcript, Walter M. Schirra Jr., interviewed by Roy Neal, San Diego, California, December 1, 1998, www.jsc.nasa.gov/history/oral_histories/SchirraWM/WMS_12-1-98.pdf.)

ETYMOLOGY / FIRST USE. The term was first applied to the *Original Seven* astronauts, who were required to appear and speak in certain home districts, suggesting the possibility that the barrel in question here is the "pork barrel"—long-established American political slang for the federal Treasury into which legislators dip to finance projects for their home districts. In 1965, columnist Jim Bishop discussed the use of astronauts for political purposes: "The puritanical Congressmen pondered how they could use the Space Men, while denying such use to anyone else. They finally came up with a scheme to ask NASA for astronauts to appear and speak in certain home districts. The seven are not permitted to decline. They say glumly: It's my week in the barrel." (*Miami Herald,* May 12, 1965, p. 7.)

weightlessness. Condition in which no acceleration within the system in question can be detected by an observer and in which gravity has no effect on a body. Also known as *zero gravity* or zero g.

Weightless Wonder. NASA flying *microgravity* laboratory, a modified C-9 aircraft that produces 25 seconds of *weightlessness* by flying in a roller-coaster path of steep climbs and free falls.

Westar. Series of commercial communications *satellites* owned and operated by Western Union Telegraph Company and launched by NASA under a contract to form the first U.S. domestic *comsat* system.

As early as 1966, Western Union petitioned the Federal Communications Commission for permission to build a domestic satellite system to relay telegraph traffic. The FCC was then making a detailed study of the need for such a system in response to requests from several organizations. When the FCC decided in 1970 to invite applications, Western Union was the first to respond, proposing a high-capacity multipurpose system to serve all 50 states.

The company won approval in January 1973 to build the first U.S. system, with authorization for three satellites. Hughes Aircraft Company was to build the comsats—or domsats (domestic satellites) as the press began to call them—and NASA signed a contract with Western Union in June 1973, agreeing to provide *launch* services, with reimbursement for the Thor-Delta *launch vehicles* and costs.

Westar 1 (Westar-A, before launch) was launched on April 13, 1974, and began commercial operation on July 16. As a new postal service, Westar 1 relayed the first satellite "Mailgrams" from New York to Los Angeles at the speed of light. Westar 2 was launched on October 10, 1974, and Westar-C was held as a spare. In synchronous *orbit,* each drum-shaped satellite could relay 12 color TV channels, up to 14,400 one-way telephone circuits, or multiple data channels. The last launch was Westar 6 aboard *STS*-41B on February 3, 1984.

ETYMOLOGY. Western Union asked its employees to suggest names for the new satellites. Westar was the one chosen, combining part of the company's name with "star," a reference to a body in space, or satellite.

SOURCES. SP-4402, pp. 79–80; Frances Shissler, Western Union Telegraph Co., McLean, Va., telephone interview, April 2, 1975; Western Union Telegraph Co., *Communicator,* Summer 1973, pp. C4–5; UPI, "Western Union Proposes Satellite Telegram System," *New York Times,* November 8, 1966, p. 15; Dow Jones News Service, "Western Union Files Domestic Satellite Plan," *Washington Evening Star,* July 31, 1970, p. A14; AP, "Satellite Relay for US Approved," *New York Times,* January 5, 1973, p. 1; "The Day of the Domsat," *Time,* April 29, 1974, p. 2; NASA program office.

Western Space and Missile Center (WSMC). Formerly USAF Pacific Missile Range and then Western Test Range. This U.S. *launch* facility is located within Vandenberg Air Force Base on the west coast of California. The site is located at 34.7 deg N, 120.6 deg W, and is primarily used for polar orbiting missions.

Western Test Range (WTR). Formerly USAF Pacific Missile Range, now *Western Space and Missile Center.*

wet-fuel rocket. Liquid-fuel *rocket* ("Glossary of Aerospace Age Terms," p. 32).

wheel. Centrifuge used to test and train astronauts. The *Original Seven* astronauts were required to spend much time on the wheel, and the term is used repeatedly in Carpenter, et al., *We Seven.*

white dwarf. A star that has exhausted most or all of its nuclear fuel and has collapsed to a very small size. Typically, a white dwarf has a radius equal to about 0.01 times that of the Sun, but it has a mass roughly equal to the Sun's. This gives a white dwarf a density about a million times that of water—roughly equivalent to the density of a soda can into which a 747 airliner has been squeezed. The Sun is expected to become a white dwarf at the end of its life. (ASP Glossary.)

White Sands Missile Range (WSMR). See *White Sands Space Harbor.*

White Sands Space Harbor (WSSH). Primary training area for *Space Shuttle* pilots flying practice approaches and landings in the shuttle-training aircraft (STA) and T-38 chase aircraft. The STA is a Gulfstream II aircraft modified to mimic the flight characteristics and instrumentation on the Shuttle and provides a realistic simulation of the Shuttle's landing from high altitudes to *touchdown.* WSSH is located approximately 30 miles (50 km) west of Alamogordo, New Mexico, on the U.S. Army White Sands Missile Range. It is operated by the White Sands Test Facility (WSTF). The facility was used during the landing of *STS*-3 in March 1982. After the STS-3 landing, WSSH became an emergency landing site, and the U.S. Congress designated the facility as the White Sands Space Harbor.

White Sands Test Facility (WSTF). See *White Sands Space Harbor.*

wild turkey. Fanciful designation used by NASA insiders, ca. 1976 and later, for a single overarching goal for the agency. In a memo that mentions the goal of "opening the space frontier to all forms of human activity," John E. Naugle added in parentheses, "To be done by Naugle if he gets sufficient Ouzo or Wild Turkey to let his imagination run wild." See also *gray mice* and *purple pigeons.* (NASA Names Files, record no. 17540, memo from John E. Naugle of November 12, 1976.)

Wilkinson Microwave Anisotropy Probe (WMAP). A mission that provided a detailed full-sky map of the microwave background radiation in the universe. The *probe* made a number of discoveries, including the determination that our universe is 13.7 billion years old, with a margin of error close to 1 percent.

Wind Mission. A *mission* to investigate the *solar wind* and its impact on the near-Earth environment.

window. (1) Interval of time within which a *spacecraft* must be launched in order to accomplish its *mission.* Short for *firing window* or *launch window.* (2) An area in the Earth's *atmosphere* through which a returning *spacecraft* must pass for successful *reentry.*

FIRST USE. "Ranger 9 is scheduled for launch during the next 'moon window' which begins March 17 and lasts for a week. A window, in NASA jargon, is the period during which the moon (or a planet) makes itself accessible to spacecraft launched from earth." ("Ranger, Cameras Head for Moon Impact," *New York Herald Tribune,* February 10, 1965, in *NASA Current News,* February 18, 1965, p. 5.)

the Worm. Nickname for the NASA logo in use from 1975 until 1992, when it was replaced with the original NASA logo, the old-school Meatball. So called because the stylized rendering of the name connected the letters N-A-S-A in a wormlike shape. See *the Meatball.* (http://history.nasa.gov/meatball.htm.)

wormhole. A hypothetical "tunnel" linking otherwise distant regions of *spacetime.*

WS. See *Wallops Station.*

WS-117L. *Reconnaissance satellite* developed by Lockheed.

WSMC. See *Western Space and Missile Center.*

WSMR. *White Sands Missile Range.*

WSSH. See *White Sands Space Harbor.*

WTR. See *Western Test Range.*

X. Designation for experimental aircraft, as in *X-15, X-20,* etc. X planes are designed to answer fundamental questions about the behavior of aircraft close to, at, or beyond the speed of sound, and to serve as prototypes for advanced aero*space vehicles.*

X-15. Joint hypersonic research program conducted by NASA with the Air Force, the Navy, and North American Aviation. The program completed its 199th and final flight on October 24, 1968, in what many consider to have been the most successful flight research effort in history. These data led to improved design tools for future hypersonic vehicles and contributed in important ways to the development of the *Space Shuttle,* including information from flights to the edge of space and back in 1961–63.

X-20. See *Dyna-Soar.*

X-30. Designation for the National Aero-Space Plane (NASP), an attempt by the United States to create a single-stage-to-*orbit* (SSTO) *spacecraft.* The project was canceled in 1993 prior to the first craft being built.

X-33. A pilotless prototype, 69 feet (21 m) long, that was to have evolved into a larger vehicle, the VentureStar, that could carry people and cargo. The VentureStar, in turn, would have had a radically new type of *engine*—the *aerospike.* Its proponents believed that the VentureStar would not only service the *International Space Station* and replace the aging Shuttle fleet but would also serve commercial customers. However, because of NASA budgetary constraints and growing concern over cost overruns, funding was withdrawn from both projects in March 2001.

X-38. Proposed crew return vehicle for the *International Space Station* that was targeted to begin operations in 2003 but was canceled in development.

Xena. Nickname for the "tenth planet" (officially cataloged as 2003 UB313 and then named Eris), photographed by NASA's *Hubble Space Telescope* on December 9–10, 2005. The photograph showed Xena to be slightly larger than Pluto. Xena's diameter is 1,490 miles (2,400 km), with an uncertainty of 60 miles (96 km); Pluto's diameter, as measured by Hubble, is 1,422 miles (2,300 km). "Hubble is the only telescope capable of getting a clean visible-light measurement of the actual diameter of Xena," said Mike Brown, planetary scientist at the California Institute of Technology and a member of the research team that discovered Xena. (NASA News Release 06-183, April 11, 2006, "Hubble Finds 'Tenth Planet' Is Slightly Larger Than Pluto.")

X-Prize. See *Ansari X Prize.*

X-ray Astronomy Explorer. See *Explorer.*

XS11. The *Excess Eleven.* Astronauts selected in mid-1967 gave themselves this nickname because it was unlikely that they would go into space. (Compton, *Where No Man Has Gone Before,* SP-4214, p. 136.)

Y. Symbol for prototype. When used as a prefix in the designation of an aerospace vehicle (YTM-61), it indicates that the vehicle is a prototype. ("Glossary of Aerospace Age Terms," p. 33.)

Yankee Clipper. *Call sign* for the *Apollo* 12 Command Module. The *mission* was crewed by an all-Navy crew of Charles Conrad Jr., Richard F. Gordon Jr., and Alan L. Bean. The *Lunar Module* for this Moon landing was the *Intrepid.*

ETYMOLOGY. Originally denoting a very fast type of multi-masted sailing ship developed by New Englanders in the mid-19th century, the name was selected from more than 2,000 suggestions submitted by employees of the manufacturers of the two *spacecraft.* Speaking of the two names, Intrepid and Yankee Clipper, Richard Gordon said, "We think the names are fitting as we sail on this new ocean of space" ("Apollo Spaceships Have Names with Salty Ring," *New York Times,* November 15, 1969, p. 24).

yestersol. Yesterday as applied to Mars by those who manage Mars missions at the *Jet Propulsion Laboratory* (JPL). The term has come into such common use—along with *sol* for the Martian day—that it is routinely used in JPL press releases and *mission* reports. For instance: "[Mars rover] Opportunity also used its . . . alpha particle x-ray spectrometer to assess the composition of the interior material of [the outcrop-rock target] 'Guadalupe' exposed yestersol by a grinding session with the rock abrasion tool." ("Mars Exploration Rovers, Daily Update—2/29/04: Guadalupe under the Microscope, Opportunity Status for sol 35," www.google.com/unclesam?q=cache:ilKX9ZkPjiAJ: www.jpl.nasa.gov/missions/mer/daily.cfm%3Fdate%3D2%26year %3D2004+yestersol&hl=en&gl=us&ct=clnk&cd=1.)

Z

Zarya (Russian for sunrise). The first element of the *International Space Station* (ISS), built in Russia under U.S. contract. During the early stages of ISS assembly, Zarya (also known as the Functional Cargo Block) provided power, communications, and attitude control functions. The module is now used primarily for storage and propulsion. It was based on the modules for the *Mir* space station. (*Reference Guide to the International Space Station,* SP-2006-557.)

Zenit (Russian for zenith). *Launch vehicle* manufactured by the Yuzhnoe Design Bureau of Ukraine and launched from Russia's *Baikonur Cosmodrome* in Kazakhstan.

zero g. *Weightlessness.* Short for *zero gravity.*

zero gravity. *Weightlessness;* the appearance that the force of gravity is absent.

ETYMOLOGY. From the tendency in aviation to describe things as absolutes when they are not. For example, in conditions of "zero visibility," there is still some visibility, though extremely limited.

FIRST USE. "Dr. Haber now with the department of space medicine US Air Force said zero gravity would be a common experience in planes traveling 2000 miles an hour at 100,000 feet" (*Los Angeles Times*, June 1, 1950, p. B10).

USAGE. "Zero gravity is a bit of a misnomer," writes Dr. Eric Christian on NASA's Cosmicopia website. "It is used to describe the condition when an object is freely falling with no resistance. You can feel zero *G* in a plane, roller coaster, or elevator. Gravity is still present, however. Something in orbit is essentially freely falling around the Earth. But oxygen and the rest of the atmosphere (mostly nitrogen) just gradually fade out and extend hundreds of miles over the surface of the Earth. Even where the shuttle and space station are (greater than 400 miles or 650 kilometers up), there is enough air resistance that there is apparent acceleration of about a millionth of that on the Earth's surface. This is why experiments there are called 'microgravity' experiments, not zero *G* experiments." (http://helios .gsfc.nasa.gov/qa_gp_gr.html#zero.)

Zero Gravity Research Facility / Zero-G Research Facility. NASA's premier facility for conducting ground-based *microgravity* research, and the largest facility of its kind within the United States. Operational since 1966, it is one of two drop towers located at NASA's *Glenn Research Center.* Microgravity can be achieved on Earth only by putting an object in a state of free fall. NASA conducts microgravity experiments on Earth by using drop towers and aircraft flying parabolic trajectories. The Zero-G Facility provides researchers the ability to obtain a near-weightless environment for a duration of 5.18 seconds, allowing an experiment to free-fall a distance of 430 feet (132 m). (http://microgravity.grc.nasa.gov/zero-g/description.htm.)

Zond (Russian for *probe*). Soviet lunar missions, 1965–70. The name was given to two series of Soviet uncrewed space missions to gather information about nearby planets and to test *spacecraft.*

Zulu time. Communications code word for Greenwich mean time (GMT) used in NASA logs and messages. Also known as z-time.

Zvezda (Russian for star). The Service Module of the *International Space Station.* The module provided the ISS's early living quarters, life support, electrical power distribution, data processing, flight control system, and propulsion system. Some of these systems were subsequently supplemented or replaced by later U.S. systems, but Zvezda remains the structural and functional center of the Russian segment of the ISS. (*Reference Guide to the International Space Station,* SP-2006-557.)

Acknowledgments

First and foremost, I would like to thank the NASA History Office for its sponsorship and support of this project: Steven J. Dick, NASA Chief Historian; Steve Garber, NASA Historian; Jane Odom, Chief Archivist; Colin Fries, John Hargenrader, and Liz Suckow, archivists; and Nadine Andreassen, program support assistant.

My thanks also to Chris Gamble and other anonymous referees for reading the entire manuscript and making numerous suggestions that greatly improved the final version.

Great indebtedness is acknowledged to the original team that compiled the original *Origins of NASA Names,* which was directed by Monte D. Wright, Director, NASA History Office, and written by Helen T. Wells, Susan H. Whiteley, and Carrie E. Karegeannes.

Reaching back to acknowledge the personal genesis of this project, in 1967 the author got a job as a full-fledged reporter for McGraw-Hill's *Electronics* magazine with a beat covering the National Aeronautics and Space Administration (NASA), the Communications Satellite Corp. (COMSAT), and the Federal Communications Commission (FCC). I had been working for the magazine in New York City as an apprentice, and the day I left for my new assignment I was handed a copy of the *Dictionary of Technical Terms for Aerospace Use* (NASA SP-7, 1965) by Sally Powell, the magazine's extraordinary copy editor. She drew a heart on the flyleaf and told me that I would need it.

She was right.

That dictionary and my experiences covering NASA during the Apollo years sowed the seeds for this work. My thanks go to Sally Powell and the reporters who taught me the language of the early Space Age: Bill Hines, Seth Payne, Ray Connolly, Joe Roberts, Louise Dick, Les Gaver, and the first Archivist of the Space Age, Lee D. Saegesser.

I would also like to thank the late Cecil Paul Means, Dr. Michael J. Neufeld (NASM), Tom Mann of the Library of Congress, and Richard D. Spencer, Reference Librarian, NASA HQ.

Also thanks to Mary V. Yates, who worked diligently to bring style and consistency to the final manuscript.

Bibliography

Acronym List—Space Station. Space Station Engineering and Integration Contractor. Reston, VA: Office of Space Station, 1992.

Adams, Frank Davis. *Aeronautical Dictionary.* TM-101286. Washington, DC: GPO, 1959. (This is a very important transition from the 1933 NACA Report 474, "Nomenclature for Aeronautics," to the early days of NASA and the emerging vocabulary of space. Adams notes in his introduction, "The border line between aeronautics and space technology is not always easy to define, and many terms are, or can be, applied in both areas; consequently, the definitions given these terms herein are worded so as to take account of applications in either field." Adams does not shy away from slang or terms that have become obsolete, and he gives many useful examples of terms actually used in a sentence. His definition of barnstorm includes this quotation from Charles Lindbergh: "I've barnstormed more than half of the forty-eight states.")

————. *The Second Aerospace Glossary.* Maxwell Air Force Base, AL: Aerospace Studies Institute, Air University, 1966. (Adams created this book semi-anonymously, signing the preface F.D.A. and identifying himself only as an employee of NASA.)

Aerospace Science and Technology Dictionary. CD-ROM. NASA STI Program. Washington, DC: NASA, 2000.

Alexander, Charles, James Grimwood, and Loyd Swenson. *This New Ocean: A History of Project Mercury.* Washington, DC: GPO, 1989.

Allen, William H. *Dictionary of Technical Terms for Aerospace Use.* SP-7. Washington, DC: NASA, 1965. Accession no. 66N10413. (The single work that did the most to stabilize the language of space exploration.)

Anderson, Frank W., Jr. *Orders of Magnitude: A History of NACA and NASA, 1915–1980.* SP-4403. Washington, DC: GPO, 1981.

Angelo, Joseph A. *The Facts on File Dictionary of Space Technology.* New York: Facts on File, 2004.

Apollo Acronyms and Abbreviations. Report no. MG404. Greenbelt, MD: Manned Flight Engineering Division, GSFC, 1968.

Apollo Program: Glossary of Acronyms and Abbreviations. Washington, DC: NASA, 1969.

Apollo Soyuz Test Project: ASTP Glossary. ASTP-20020.1. Washington, DC: NASA, 1975.

Apollo Terminology. SP-6001. Washington, DC: NASA, August 1963. (In view of the inclination of even popular accounts of the program to use acronyms and technical terms, this is a virtually indispensable reference work for those not already familiar with the terminology. It is also the only official NASA glossary to contain colloquialisms. Even though it is part of the SP series, it is a difficult work to find.)

Armstrong, Neil, Michael Collins, and Gene Farmer. *First on the Moon.* Boston: Little, Brown, 1970.

Associated Press. "NASA Speak Befuddles New Reporters." *New Haven Register,* October 1, 1988, p.18.

Astronyms: Acronyms and Space Terms Used in Spacecraft Communication. Houston: Philco, 1968. (The most valuable element of this 128-page glossary is a special addendum on terms that came into play in air-to-ground communications during the first four Apollo missions. It is a valuable linguistic snapshot of that time.)

Ballingrud, David. "Glossary for the Armchair Star Explorer." *St. Petersburg Times,* April 10, 1990, p. A8.

Barrett, George. "Visit to 'Earthship No. 1.'" *New York Times Magazine,* September 8, 1957, SM 13+. (This article served as a primer on the emerging language of space.)

Baughman, Harold. *Aviation Dictionary and Reference Guide: Aero-Thesaurus.* Glendale, CA: Aero, 1940–51.

Benson, Charles D., and William Barnaby Faherty. *Moonport: A History of Apollo Launch Facilities and Operation.* SP-4204. NASA History Series. Washington, DC: NASA, 1978. (The University Press of Florida has republished the book in two volumes, *Gateway to the Moon* and *Moon Launch.*)

Bilstein, Roger E. *Orders of Magnitude: A History of the NACA and NASA, 1915–1990.* SP-4406. Washington, DC: GPO, 1989.

———. *Stages to Saturn: A Technological History of the Apollo/Saturn Launch.* SP-4206. Washington, DC: GPO, 1989.

Bishop, Jerry. "Themes and Variations: Memo on Space Semantics." *Wall Street Journal,* September 9, 1965, p. 12. (How a great newspaper deals with style issues brought on by the Space Age.)

Borenstein, Seth. "Translation Please: NASAspeak." *Orlando Sentinel,* October 12, 1997, p. 5.

Bower, Tom. *The Paperclip Conspiracy.* Boston: Little, Brown, 1986.

Bridges, Andrew. "Mars Mission Spawns Its Own Unworldly Lingo." *Chico (CA) Enterprise Record,* February 22, 2004.

Bright, Charles D., ed. *Historical Dictionary of the US Air Force.* Westport, CT: Greenwood, 1992.

Brooks, Courtney G., James M. Grimwood, and Loyd S. Swenson. *Chariots for Apollo.* SP-4205. Washington, DC: GPO, 1989.

Bullock, Gilbert D. *Spanish Language Equivalents for a Glossary of Terms Used in the Field of Space Exploration.* TM-86211. Washington, DC: NASA, 1985.

Burrows, William E. *This New Ocean.* New York: Random House, 1998.

Byrnes, Mark E. *Politics and Space: Image Making by NASA.* Westport, CT: Praeger, 1994. (This work argues, among other things, that NASA has used euphemistic language to make itself look good and obfuscate the truth in times of crisis or tragedy. Byrnes makes some valid points; but some of his arguments are based on the belief that NASA is a monolith that speaks with a single voice.)

Caidin, Martin. *The Man-in-Space Dictionary: A Modern Glossary.* New York: Dutton, 1963.

Carpenter, M. Scott, et al. *We Seven, by the Astronauts Themselves.* New York: Simon and Schuster, 1962.

Carpenter, Scott, with Kris Stoever. *For Spacious Skies: The Uncommon Journey of a Mercury Astronaut.* New York: New American Library, 2003.

Cartier, Arthur T. *Missile Technology Abbreviations and Acronyms.* New York: Hayden, 1965.

Chaikin, Andrew. *A Man on the Moon: The Voyages of the Apollo Astronauts.* New York: Viking Penguin, 1994.

Chaisson, Eric. *The Hubble Wars.* New York: HarperCollins, 1994.

Chui, Glennda. "Remembering a Space Pioneer—Columbia, Nation's First Shuttle, Proved It Could Be Done, Then Did It for 22 Years." *San Jose Mercury News,* February 11, 2003, p. A1.

Clarke, Arthur C., ed. *The Coming of the Space Age: Famous Accounts of Man's Probing of the Universe.* New York: Meredith, 1967.

———. *The Exploration of Space.* New York: Pocket Books (Cardinal ed.), 1954.

———. *The Making of a Moon.* New York: Harper and Brothers, 1957.

———. *The Promise of Space.* New York: Harper and Row, 1968.

Collins, John M. *Military Space Forces.* Washington, DC: Pergamon-Brassey's, 1989.

Collins, Michael. *Carrying the Fire: An Astronaut's Journeys.* New York: Farrar, Straus and Giroux, 1974.

Compton, William David. *Where No Man Has Gone Before: A History of Apollo Lunar Exploration Missions.* SP-4214. Washington, DC: NASA, 1989.

Crouch, Tom D. *Aiming for the Stars: The Dreamers and Doers of the Space Age.* Washington, DC: Smithsonian Institution Press, 1999.

Daniels, Lynne. *Statement of Prominent Americans at the Beginning of the Space Age.* NASA Historical Note 22. Washington, DC: NASA Headquarters, 1965.

Date, Shirish. "'Ok, I Go Ready to Copy': Ground Crews Have a Hard Time Following NASA Jargon As Said with German Phrasing." *Orlando Sentinel,* April 30, 1993, p. A5.

Davies, Merton E., and William R. Harris. *RAND's Role in the Evolution of Balloon and Satellite Observation Systems and Related US Space Technology.* R-3692-RC. Santa Monica: RAND Corporation, September 1988.

Davis, Deane. "The Talking Satellite: A Reminiscence of Project SCORE." *Journal of the British Interplanetary Society* 52 (1999).

Davis, Kenneth S., "Father of Rocketry." *New York Times,* October 23, 1960.

Day, Dwayne A., John M. Logsdon, and Brian Latell. *Eye in the Sky: The Story of the Corona Spy Satellite.* Smithsonian History of Aviation Series. Washington, DC: Smithsonian Institution Press, 1998.

Dethloff, Henry C. *"Suddenly, Tomorrow Came . . .": A History of the Johnson Space Center.* SP-4307. Houston: NASA, Lyndon B. Johnson Space Center, 1993.

Dick, Steven J., and James E. Strick. *The Living Universe: NASA and the Development of Astrobiology.* New Brunswick, NJ: Rutgers University Press, 2004.

Dickson, Paul. *Sputnik: The Shock of the Century.* New York: Walker, 2001.

Dryden, Hugh L. "The IGY: Man's Most Ambitious Study of His Environment." *National Geographic,* February 1956.

Earth Observing System: EOS: Acronym List. Long Beach, CA: McDonnell Douglas Corp., 1992.

Earth Observing System (EOS): Glossary and List of Acronyms/Abbreviations. Greenbelt, MD: EOS Project Science Office, GSFC, 1990.

Emme, Eugene M., comp. *Aeronautics and Astronautics: An American Chronology of Science and Technology in the Exploration of Space, 1915–1960.* Washington, DC: NASA, 1961.

———, ed. *Two Hundred Years of Flight in America: A Bicentennial Survey.* San Diego: American Astronautical Society, 1977.

EOS Data Glossary. Greenbelt, MD: GSFC, NASA, 1990.

Evans, Bergen. "New World, New Words." *New York Times Magazine,* April 9, 1961, pp. 62–65.

Ezell, Edward Clinton, and Linda Neuman Ezell. *The Partnership: A History of the Apollo-Soyuz Test Project.* SP-4209. Washington, DC: GPO, 1978.

Fink, Ken. "Moon Trips Gave World of Jargon." *Biloxi Sun Herald,* July 16, 1994, p. 4.

Fleming, Thomas J. "Space Slanguage." *American Weekly,* a news supplement to the *Washington Post,* August 30, 1959, p. 2.

"For the Space Age Dictionaries." *Newport News (VA) Times-Herald,* July 31, 1961. In *NASA Current News,* August 3, 1961, p. 2.

Gall, Sarah L., and Joseph T. Pramberger. *NASA Spinoffs: 30 Year Commemorative Edition.* Washington, DC: NASA, 1992.

Garber, Stephen J., comp. *Research in NASA History: A Guide to the NASA History Program.* HHR-64. Washington, DC: NASA, 1997.

Gaynor, Frank. *Aerospace Dictionary.* New York: Philosophical Library, 1960.

———. *Dictionary of Aeronautical Engineering.* New York: Philosophical Library, 1959.

Gentle, Ernest J., ed. *Aviation and Space Dictionary.* Fallbrook, CA: Aero, 1961, 1974.

Glenn, John, with Nick Taylor. *John Glenn: A Memoir.* New York: Bantam Books, 1999.

Glennan, T. Keith. *The Birth of NASA: The Diary of T. Keith Glennan.* Edited by J. D. Hunley. SP-4105. Washington, DC: GPO, 1993.

"Glossary of Aerospace Age Terms." In "Can You Talk the Language of the Aerospace Age?" Brochure published by the U.S. Air Force Recruiting Service, Wright Patterson Air Force Base, Ohio, 1963. (This is a recruiting glossary aimed at getting students interested in aerospace. It serves as a popular vehicle for passing along terms like scrub, go/no-go, and soft landing.)

"Glossary of Missile Talk." *New York Herald Tribune,* March 21, 1960. In *NASA Current News,* March 21, 1960, p. 3.

Glossary of Personnel Terms. Washington, DC: NASA Headquarters, 1979.

"A Glossary of Space Age Terminology." *New York Times,* October 8, 1961, p. M7.

"Glossary of Space Talk." *Popular Mechanics,* March 1959, p. 72.

Glossary of Technical Terms and Abbreviations (Including Acronyms and Symbols) for the NASA Project and Program Management Training Course. Houston: DEF Enterprises, 1991.

"Glossary of Technical Terms in Flight of Apollo 10." *New York Times,* May 18, 1969, p. 69.

"Glossary of Terms Used in Space Flight." *New York Times,* March 4, 1969, p. 15.

Glossary of Terms Used in the Exploration of Space. X-202-63-140, TM-X-51299. Greenbelt, MD: GSFC, NASA, 1963.

Goddard Space Flight Center. *Encyclopedia: Satellites and Sounding Rockets of Goddard Space Flight Center, 1959–1969.* Greenbelt, MD: GSFC, 1970.

———. *The United States Sounding Rocket Program.* 47X-740-71-337 preprint. Greenbelt, MD: GSFC, July 1971.

Goldstein, Stanley H. *Reaching for the Stars: The Story of Astronaut Training and the Lunar Landing.* New York: Praeger, 1987. (Contains a glossary.)

Gor'kov, V., and Avdeev Yu. *An A–Z of Cosmosnautics.* Moscow: Mir, 1989.

Green, Constance McLaughlin, and Milton Lomask. *Vanguard: A History.* Washington, DC: Smithsonian Institution Press, 1971.

Green, Jonathan. *Tuttle Dictionary of New Words since 1960.* Boston/Rutland, VT: Charles E. Tuttle, 1991.

Guide to Terminology for Space Launch Systems. Washington, DC: American Institute of Aeronautics and Astronautics, 1994.

Gunston, Bill. *The Cambridge Aerospace Dictionary.* New York: Cambridge University Press, 2004.

Guy, Andrew Jr., and Lana Berkowitz, "It Doesn't Take a Rocket Scientist to Figure out What We at NASA Are Talking About—Well, Actually, Maybe It Does. Bringing Jargon Down to Earth." *Houston Chronicle,* August 2, 2005, p. 1.

Hacker, Barton C., and James M. Grimwood. *On the Shoulders of Titans: A History of Project Gemini.* SP-4203. Washington, DC: NASA, 1977. Reprinted 2002.

Hale, Edward E. "The Brick Moon." In *Exploring the Unknown.* Vol. 1. Ed. John M. Logsdon with Linda Lear et al. SP-4407. Washington, DC: NASA, 1995.

Hale, Leon. "NASAese As a Taxi Driver Might Speak It, Affirmative." *Houston Post,* February 22, 1963. In *NASA Current News,* March 18, 1963, p. 7.

————. "NASAese Turns Texas Talk to Mulitiple Use Facility." *Houston Post,* February 20, 1963. In *NASA Current News,* March 15, 1963, p. 8.

Hall, R. Cargill, and Jacob Neufeld. *The US Air Force in Space, 1945 to the Twenty-first Century.* Washington, DC: USAF History and Museums Program, 1998.

Harper, Harry. *Dawn of the Space Age.* London: Samson, Low, Marston and Co., n.d.

Harwood, William B. *Raise Heaven and Earth: The Story of Martin Marietta People and Their Pioneering Achievements.* New York: Simon and Schuster, 1993.

Heflin, Woodford Agee. *The Aerospace Glossary.* Maxwell Air Force Base, AL: Research Studies Institute, Air University, September 1959.

————. *Interim Glossary Aero-Space Terms.* Maxwell Air Force Base, AL: Aerospace Studies Institute, Air University, 1958. (Significant for a number of reasons, not the least of which is that it marks the lexical debut of the term aerospace, albeit as a hyphenated term. It should also be noted that Heflin's 1959 *Glossary* and the *NASA Aeronautical Dictionary* have done much to stabilize the language of the Space Age. Both were widely used by government agencies and outside organizations, and both have been used as sources for other dictionaries.)

————. *The Second Aerospace Glossary.* Maxwell Air Force Base, AL: Documentary Research Division, Aerospace Studies Institute, Air University, 1966.

Heppenheimer, T. A. *Countdown: A History of Space Flight.* New York: John Wiley and Sons, 1997.

————. *History of the Space Shuttle.* 2 vols. Washington, DC: Smithsonian Institution Press, 2002.

————. *The Space Shuttle Decision: NASA's Search for a Reusable Space Vehicle.* SP-4221. Washington, DC: NASA, 1999.

"Here's Your Glossary of Spacemen's Terms." *Washington Star,* May 24, 1962. In *NASA Current News,* May 25, 1962, p. 11.

Herring, Mack R. *Way Station to Space: A History of the John C. Stennis Space Center.* SP-4310. Washington, DC: NASA, 1997.

Hines, William. "The Wrong Stuff." *The Progressive,* July 1994, p. 18+. (Hines, who covered space from 1955 through 1989 for the *Washington Evening Star* and the *Chicago Sun Times,* was perhaps the single toughest critic of human spaceflight and in that capacity had the ability to generate a countervocabulary to the official one. Even in this piece, which is titled in opposition to *The Right Stuff,* he speaks of the Incredible Shrinking Space Station and resurrects the 1960s interpretation that the real meaning of NASA is "Never a Straight Answer.")

Hopkins, Jeanne. *Glossary of Astronomy and Astrophysics.* Chicago: University of Chicago Press, 1976.

Hubble Space Telescope: Acronyms and Abbreviations. Sunnyvale, CA: Lockheed Missiles and Space Co., Space Systems Division, 1984.

Hunsaker, Jerome C. *Forty Years of Aeronautical Research.* Smithsonian

Publication no. 4237. Reprinted from *Smithsonian Report for 1955*. Washington, DC: Smithsonian Institution, 1956. (This work contains detailed information on the naming and founding of the components of NACA that became part of NASA. It is often cited in *Origins of NASA Names*, SP-4402.)

Jane's Aerospace Dictionary. London: Jane's, 1988.

Jastrow, Robert. *Red Giants and White Dwarfs*. New York: W. W. Norton, 1979.

Jean, Charlie. "NASA 'Go' for Anomaly in Doublespeak Mode." *Orlando Sentinel*, November 22, 1986, p. A1.

Kaplan, Joseph, Wernher von Braun, et al. *Across the Space Frontier*. New York: Viking, 1952. (Book form of the series of *Collier's* magazine articles that first confronted the U.S. public with the imminent possibility of human spaceflight.)

Keller, Russell B., ed. *Glossary of Terms and Table of Conversion Factors Used in Design of Chemical Propulsion Systems*. SP-8126. Washington, DC: Lewis Research Center, NASA, 1979.

Klingaman, William A. *APL—Fifty Years of Service to the Nation: A History of the Johns Hopkins University Applied Physics Laboratory*. Laurel, MD: APL, 1993.

Koller, A. M., Jr. *Glossary, Acronyms, Abbreviations: Space Transportation System and Associated Payloads*. Kennedy Space Center, FL: NASA, 1977.

Kranz, Gene. *Failure Is Not an Option*. New York: Simon and Schuster, 2000. (Nice straightforward glossary from America's most famous flight director.)

Lambright, W. Henry. *Governing Science and Technology*. New York: Oxford University Press, 1976.

———. *Powering Apollo: James E. Webb of NASA*. Baltimore: Johns Hopkins University Press, 1995.

"Language of the Spaceman." *Newsweek*, July 11, 1960, p. 58.

Lashmar, Paul. *Spy Flights of the Cold War*. Annapolis: Naval Institute Press, 1996.

Lattimer, Dick, as told to by the Astronauts. *All We Did Was Fly to the Moon*. Gainesville, FL: Whispering Eagle Press, n.d. (Invaluable resource as Lattimer takes first-person accounts from the astronauts themselves as to how they gave call signs—hence names—to their spacecraft. This work will have increasing value over time as the only coherent attempt to collect this information while it was still fresh. The back of this book contains many solid endorsements—Arthur C. Clarke, Isaac Asimov, Alan B. Shepard—but the most impressive may be from KSC Librarian Marian Rawls: "It is a gem! . . . one of those little treasures that makes a reference librarian wonder how she ever lived without it.")

Lauer, Carl. *Acronyms, Abbreviations, and Initialisms*. MDC-0001. St. Louis, MO: McDonnell Douglas Library, 1989.

Launius, Roger D. "Eisenhower, Sputnik, and the Creation of NASA: Technological Elites and the Public Policy Agenda." *Prologue: Quarterly of the National Archives and Records Administration*, Summer 1996.

———. *Frontiers of Space Exploration*. Westport, CT: Greenwood, 1998.

Launius, Roger D., and Aaron K. Gillette, comps. *Toward a History of the Space Shuttle: An Annotated Bibliography.* Monographs in Aerospace History no. 1. Washington, DC: NASA History Office, 1992.

Launius, Roger D., and J. D. Hunley, comps. *An Annotated Bibliography of the Apollo Program.* Monographs in Aerospace History no. 2. Washington, DC: NASA History Office, 1994.

Launius, Roger D., and Dennis R. Jenkins, eds. *To Reach the High Frontier: A History of U.S. Launch Vehicles.* Lexington: University Press of Kentucky, 2002.

Launius, Roger D., and Bertram Ulrich. *NASA and the Exploration of Space.* New York: Stewart, Tabori and Chang, 1998.

Linstedt, Sharon. "Think Outside the Box and Skip Jargon 24/7." *Buffalo News,* January 3, 2005, p. B6.

Logsdon, John M. *The Decision to Go to the Moon: Project Apollo and the National Interest.* Chicago: University of Chicago Press, 1970.

Logsdon, John M., ed, with Dwayne A. Day and Roger D. Launius. *Exploring the Unknown: Selected Documents in the History of the U.S. Civil Space Program.* Vol. 2, *External Relationships.* SP-4407. Washington, DC: GPO, 1996.

Logsdon, John M., ed, with Roger D. Launius, David H. Onkst, and Stephen J. Garber. *Exploring the Unknown: Selected Documents in the History of the U.S. Civil Space Program.* Vol. 3, *Using Space.* SP-4407. Washington, DC: GPO, 1998.

Logsdon, John M., ed., with Linda J. Lear et al. *Exploring the Unknown: Selected Documents in the History of the U.S. Civil Space Program.* Vol. 1, *Organizing for Exploration.* SP-4407. Washington, DC: GPO, 1995.

Logsdon, John M., ed., with Ray A. Williamson et al. *Exploring the Unknown: Selected Documents in the History of the U.S. Civil Space Program.* Vol. 4, *Accessing Space.* SP-4407. Washington, DC: GPO, 1999.

Looking at Earth from Space: Glossary of Terms. EP-302. Washington, DC: NASA, Office of Mission to Planet Earth, 1994.

Lovell, Jim, and Jeffrey Kluger. *Lost Moon: The Perilous Voyage of Apollo 13.* Boston: Houghton Mifflin, 1994.

Lule, Jack. "Roots of the Space Race: Sputnik and the Language of US News in 1957." *Journalism Quarterly* 68 (1991): 76–86.

McBee, Susan. "Space Age Brings New Language That Starts with Sixth Graders." *Washington Post,* October 29, 1962. In *NASA Current News,* October 29, 1962, p. 2.

McCurdy, Howard E. *Space and the American Imagination.* Smithsonian History of Aviation Series. Washington, DC: Smithsonian Institution Press, 1997.

McDougall, Walter A. *The Heavens and the Earth: A Political History of the Space Age.* New York: Basic Books, 1985.

McLaughlin, Charles. *Space Age Dictionary.* Princeton, NJ: Van Nostrand, 1959. 2nd ed. 1963. (Clear, concise, and detailed, the second edition captures the urgency and informality of the early days of the human spaceflight program.)

Mailer, Norman. *Of a Fire on the Moon.* New York: New American Library, 1970.

Makkai, Adam, ed. *A Dictionary of Space English.* Chicago: English-Language Institute of America, 1973.

Mallan, Lloyd. "The Big Red Lie." *True: The Man's Magazine,* May 1959.

———. *Peace Is a Three-Edged Sword.* New York: Prentice-Hall, 1964.

Marks, Robert W., ed. *The New Dictionary and Handbook of Aerospace: With Special Sections on the Moon and Lunar Flight.* New York: Praeger, 1969.

Martin, M. J. *STS and Cargo Glossary, Acronyms and Abbreviations.* GP-1052, rev. 13. TM-84707. Kennedy Space Center, FL: NASA, 1982.

Matson, Wayne R., ed. *Cosmonautics: A Colorful History of the Soviet-Russian Space Programs.* Washington, DC: Cosmos Books, 1994.

Merrill, Grayson, ed. *Dictionary of Guided Missiles and Space Flight.* Princeton: Van Nostrand, 1959.

Merzer, Martin. "When NASA and Astronauts Communicate, They're Speaking in Tongues." *Akron Beacon Journal,* January 13, 1990.

Minton, Arthur. "Sputnik and Some of Its Offshootniks." *Names,* June 1958.

"A Missile and Space Glossary." *Air Force/Space Digest,* April 1962, pp. 143–59.

Moore, Dianne F. *The HarperCollins Dictionary of Astronomy and Space Science.* New York: HarperPerennial, 1992.

Moore, Patrick, ed. *The International Encyclopedia of Astronomy.* New York: Orion Books, 1987.

Morinigo, Fernando B., comp. *Aerospace and Defense Acronyms.* Washington, DC: American Institute of Aeronautics and Astronautics, 1992.

Moser, Reta C. *Space-Age Acronyms: Abbreviations and Designations.* New York: IFI/Plenum, 1969.

MSFC Space Station Program. *Commonly Used Acronyms and Abbreviations.* MHR-17. Marshall Space Flight Center, AL: NASA, 1988.

Murray, Charles, and Catherine Bly Cox. *Apollo—The Race to the Moon.* New York: Simon and Schuster, 1989.

Myler, Joseph L. "New Space Glossary Shows NASA 'Way Out.'" *Orlando Sentinel,* June 24, 1962. In *NASA Current News,* July 6, 1962, p. 9.

NASA Acronym Dictionary. Accession no. NASA CR-193218. Linthicum Heights, MD: Center for Aerospace Information, 1993.

NASA Project Names: A Listing of Names and Code Words Associated with NASA Programs. 62N11673. Washington, DC: NASA, 1962.

National Aeronautics and Space Administration. *Glossary of Terms Used in Connection with Congressional Budget Submission,* February 4, 1963.

———. *The Voyager Neptune Travel Guide.* Pasadena: Jet Propulsion Laboratory, 1989.

National Commission on Space. *Pioneering the Space Frontier: Report of the National Commission on Space.* New York: Bantam, 1986.

Naugle, John E. *First among Equals—The Selection of NASA Space Science Experiments.* SP-4215. Washington, DC: GPO, 1991.

Nayler, Joseph Lawrence. *A Dictionary of Astronautics.* New York: Hart, 1964.

Neufeld, Michael J. *The Rocket and the Reich: Peenemünde and the Coming of the Ballistic Missile Era*. New York: Free Press, 1995.

Newell, Homer E. *Beyond the Atmosphere: Early Years of Space Science*. SP-4211. Washington, DC: GPO, 1980.

———. *Express to the Stars*. New York, McGraw-Hill, 1961. (This book contains a basic glossary that has an elegant simplicity: "Astronautics" is defined as "the science and practice of flight through space.")

———, ed. *Sounding Rockets*. New York: McGraw-Hill, 1959.

Newlon, Clarke. *The Aerospace Age Dictionary*. New York: F. Watts, 1965.

Oberg, James E. *Red Star in Orbit*. New York: Random House, 1981.

Ordway, Frank, and Mitchell Sharpe. *The Rocket Team*. New York: Crowell, 1979.

Pate, Nancy. "'What Is the Right Stuff'—Name for Space Heroism Has Become Part of American Lexicon." *Newport News (VA) Daily Press*, May 10, 2003, p. D1.

Pearce, T. M. "The Names of Objects in Aerospace." *Names*, March 1962.

Pendray, G. Edward. *The Coming Age of Rocket Power*. New York: Harper and Brothers, 1945. (Contains a "Lexicon of Rocket Power" establishing early space language—"shot" for a rocket launch, etc.)

Pitts, John A. *The Human Factor: Biomedicine in the Manned Space Program to 1980*. SP-4213. Washington, DC: GPO, 1985. (See pp. 254–65 for selected biomedical terms of aerospace interest.)

Plant, Malcolm. *Dictionary of Space*. Harlow, Essex, England: Longman, 1986. (Plant's definitions are often encyclopedic rather than purely definitional. His definition of V-2 tells us that "Von Braun's V-2s killed 2,511 English people and seriously wounded 5,869 others.")

Portee, David S. F. *NASA's Origins and the Dawn of the Space Age*. Monographs in Aerospace History no. 10. Washington, DC: GPO, 1998.

Reference Guide to the International Space Station. SP-2006–557. Washington, DC: NASA, 2006.

Reithmaier, L[awrence]. W. *The Aviation/Space Dictionary*. Blue Ridge Summit, PA: Aero, 1990.

Rendon, Ruth. "NASA Lingo a Real Challenge." *Houston Chronicle*, October 15, 1994, p. 33.

Rice, Jack. "Quick Course in Space Jargon." *St. Louis Post-Dispatch*, September 10, 1965. In *NASA Current News*, September 22, 1965, pp. 8–9. (Valuable in that a reporter covering Gemini reacted to the vocabulary of those working on the project at NASA and at McDonnell Aircraft rather than official glossaries.)

Ridpath, Ian. *A Dictionary of Astronomy*. New York: Oxford University Press, 1998.

———. *Longman Illustrated Dictionary of Astronomy and Astronautics: The Terminology of Space*. Harlow, Essex, England: Longman, 1987.

Roland, Alex. *A Space-Faring People: Perspectives on Early Spaceflight*. SP-4405. Washington, DC: GPO, 1985.

Rosen, Milton. *The Viking Rocket Story.* New York: Harper, 1955.

Rosenfield, Loyd. "Space Glossary for Unspacified People." *Saturday Evening Post,* March 2, 1963. In *NASA Current News,* March 26, 1963, p. 11.

Rosenthal, Alfred. *Venture into Space: Early Years of Goddard Space Flight Center.* SP-4301. Washington, DC: NASA, 1968.

Rosenthal, Harry F., "Rogers Cut through Witnesses' Jargon." *Lexington (KY) Herald-Leader,* June 10, 1986, p. A4.

Rosholt, Robert L. *An Administrative History of NASA, 1955–1965.* SP-4101. Washington, DC: GPO, 1966.

Ruffner, Frederick G. *Code Names Dictionary: A Guide to Code Names, Slang, Nicknames, Journalese, and Similar Terms: Aviation, Rockets and Missiles, Military, Aerospace, Meteorology, Atomic Energy, Communications, and Others.* Detroit: Gale Research, 1963.

Ruffner, Kevin. *Corona: America's First Satellite Program.* Washington, DC: Center for the Study of Intelligence, 1995.

Rumerman, Judy. *US Human Spaceflight; A Record of Achievement, 1961–1998.* Monographs in Aerospace History no. 9. Washington, DC: NASA History Office, July 1998.

Saturn V Glossary: Apollo/Saturn V Space Terms and Abbreviations. Huntsville, AL: Saturn V Program Control Office, MSFC, 1965.

Schachter, Ken. "A Glossary Of Shuttle Terms." *Miami Herald,* January 30, 1986.

Seamans, Robert C., Jr. *Aiming at Targets: The Autobiography of Robert C. Seamans, Jr.* SP-4106. Washington, DC: NASA, 1996.

Sears, Donald A., and Henry A. Smith, "A Linguistic Look at Aerospace English." *Air Force Space Digest,* December 1969.

Shelton, William. *Soviet Space Exploration.* New York: Washington Square, 1968.

Shelton, William Roy. *Countdown: The Story of Cape Canaveral.* Boston: Little, Brown, 1960. (This book written for young readers has an impressive glossary attached to it entitled "The Language of the Missile Men.")

Shepard, Alan, and Deke Slayton. *Moon Shot: The Inside Story of America's Race to the Moon.* Atlanta: Turner, 1994.

Short Glossary of Space Terms. SP-1. Washington, DC: NASA, 1962.

Simon, Seymour. *Space Words: A Dictionary.* New York: HarperCollins, 1991.

Skylab Astronyms: Acronyms and Space Terms Used in Air-to-Ground Communication. Houston: Philco-Ford, Aerospace and Defense Systems Operations, WDL Division, Houston Operation, 1973.

Slayton, Donald K. *Deke! An Autobiography.* New York: Forge, 1994.

Smal-Stocki, Roman. *The Impact of the "Sputnik" on the English Language of the USA.* Chicago: Shevchenko Scientific Society Study Center, 1958.

Smart, Tim. "Panel Hears Little New on Tragedy—Shuttle Hearing Deals in Jargon Generalities." *Orlando Sentinel,* February 7, 1986, p. A1.

Smith, Marcia M. *CRS Report for Congress: Commonly Used Acronyms and Program Names in the Space Program.* Washington, DC: Congressional Research Service, Library of Congress, 1997.

"Space-Age Language." *The Times* (London), March 9, 1959, p. 13.

"Space Age Slang." *Time* magazine, August 10, 1962, p. 12.

The Space Encyclopedia. New York: E. P. Dutton, 1964. (Compiled with the help of a number of experts from the United States and United Kingdom, including Sir Bernard Lovell and NASA's Homer Newell.)

Space Lexicon: Commonly Used Terms in Space Operations. Peterson Air Force Base, CO: United States Space Command, 1988.

Space Station Freedom: List of Acronyms. Sunnyvale, CA: Lockheed Missiles and Space Co., 1988.

Space Station Program Glossary, Acronyms and Abbreviations. JSC-30235. Houston: Space Station Program Office, Lyndon B. Johnson Space Center, 1986.

Space: The New Frontier. EP-6. Washington, DC: NASA, 1966.

Space Transportation System and Associated Payloads: Glossary, Acronyms, and Abbreviations. RP-1059. Washington, DC: NASA, 1981.

Space Transportation System and Associated Payloads: Glossary, Acronyms, and Abbreviations. RP-1059, rev. ed. Washington, DC: NASA, 1985.

Space Transportation System and Associated Payloads: Glossary, Acronyms, and Abbreviations. TM-103575. Marshall Space Flight Center, AL: NASA, 1992.

Spitz, Armand, and Frank Gaynor. *Dictionary of Astronomy and Astronautics.* New York: Philosophical Library, 1959.

Stapleton, William B. "Space Age Sign Language." *Houston Post,* July 12, 1964, pp. 5–6.

Stover, Dawn. "Anomaly = Disaster, and Other Handy NASA Euphemisms." *Popular Science,* February 2004, p. 96. (One of the stronger critical articles on NASAese. Stover begins her examination, "The US space agency has a language all its own. NASA uses so many acronyms that the agency issues a book to its employees to keep track of them. And even when NASA uses ordinary words, they're often imbued with special meaning, generally designed to take the edge off graphic situations.")

Swanson, Glen E., ed. *"Before This Decade Is Out . . .": Personal Reflections on the Apollo Program.* SP-4223. Washington, DC: NASA, 1999.

Swenson, Loyd S., Jr., James M. Grimwood, and Charles C. Alexander. *This New Ocean: A History of Project Mercury.* SP-4201. Washington, DC: GPO, 1966.

Thomas, Shirley. *Men of Space.* 7 vols. Philadelphia: Chilton, 1960–65.

Tikhonravov, M. K. "The Creation of the First Artificial Earth Satellite: Some Historical Details." *Journal of the British Interplanetary Society* 47, no. 5 (May 1994): 191–94.

Tomsic, Joan L. *SAE Dictionary of Aerospace Engineering.* Warrendale, PA: Society of Automotive Engineers, 1998.

Turnill, Reginald. *The Language of Space: A Dictionary of Astronautics.* New York: Day, 1971.

Tver, David F., with the assistance of Lloyd Motz and William K. Hartmann. *Dictionary of Astronomy, Space, and Atmospheric Phenomena.* New York: Van Nostrand Reinhold, 1979.

Volgenau, Gerald. "NASA Uses a Language All Its Own," *Detroit Free Press,* September 29, 1988, p. A14.

Von Braun, Wernher von, and Frederick I. Ordway III. *History of Rocketry and Space Travel.* New York: Crowell, 1969.

—————— *History of Rocketry and Space Travel.* 3rd rev. ed. New York: Crowell, 1975.

Wagener, Leon. *One Giant Leap.* New York: Forge, 2004.

Walker, P. M. B. *Cambridge Air and Space Dictionary.* Cambridge: Cambridge University Press, 1990.

Wallace, Lane E. *Dreams, Hopes, Realities: NASA's Goddard Space Flight Center, the First Forty Years.* SP-4312. Washington, DC: GPO, 1999.

Webb, James E. *Space Age Management.* New York: McGraw-Hill, 1962.

Weintraub, Jessica. "From A-OK to Oz: The Historical Dictionary Of American Slang." *Humanities,* March–April 2004.

Weiss, Jeffrey. "Learning the Lingo of the Final Frontier." *Dallas Morning News,* September 5, 1988.

Wells, Helen T., Susan B. Whiteley, and Carrie Karegeannes. *Origins of NASA Names.* SP-4402. Washington, DC: GPO, 1976.

Williamson, Mark. *The Cambridge Dictionary of Space Technology.* Cambridge: Cambridge University Press, 2001.

—————— *Dictionary of Space Technology.* New York: Adam Hilger, 1990.

Wilson, Glen P. "The Legislative Origins of NASA: The Role of Lyndon B. Johnson." *Prologue: Quarterly of the National Archives,* Winter 1993.

Winter, Frank H. *Rockets into Space.* Cambridge, MA: Harvard University Press, 1990.

Withers, A. M. "Words and the Space Age." *Word Study* (Merriam-Webster), February 1962, pp. 1–3. (Important cataloging of the influence of Greek and Latin on the language of the Space Age.)

Wolfe, Tom. *The Right Stuff.* New York: Farrar, Straus and Giroux, 1979.

Wragg, David W. *A Dictionary of Aviation.* New York: F. Fell, 1974.

Yeager, Gen. Chuck, and Leo Janos. *Yeager.* New York: Bantam, 1985.

Zimmerman, Robert. *Genesis: The Story of Apollo 8, the First Manned Flight to Another World.* New York: Four Walls Eight Windows, 1998.

Internet Resources

In addition to the sources listed below, NASA Headquarters library maintains its own listing of space and aerospace dictionaries at www.hq.NASA.gov/office/hqlibrary/pathfinders/aerodic.htm.

Astronomical Society of the Pacific. "Introductory Astronomy Glossary." *The Universe in the Classroom,* nos. 14 and 15 (Spring 1990). www.astrosociety .org/education/publications/tnl/14/14.html.

Banholzer, Pete. *NASA Acronyms* [cited June 28, 2004]. http://library.gsfc.NASA
.gov/Databases/Acronym/acronym.html.

Doody, Dave. *Basics of Space Flight Glossary* [cited June 28, 2004]. www2.jpl
.NASA.gov/basics/bsfgloss.htm.

Dumoulin, Jim. *NASA/KSC Acronym List.* October 27, 1996 [cited June 28, 2004].
www.ksc.NASA.gov/facts/acronyms.html.

Glover, Daniel R. *Dictionary of Technical Terms for Aerospace Use.* November 15,
2001 [cited June 28, 2004]. http://vesuvius.jsc.NASA.gov/er/seh/menu.html.

JSC Aerospace Scholars Glossary. http://aerospacescholars.jsc.nasa.gov/HAS/
cirr/glossary.cfm#top.

Kennedy, Garry, comp. "Glossary" (November 18, 2007) appended to NASA's
Apollo Lunar Surface Journal. http://history.nasa.gov/alsj.

NASA Academy of Program and Project Leadership. *Lexicon.* September 15,
2003 [cited June 28, 2004]. http://appl.NASA.gov/resources/lexicon/index
.html.

Olsen, Lola. *Global Change Master Directory: List of Earth Science Acronyms,
Glossaries and Gazetteers* [cited June 28, 2004]. http://globalchange.NASA
.gov/Resources/FAQs/acronyms.html.

Shaw, Robert J. *Ultra-Efficient Engine Technology Engine Glossary* [cited June 28,
2004]. www.ueet.NASA.gov/glossaries/EngineGlossary.html.

Wilton, Dave. "Moonshot Terms." *A Way with Words: The Weekly Newsletter of
Wordorigins.org* 4, no. 17 (July 22, 2005). www.wordorigins.org.